The Role of Neutrinos, Strings, Gravity, and Variable Cosmological Constant in Elementary Particle Physics

The Role of Neutrinos, Strings, Gravity, and Variable Cosmological Constant in Elementary Particle Physics

Edited by

Behram N. Kursunoglu
Global Foundation, Inc.
Coral Gables, Florida

Stephan L. Mintz
Florida International University
Miami, Florida

and

Arnold Perlmutter
University of Miami
Coral Gables, Florida

Springer Science+Business Media, LLC

Proceedings of an International Conference on the Role of Neutrinos, Strings, Gravity, and Variable Cosmological Constant in Elementary Particle Physics, held December 14–17, 2000, in Fort Lauderdale, Florida
ISBN 978-1-4757-8700-9 ISBN 978-0-306-47116-2 (eBook)
DOI 10.1007/978-0-306-47116-2

Originally published by Kluwer Academic / Plenum Publishers, New York in 2001
Softcover reprint of the hardcover 1st edition 2001

http://www.wkap.nl

10 9 8 7 6 5 4 3 2 1

A C.I.P. record for this book is available from the Library of Congress.

PREFACE

The 29th International Conference was held as the first one of the millennium at its Fort Lauderdale venue. These conferences began, with High Energy Physics being the main topic, by introducing gradually cosmology into its programs. These proceedings of the 2000 conference reflect the variety of topics and ideas discussed. Our future conferences will be designed somewhat akin to the early Coral Gables Conferences where we shall seek some convergence of ideas. For this reason various committees have been formed from among the participating physicists. The committees and their memberships are listed in these proceedings. We further decided for the first time to include some graduate student participants in our future meetings for which also a committee has already been established. The topics will demonstrate a more *activist structure* of the Coral Gables Conferences, for example the *duality of the gravitational forces* and expansion of the universe will be discussed from this point of view since it conveys a convergence to the ideas of quintessence versus the ordinary theory, which are considered as the cause of the expansion of the universe.

We further wish to announce that the future conferences will assume a collective organization where several committees as listed in these proceedings will have their input into the conference. We have now introduced new topics and ideas, which referred especially to the attractive and repulsive nature of the gravitational force. These proceedings of the conference contain a variety of topics and ideas. The meeting for 2001 next December will represent these anticipated changes and hopefully will follow the established early tradition of these series of conferences.

The Chairman and Trustees of the Global Foundation, Inc. wish to gratefully acknowledge Lady Blanka Rosenstiel, Founder and President of the American Institute of Polish Culture, Chopin Foundation and Honorary Consul of the Republic of Poland in Miami, and to Dr. and Mrs. Edward Bacinich of Palm Beach, Florida for their continued generous annual support.

Behram N. Kursunoglu
Stephan L. Mintz
Arnold Perlmutter
Coral Gables, Florida
February 2001

About the Global Foundation, Inc.

The Global Foundation, Inc., which was established in 1977, utilizes the world's most important resource... people. The Foundation consists of distinguished men and women of science and learning, and of outstanding achievers and entrepreneurs from industry, governments, and international organizations, along with promising and enthusiastic young people. These people convene to form a unique and distinguished interdisciplinary entity to address global issues requiring global solutions and to work on the frontier problems of science.

Global Foundation Board of Trustees

Former Trustees

GLOBAL FOUNDATION'S RECENT CONFERENCE PROCEEDINGS

Making the Market Right for the Efficient Use of Energy
Edited by: Behram N. Kursunoglu
Nova Science Publishers, Inc., New York, 1992

Unified Symmetry in the Small and in the Large
Edited by: Behram N. Kursunoglu, and Arnold Perlmutter
Nova Science Publishers, Inc., New York, 1993

Unified Symmetry in the Small and in the Large - 1
Edited by: Behram N. Kursunoglu, Stephen Mintz, and Arnold Perlmutter
Plenum Press, 1994

Unified Symmetry in the Small and in the Large - 2
Edited by: Behram N. Kursunoglu, Stephen Mintz, and Arnold Perlmutter
Plenum Press, 1995

Global Energy Demand in Transition: The New Role of Electricity
Edited by: Behram N. Kursunoglu, Stephen Mintz, and Arnold Perlmutter
Plenum Press, 1996

Economics and Politics of Energy
Edited by: Behram N. Kursunoglu, Stephen Mintz, and Arnold Perlmutter
Plenum Press, 1996

Neutrino Mass, Dark Matter, Gravitational Waves, Condensation of Atoms and Monopoles, Light Cone Quantization
Edited by: Behram N. Kursunoglu, Stephen Mintz, and Arnold Perlmutter
Plenum Press, 1996

Technology for the Global Economic, Environmental Survival and Prosperity
Edited by: Behram N. Kursunoglu, Stephen Mintz, and Arnold Perlmutter
Plenum Press, 1997

25th Coral Gables Conference on High Energy Physics and Cosmology
Edited by: Behram N. Kursunoglu, Stephen Mintz, and Arnold Perlmutter
Plenum Press, 1997

Environment and Nuclear Energy
Edited by: Behram N. Kursunoglu, Stephan Mintz, and Arnold Perlmutter
Plenum Press, 1998

Physics of Mass
Edited by: Behram N. Kursunoglu, Stephan Mintz, and Arnold Perlmutter
Plenum Press, 1999

Preparing the Ground for Renewal of Nuclear Power
Edited by: Behram N. Kursunoglu, Stephan Mintz, and Arnold Perlmutter
Plenum Press, 2000

Confluence of Cosmology, Massive Neutrinos, Elementary Particles & Gravitation
Edited by: Behram N. Kursunoglu, Stephan Mintz, and Arnold Perlmutter
Plenum Press, 2000

Global Warming and Energy Policy
Edited by: Behram N. Kursunoglu, Stephan Mintz, and Arnold Perlmutter
Plenum Press, 2001

The Role of Neutrinos, Strings, Gravity, and Variable Cosmological Constant in Elementary Particle Physics
Edited by: Behram N. Kursunoglu, Stephan Mintz, and Arnold Perlmutter
Plenum Press, 2001

Global Foundation, Inc.

A Nonprofit Organization for Global Issues Requiring Global Solutions, and forProblems on the Frontiers of Science

Center for Theoretical Studies

MILLENIUM'S FIRST
INTERNATIONAL CONFERENCE
ON
ORBIS SCIENTIAE 2000

THE ROLE OF ATTRACTIVE AND REPULSIVE GRAVITATIONAL FORCES IN COSMIC ACCELERATION OF PARTICLES

The Origin of the Cosmic Gamma Ray Bursts

(29th Conference on High Energy Physics and Cosmology Since 1964)

December 14 - 17, 2000
Lago Mar Resort
Fort Lauderdale, Florida

This conference is supported in part by Alpha Omega Research Foundation and Lady Blanka Rosenstiel

Sponsored by:
Global Foundation Inc.

P. O. Box 249055
Coral Gables, Florida 33124-9055
Phone: (305) 669-9411
Fax: (305) 669-9464
E-mail: kursun@globalfoundationinc.org
http://www.globalfoundationinc.org

Conference Hotel:
Lago Mar Resort

1700 South Ocean Lane
Fort Lauderdale, Florida 33316
Ph: 1-800-255-5246
Fax: (954) 524-6627
Deadline: October 15, 2000
Group rate: $125.00/night

Précis

This year's conference will take off with some new ideas giving special prominence to the role of very large, both attractive and repulsive, gravitational forces in cosmic regions of the universe. There exist charged particle accelerations in the presence of tremendous gravitational power of supermassive black hole, which can accelerate for example the electrons to acquire energies far above the capability of any accelerator we can build on the earth. These particles will emit high-energy gamma rays and will emerge in various directions depending on the black hole's location. It thus appears that very high gravitational fields can explicitly act as the source of an electromagnetic field.

This précis is based on the prediction by unified field theory with quintessential matter (infinite number of positive and negative magnetic charges of different magnitudes) permeating the universe, which could alternate between gravitational attraction and repulsion modes.

NASA in collaboration with the department of energy is proposing to study such gamma rays in a project called GLAST (the gamma ray large area space telescope). Already there are preparations to deploy this particular telescope by 2005. The participants of this research project include physicists from various universities. If the attractive and repulsive gravity is the main accelerating mechanism then they will find not only directional arrival of gamma rays but also indirectly the presence of these gamma rays will demonstrate that gravity can be a source of electromagnetism. We already know that electromagnetism, as described in Einstein's general theory of relativity, is a source of gravity. Thus have we come to experimental verification of a unified field theory of gravity and electromagnetism at high energy?

The Conference will further include the topics of superstring theory based unification, gravitational waves, current understanding of dark matter, proton spin, neutrino cosmology and quantum gravity.

Dedication

The trustees of the Global Foundation and members of the 29th Orbis Scientiae 2000, dedicate this conference to Dr. Fred Zachariasen of CALTECH. The late Professor Zachariasen was a loyal and active member of this series of conferences on the frontiers of physics since 1964. He also served as a member of the Conferences' International Advisory Committee. We shall all miss Fred. We extend our deepest condolences to his wife Nancy Zachariasen, his two daughters Kerry and Judy, his two grandchildren and all the other members of his family.

--NOTES--

1. Each presentation is allotted a maximum of 25 minutes and an additional 5 minutes for questions and answers.

2. Moderators are requested not to exceed the time allotted for their sessions.

Moderator: Presides over a session. Delivers a paper in own session, if desired, or makes general opening remarks.

Dissertator: Presents a paper and submits it for publication in the conference proceedings at the conclusion of the conference.

Annotator: Comments on the dissertator's presentation or asks questions about it upon invitation by the moderator.

CONFERENCE PROCEEDINGS

1. Instructions to the authors from the publisher for preparing typescripts for the conference proceedings will be sent to you via e-mail.

2. Papers must be received at the Global Foundation by February 15, 2001.

An edited Conference Proceedings will be submitted to the Publisher by March 14, 2001.

FORMER TRUSTEES

INTERNATIONAL ADVISORY COMMITTEE

ORBIS SCIENTIAE 2000
PROGRAM OUTLINE

Thursday, December 14, 2000
Lakeview Room

8:00 AM - 12:00 PM **Registration**

9:00 AM—11:30 AM **International Advisory Committee Meeting,**
Director's Room

12:00 PM		**Committee Members Lunch**, Director's Room
1:30 PM	SESSION I:	**Cosmological Constant is Not Really a Constant**
	Moderator:	**Behram N. Kursunoglu**, Global Foundation, Inc.
	Dissertators:	**Edward Teller**, Lawrence Livermore Laboratory
		Behram N. Kursunoglu "The Quartet of Attractive and Repulsive Gravity Forces"
		Paul Frampton, University of North Carolina "Quintessence and Cosmic Microwave Background"
	Annotators:	**Thomas W. Kephart,** Vanderbilt University
	Session Organizer:	**Behram N. Kursunoglu**
3:00 PM	Coffee Break	
3:30 PM	SESSION II:	**Beyond the Standard Model**
	Moderator:	**Don Colladay,** University of South Florida
	Dissertators:	**Alan Kostelecky,** Indiana University "Violation of Spacetime Symmetries"
		Pierre Sikivie, University of Florida "Dark Matter, Axions and Caustics"
		Tom Weiler, Vanderbilt University "Neutrino Oscillations from Broken Lorentz Invariance"
	Annotator:	**Behram N. Kursunoglu**, Global Foundation, Inc.
	Session Organizer:	**Alan Kostelecky**
5:00 PM	SESSION III:	**Neutrino Cosmology**
	Moderator:	**George Fuller**, UCSD
	Dissertators:	**George Fuller** "Neutrino Oscillations in Big Bang Cosmology"
		David Cline, UCLA "Measurement of Neutrino Mass with SN II"
		Randell Mills , CEO of Black Light Power, Princeton , NJ

"The Grand Unified Theory of Classical Quantum Mechanics"

Annotator: **Stephan L. Mintz**, Florida International University

Session Organizer: **David Cline**

6:30 PM **Orbis Scientiae adjourns for the day**

Friday, December 15, 2000

8:30 AM SESSION IV: **The Role of Neutrinos in Super String Theory Based Unification**

Moderator: **Stephan L. Mintz**, Florida International University

Dissertators: **A.J. Meyer**, International Scientifics Projects, Inc.
"On CPT Asymmetry as Predicted by the Super Sp Model"

Osher Doctorow, Culver City, CA
"Light Speed Energy Dependence"

Gabor Domokos & Susan Kovesi-Domokos, Johns Hopkins
"A Neutrino Component of Ultra High Energy Cosmic Rays?"

Ali R. Fazely, Southern University, Baton Rouge, LA
"Probing the Nature of Neutrinos at Pion Factories"

Annotators: **Freydoon Mansouri**, University of Cincinnati

Session Organizer: **Stephan L. Mintz**

10:00 AM Coffee Break

10:30 AM SESSION V: **Recent Progress on Old and New Ideas I**

Moderator: **Paul Frampton**, University of North Carolina

Dissertators: **Stephan L. Mintz,** Florida International University
"Neutrino Reactions in Nuclei at Higher Energies"

Thomas W. Kephart, Vanderbilt University
"AdS/ CFT Phenomenology"

Thomas L. Curtright, University of Miami

"Duality and Time"

Annotator: **Alan Kostelecky**, Indiana University

Session Organizer: **Paul Frampton**

12:00 Noon: Lunch Break

1:30 PM SESSION VI: **New Directions In Gravitational Research**

Moderator: **Arnold Perlmutter**, University of Miami

Dissertators: **Philip Mannheim**, University of Connecticut
"Attractive and Repulsive Gravity—Cosmic Acceleration as the Solution to the Cosmological Constant Problem"

Richard P. Woodard, University of Florida
"Back-reaction is for real"

Bhaskar Dutta, Texas A&M University
"Phenomenology and Horava-Witten M Theory"

Annotator: **Don Lichtenberg**, Indiana University

Session Organizer: **Arnold Perlmutter**

3:00 PM Coffee Break

3:30 PM SESSION VII: **Proton Spin**

Moderator: **Alan Krisch**, University of Michigan

Dissertator: **Alan Krisch**
"New Experiment on Very Violent Collisions of Spinning Protons"

D.W. Sivers, Michigan-Portland University
"Single Spin Asymmetry; A Probe of Hadron Structure"

Annotator: **Arnold Perlmutter**, University of Miami

Session Organizer: **Alan Krisch**

5:00 PM **Orbis Scientiae adjourns for the day**

6:30 PM WELCOMING COCKTAILS, FOUNTAINVIEW LOBBY

Courtesy of Lago Mar Resort

7:30 PM CONFERENCE BANQUET, FOUNTAINVIEW LOBBY
Courtesy of Edward and Maria Bacinich

Saturday, December 16, 2000

8:30 AM SESSION VIII: **Nature's Highest Energy Particle Accelerators**

Moderator: **Elliott Bloom**, SLAC/Stanford University

Dissertators: **Elliott Bloom**
"Nature's highest energy accelerators"

Steve Ritz, Goddard Space Flight Center
"Exploring the extreme universe with GLAST"

Stirling Colgate, LANL
"The Free Energy of the Universe and the Acceleration of Ultra High-Energy Cosmic Rays"

Annotators: **Grzegorz Madejski,** SLAC, Stanford University

Session Organizer: **Elliott Bloom**

10:00 AM Coffee Break

10:30 AM SESSION IX: **Gamma Ray Bursts**

Moderator: **Brenda L. Dingus,** University of Wisconsin

Dissertators: **Brenda L. Dingus**
"Highest Energy Gamma Ray from Gamma Ray Bursts"

Donald Q. Lamb, University of Chicago
"Implications of Gamma-Ray Bursts on the Early Universe

Grant Mathews, Notre Dame University
"Models of the Engines of Gamma-Ray Bursts"

Annotator: **David Cline**, UCLA
"Primordial Black Holes and Short Gamma-Ray Bursts"

Session Organizer: **Brenda L. Dingus**

12:00 Noon: **Orbis Scientiae adjourns for the day**

Sunday, December 17, 2000

9:00 AM SESSION X: **Recent Progress on Old and New Ideas II**

Moderator: **Don Lichtenberg**, Indiana University

Dissertators: **Don Lichtenberg**, Indiana University
"Comments on Supersymmetry, extra dimensions, and the accelerating universe"

Freydoon Mansouri, University of Cincinnati
"Twisted Affine Kac-Moody Algebras and the Black Hole Entropy"

Uwe Trittmann, Ohio State University
"Testing the Maldacena conjecture with SDLCQ"

Annotator: **Stephen Pinsky**, Ohio State University

Session Organizer: **Don Lichtenberg**

10:30 AM Coffee Break

11:00 AM SESSION XI: **The Latest Developments In High Energy Physics and Cosmology**

Moderator: **Sydney Meshkov**, CALTECH

Dissertators: **Carsten Hast**, SLAC / Stanford University
"CP-Violation In B-Decays : Status and Prospects with BABAR"

Steve Pinsky, Ohio State University
"Quarks in Super symmetric discrete light-cone quantization (SDLCQ)"

Keith Dienes, University of Arizona
"The Role of Extra Dimensions"

Annotator: **Alan Krisch**, University of Michigan

Session Organizer: **Sydney Meshkov**

12:30 PM **Orbis Scientiae 2000 adjourns**

CONTENTS

THE COSMOLOGICAL CONSTANT IS NOT REALLY A CONSTANT

BEYOND THE STANDARD MODEL

NEUTRINO PHYSICS AND COSMOLOGY

NATURE'S HIGHEST ENERGY PARTICLE ACCELERATORS

THE LATEST DEVELOPMENTS IN HIGH ENERGY PHYSICS AND COSMOLOGY

THE COSMOLOGICAL CONSTANT IS NOT REALLY A CONSTANT

RECENT DEVELOPMENTS IN GRAVITATIONAL THEORY AND AN INTRINSIC COSMOLOGICAL PARAMETER

Behram N. Kursunoglu
Global Foundation, Inc., P.O. Box 249055
Coral Gables, Florida 33124-9055

INTRODUCTION

General relativistic theory of gravitation has made it possible to study gravitation beyond solar system and test some new predictions like, for example, gravitational collapse of very massive stars some of which containing interior cosmic fires initiated and maintained by nuclear reactions. The existence of black holes has been observationally established. The recent observations have further revealed the increasing acceleration of the expanding universe. To explain this fact we also need to ask why the universe is expanding? In order to answer this question and the reasons for the expansion's increasing acceleration we need to look more carefully into the nature of the gravitational force. General relativity does not directly predict the existence of anti-gravity or, more precisely, the fact that gravity is either an attractive or a repulsive force. If the existence of repulsive gravitational force is recognized then we have established a kinship with electrodynamics.

Now, I would like to state that these entirely novel results are obtained from the author's version of the unified field theory of gravitation and electromagnetism. A unified field theory in physics should comprise the entire physical reality from the innermost structure of matter – the ultimate constituents – to the outermost structure of the boundless universe. All forms of matter or energy experience gravitational interactions. Therefore, we may expect the gravitational field to be the basis to unify all interactions.

The place of gravitation in physics in general and its role in elementary particle physics in particular have attracted the attention of the physicists of this and the passed century. They have worked largely on the elementary particle aspects of gravitation in terms of direct quantization of the gravitational field, although these efforts have been mainly unsuccessful, some theorists are still hopeful to obtain results.

The Role of Neutrinos, Strings, Gravity, and Variable Cosmological Constant in Elementary Particle Physics, Edited by Kursunoglu *et al.*, Kluwer Academic/Plenum Publishers, New York, 2001

For the gravitational field, the coupling with the matter field is expressed in terms of a dimensional constant, and also there exist no global gauge theoretical formulation of general relativity. However, a unified theory based on a generalization of Einstein's general relativity has a unique global gauge-invariance property. The work in this paper is based on the classical unification of the two long-range fields of gravitation and electromagnetism or, in other words, unification of the electromagnetic interaction with the weakest interaction, the gravitational force. Independently of the quantum behavior of matter in the small, it is of great interest to see how and if a generalized theory of gravitation will lead to new results of the unified theory other than gravitation, as formulated within the concepts of general relativity. In other words, in place of quantization of the gravitational field, we would like to consider general relativization of electromagnetism along with gravitation itself.

MATHEMATICAL BASIS OF THE THEORY

The success of the geometrization of gravity by the general theory of relativity had led Einstein and others, *ipso facto*, to further extend the geometrical concepts to describe other force fields of physics. The contents of this paper are based on my earlier work, which was begun in 1950 in Cambridge where I was a graduate student. My work in the past few years has enabled me to obtain some new results and to further understand the physics of the theory. The simplest formulation of a unified theory according to Einstein's proposal in 1950 can be based on a reducible, with respect to coordinate transformations, non-symmetric 16 components tensor,

$$\hat{g}_{\mu\nu} = \hat{g}_{\underline{\mu\nu}} + \hat{g}_{\underset{\vee}{\mu\nu}} \tag{1}$$

where the underlined subscripts imply $\hat{g}_{\underline{\mu\nu}}$ is the symmetric part of the non-symmetric tensor $\hat{g}_{\mu\nu}$, and the symbol (ˇ) refers to the anti-symmetric part of $\hat{g}_{\mu\nu}$. The fundamental Einstein tensor $\hat{g}_{\mu\nu}$ is with respect to coordinate transformations reducible where the symmetric and anti-symmetric parts transform independently. Thus no unification is manifestly shown. Einstein's answer to this very serious objection was the fact that in the inverse of $\hat{g}_{\mu\nu}$ the symmetric and anti-symmetric parts consist of mixed terms. Therefore, the reducibility of the covariant tensor does no harm! This Einstein argument was unacceptable since a theory based half on reducible and the other half on irreducible field variables is self-contradictory. Actually, the problem of reducibility will cease to be of importance as soon as we recognize the gauge invariance of the fundamental field variables $\hat{g}_{\mu\nu}$. For real tensor $\hat{g}_{\mu\nu}$ the gauge invariance group is represented by SO (3,1) x SO (2), while for the complex $\hat{g}_{\mu\nu} = \hat{g}_{\underline{\mu\nu}} + i\, \hat{g}_{\underset{\vee}{\mu\nu}}$ the gauge invariance group is represented by SU(3,1) x U(1). Details can be found in my paper[(1)] [Behram N. Kursunoglu, Journal of Physics Essays, Vol.4, No.4 , pp439-518,1991].

The next important issue is the physical interpretation of the non-symmetric tensor $\hat{g}_{\mu\nu}$ where both Einstein and Schrödinger proposals were less than satisfactory. There was no way to compare the symmetric and anti-symmetric parts of fundamental tensor $\hat{g}_{\mu\nu}$, which, for distances large compared to *Planck Length,* refer to gravitation and electromagnetism, respectively. However, we can introduce a parameter q to make the tensor $\hat{g}_{\mu\nu}$ a dimensionless quantity and record the actual field variables in the form

$$\hat{g}_{\mu\nu} = \hat{g}_{\underline{\mu\nu}} + q^{-1}\, \hat{g}_{\underset{\vee}{\mu\nu}}, \tag{2}$$

where q has the dimensions of an electric field which can be calculated as a function of the electric and magnetic charges. All the field elements are now functions of q, which helps to link symmetric and anti-symmetric parts of the total field. The basic functions of the theory include the non-symmetric curvature tensor, $\hat{R}_{\mu\nu}$ which is a function of the 64 component non-symmetric affine connection $\Gamma^{\rho}_{\mu\nu}$. The laws of the unified field theory can be based on a relationship between the fundamental field variables $\hat{g}_{\mu\nu}$ and the curvature tensor $\hat{R}_{\mu\nu}$. The tensor $\hat{R}_{\mu\nu}$ has the dimension of the square of an inverse length. Thus the proposed linkage between $\hat{g}_{\mu\nu}$ and $\hat{R}_{\mu\nu}$ will automatically introduce a fundamental length r_o. This r_o is the *intrinsic cosmological parameter.* There are just two more steps to complete the origin and the history of the fundamental length r_o.

BIANCHI-EINSTEIN IDENTITIES

The original derivation in 1951 of the third version of the non-symmetric unified field theory was based by this author on Bianchi-Einstein differential identities

$$[\, \mathfrak{g}^{\underline{\mu\nu}} \hat{R}_{\underline{\mu\rho}} - \frac{1}{2}\delta^{\nu}_{\rho}\mathfrak{g}^{\underline{\mu\sigma}} \hat{R}_{\underline{\mu\sigma}}\,]_{\|\nu} = \frac{1}{2}\mathfrak{g}^{\overset{\mu\nu}{\vee}} (\hat{R}_{\underset{\vee}{\mu\nu},\rho} + \hat{R}_{\underset{\vee}{\nu\rho},\mu} + \hat{R}_{\underset{\vee}{\rho\mu},\nu}), \qquad (3)$$

where the parallel lines ($\|$) indicate covariant differentiation with respect to the tensor $b_{\mu\nu}$ defined by

$$b_{\mu\nu} = \frac{(1+\frac{1}{2}\Omega)g^{\mu\nu} + T^{\mu\nu}}{(1+\Omega-\Lambda^2)^{\frac{1}{2}}}, \qquad (4)$$

for more details, the reader should consult my paper in reference (1). I spend some time to succeed to write Bianchi-Einstein identities in the form of the divergence of a symmetric tensor yielding, as in equation (3), familiar in electrodynamics, a force density on the right hand side. This was crucial for the eventual discovery of the field equations for the unified field theory. I also observed that the same identities (4) are satisfied by the fundamental tensor

$$\hat{g}_{\mu\nu} = g_{\mu\nu} + q^{-1}\Phi_{\mu\nu}, \qquad (5)$$

where we made slight change of notation $g_{\mu\nu} = \hat{g}_{\underline{\mu\nu}}$, $\Phi_{\mu\nu} = \hat{g}_{\underset{\vee}{\mu\nu}}$. The fundamental tensor $\hat{g}_{\mu\nu}$ also satisfies the differential identity (4) above,

$$[\mathfrak{g}^{\underline{\mu\nu}} g_{\mu\rho} - \frac{1}{2}\delta^{\nu}_{\rho}\mathfrak{g}^{\underline{\mu\sigma}} g_{\mu\sigma}]_{\|\nu} = \frac{1}{2}\mathfrak{g}^{\overset{\mu\nu}{\vee}} (\Phi_{\mu\nu,\rho} + \Phi_{\nu\rho,\mu} + \Phi_{\rho\mu,\nu}), \qquad (6)$$

In general the Einstein-Bianchi identities are satisfied by the tensor

$$\hat{Q}_{\mu\nu} = \hat{R}_{\mu\nu} + \lambda \hat{g}_{\mu\nu} + \gamma \hat{b}_{\mu\nu}, \tag{7}$$

where

$$\hat{b}_{\mu\nu} = b_{\mu\nu} + \mathrm{q}^{-1} F_{\mu\nu} \tag{8}$$

$$F_{\mu\nu} = \partial_\mu A_\nu - \partial_\nu A_\mu \tag{9}$$

and where lambda and gamma are arbitrary constants with the dimension of inverse square of a length. In 1950 when Einstein announced the discovery of his theory his proposals for the field equations included the four equations below

$$\hat{R}_{\underset{\vee}{\mu\nu,\rho}} + \hat{R}_{\underset{\vee}{\nu\rho,\mu}} + \hat{R}_{\underset{\vee}{\rho\mu,\nu}} = 0. \tag{10}$$

If we set $\Phi_{\mu\nu} = 0$, the identities (3) reduce to the Bianchi identities of the symmetric theory

$$[\mathfrak{g}^{\mu\nu} R_{\mu\rho} - \frac{1}{2} \delta^\nu_\rho \mathfrak{g}^{\mu\sigma} R_{\mu\sigma}]_{|\nu} = 0 , \tag{11}$$

where

$$\mathfrak{g}^{\mu\nu} = \sqrt{-g}\, g^{\mu\nu} , \tag{12}$$

and where in (9) $R_{\mu\nu}$ is the curvature tensor of general relativity. The four identities (3) are also satisfied by $\hat{b}_{\mu\nu} = b_{\mu\nu} + q^{-1} F_{\mu\nu}$. The right-hand side of (3) in analogy with general relativity, looks formally like a force density. Einstein used the identities (3) alone to derive his field equations, even though the same identities were also satisfied by the field tensor $\hat{g}_{\mu\nu}$, as in (6) and also by $\hat{b}_{\mu\nu}$, where the latter, for an arbitrary tensor $F_{\mu\nu} = \partial_\mu A_\nu - \partial_\nu A_\mu$, corresponds to a *special kind of gauge invariance*. Thus replacements in (3) of $\hat{R}_{\underline{\mu\nu}}$ and $\hat{R}_{\underset{\vee}{\mu\nu}}$ by $\hat{R}_{\underline{\mu\nu}} + \lambda b_{\mu\nu}$ and $\hat{R}_{\underset{\vee}{\mu\nu}} + \lambda F_{\mu\nu}$, respectively leave the identities unchanged. Therefore, in general, the four identities (3) are satisfied by the tensor

$$\hat{Q}_{\mu\nu} = \hat{R}_{\mu\nu} + \lambda \hat{g}_{\mu\nu} + \gamma \hat{b}_{\mu\nu} \tag{13}$$

Thus the most general possible form of the field equations is contained in the statement

$$\hat{R}_{\mu\nu} + \lambda \hat{g}_{\mu\nu} + \gamma\left(b_{\mu\nu} + F_{\mu\nu}\right) = 0. \tag{14}$$

Hence we obtain

$$\hat{R}_{\underline{\mu\nu}} = -\lambda g_{\mu\nu} - \gamma b_{\mu\nu}, \quad \hat{R}_{\underset{\vee}{\mu\nu},\rho} + \hat{R}_{\underset{\vee}{\nu\rho},\mu} + \hat{R}_{\underset{\vee}{\rho\mu},\nu} = -\lambda I_{\mu\nu\rho}. \tag{15}$$

If we impose the requirement that for $\Phi_{\mu\nu} = 0$ the field equations must reduce to the field equations of a pure gravitational field without a cosmological constant, we obtain $\gamma = -\lambda\left(= -r_o^{-2}\right)$, and hence we obtain the ten field equations

$$\hat{R}_{\underline{\mu\nu}} = r_o^{-2}\left(b_{\mu\nu} - g_{\mu\nu}\right), \tag{16}$$

The remaining field equations are

$$\hat{R}_{\underset{\vee}{\mu\nu},\rho} + \hat{R}_{\underset{\vee}{\nu\rho},\mu} + \hat{R}_{\underset{\vee}{\rho\mu},\nu} + r_o^{-2}\, I_{\mu\nu\rho} = 0 \tag{17}$$

$$\overset{\mu\nu}{\underset{\vee}{g}}{}_{,\nu} = 0 \tag{18}$$

where the magnetic current

$$I_{\mu\nu\rho} = \Phi_{\mu\nu,\rho} + \Phi_{\nu\rho,\mu} + \Phi_{\rho\mu,\nu} \tag{19}$$

The intrinsic cosmological parameter r_o is the basic reason for the unification of the electromagnetic and the gravitational forces. For the infinitely large values of r_o the field equations reduce to equations for the flat space-time and therefore we can assume that the radius of the universe can be represented by r_o. However, in the limit of r_o tending to zero the field equations reduce to the field equations of general relativity. This latter limiting process is just the *correspondence principle* of the theory and it is represented by the fact that the parameters r_o and q are related by

$$r_o^2 q^2 = \frac{c^4}{2G}, \tag{20}$$

which was first derived in 1950 when I was a graduate student in Cambridge.

ATTRACTIVE AND REPULSIVE GRAVITATIONAL FORCES AS A DUALITY OF THE GRAVITATIONAL FIELD

All the early attempts (1918-1950) sought a unification of electromagnetism and gravitation. The reasons for attempting to unify only the two long-range fields were based on the then known facts that either the scantily understood short-range weak and strong interactions were not amenable to the proposed schemes of unification of force fields, or the latter forces could result from the unification of gravitation and electromagnetism. In fact failure was the common fate of all these past attempts. In view of these failures to reach an acceptable unified field theory, the physicists assume *ab initio* that an approach based on a generalization of the relativistic theory of gravitation is somewhat solitary; thus this kind of approach may not be appreciated by some of the experts in current theories of elementary particle physics. This work is an effort to start a new path by casting the author's nonsymmetric generalization of the gravitation theory (1950-1952), versus those of Einstein and Schrödinger, within some of the mathematical methods and techniques of elementary particle physics.

In a unified theory the electromagnetic field that carries energy and momentum acts, therefore as a source of the gravitational field. Experimentally in some cosmic distances there exist charged particle acceleration in the presence of a super massive black hole powerful gravity force far above the capability of any accelerator we can build on Earth. These particles will emit high-energy gamma rays bursts, which will emerge in various directions depending on the location of the black hole. It thus appears that powerful gravity forces can explicitly act as the source of an electromagnetic field. This is a good example of unification of gravitation and electromagnetism. Thus electromagnetism and gravitation can, especially at high energies, act as sources to each other.

The present approach intends to analyze for example, the place of electric charge in the theory as compared to the role of say, magnetic charge. What is the role of a free electron? A free electron as may be expected to interact in such a way that it allows condensation of magnetic charges around the electron and confine it in some configuration of magnetic monopoles. At this point it seems quite natural to place the entire cosmology on a *dualist philosophy* where matter and quintessence, the latter sourced by the presence of magnetic charges, constitute a *dual system* generating attractive and repulsive gravitational forces, respectively. Furthermore, it is the difference between opposing gravitational forces of the dual system that governs the expansion of the universe.

The five most important advances, amongst others, in cosmology in the past six decades after Hubble's observation in 1929 of the expanding universe, and after George Gamow's theory of the Big-Bang creation of the universe, include, in chronological order: (1) Ralph A. Alpher and Robert Herman's theoretical prediction in the 1940's of the cosmic microwave background radiation (CMBR) left over from the Big-Bang; (2) Arno A. Penzias and Robert W. Wilson's observation in 1964 in the residual heat detected as the CMBR; (3) Alan H. Guth's hypothesis in 1979 of an inflationary universe; (4) the observation in 1992 of microwave anisotropies in the CMBR as seen through COBE by physicists and cosmologists led by George Smoot; (5) The recent data on distant light from stars, exploded before the sun was born, gives support to the idea of accelerated expansion of the universe which may be destined to continue forever.

A further advancement in the physics of gravity refers to the long time unrealized fact that gravity, like electrodynamics, is both an attractive and a repulsive force. This property of gravity plays, as was demonstrated earlier, the most important role in the expansion of the universe. Furthermore gravity manifests itself not only in attraction and repulsion pertaining to long-range coupling, but it has short-range components of the two short-range gravitational interaction, one is oscillatory. All these facts about gravity are revealed and explained as a result of the existence of a length parameter contained in my version of the generalized theory of gravitation [1]. The Expansion of the universe with increasing acceleration is explained by recalling Einstein's old cosmological constant that he had regretted for introducing it into general theory of relativity. The result of the unified field theory leads to the existence of quintessential matter or just quintessence, an ephemeral substance, which is created in the form of positive and negative magnetic charges and its gravitational interaction, is repulsive.

(1) Behram N. Kursunoglu, "After Einstein and Schrödinger – A New Unified Field Theory", Journal of Physics Essays, Vol. 4, No. 4, pp. 439-518, 1991. See also other references in the same paper.

EQUATION OF STATE AND A QUARTET OF GRAVITATIONAL FORCES

It was a serendipitous observation of the physical significance of a relation, which is now more than forty years old,

$$q_o^2 r_o^2 = \frac{c^4}{2G}, \tag{21}$$

that inspired the writing of this paper, where c and G represent speed of light and gravitational constant, respectively, and where r_o is a "fundamental length" and q_o^2 is the energy density, with q_o having the dimensions of an electric field. The relation (21) was the result of correspondence requirement, where in the r_o=0 limit the generalized theory of gravitation reduces to general relativity in the presence of the electromagnetic field without explicit appearance of the electric and magnetic charges. The energy density q_o^2 in the region of size r_o (Planck-length size) is, at the instant of Big Bang, of the order of 10^{114} ergs.cm^{-3} and it represents the energy density at the beginning of time when the size of the universe was of the order of r_o.

The entire evolution of the birth of the universe and its continuous expansion is governed by this intrinsic cosmological parameter r_o. The presence of r_o has made it possible to answer the question "Why is the Universe expanding?" The fundamental relation (21) does in fact govern the general behavior of the universe including the process of nucleosynthesis and the elementary particles as described by the field equations of the theory[6] for the 16 non-symmetric hermitian field variables $\hat{g}_{\mu\nu} = g_{\mu\nu} + iq_o^{-1}\Phi_{\mu\nu}$ where $g_{\mu\nu}$ and $\Phi_{\mu\nu}$ represent generalized gravitational and generalized electromagnetic fields, respectively. The theory includes also the ***supersymmetric*** transition of the hermitian field variables into the 16 non-symmetric non-hermitian field variables $\hat{g}_{\mu\nu} = g_{\mu\nu} + q_o^{-1}\Phi_{\mu\nu}$, both sets of corresponding field equations reducing in the limit $r_o = 0$, to general relativity.

The current discussion of cosmology, especially to include in general relativity a repulsive gravity, there is almost a religious acceptance of Einstein's cosmological constant. Its adoption blindly is so intense that it now plays the role of an obstacle to further progress in cosmology. It is a constant and it affects the unfolding of gravity only say at one point and it does not embrace the entire evolution of an expanding universe. In this paper I would like to point out that there exists a cosmological intrinsic parameter r_o

which does everything that Einstein's cosmological constant does and it also covers the evolution of the universe from the Big Bang to the present time. The invariant length parameter r_o is the result of the unification of gravitational field with the electromagnetic field. In connection with the unified field theory it yields the results:

(1) The mass of everything from the elementary particles all the way to the mass of the universe, is given by

$$m = \left(\frac{c^2}{2G}\right) r_o \tag{22}$$

(2) In the limit of $r_o = 0$ the field equations of the unified field theory reduces the field equations of the general theory of relativity.

(3) In the limit of $r_o \to \infty$ the field equations correspond to that of flat space-time.

(4) If r_o is interpreted as a Schwarzschild radius then for the mass of the proton we obtain $r_o \sim 10^{-54}$cm while for the entire universe Schwarzschild radius is of the order of 10^{10} light years we obtain for the total mass the value $M_u \sim 10^{33}$ solar masses. For the case of gravitational attraction (G positive) the relation

$$q_o^2\, r_o^2 = \frac{c^4}{2G} \tag{23}$$

shows that G is indeed positive provided r_o and q_o are real. We can however replace them by their imaginary forms ir_o and iq_o, respectively. We still obtain G as a positive quantity thereby no change of force takes place. However the alternative of replacing in the equation (23) the imaginary versions of these quantities by one of them at a time in which case the equation (23) says that G is negative and corresponding gravitational interaction is a repulsive force thus we have obtained four different forces partaking in the action of gravity. *This quartet of repulsive and attractive gravitational forces yield, as can be seen from the field equations of the unified field theory, long range attractive and repulsive gravity forces which are accompanied with short- range attractive or repulsive gravitational forces.*

WHY THE UNIVERSE IS EXPANDING

(a) Some Preliminary Considerations.

In 1974, I was giving a lecture on the "Birth of the Universe" to a joint colloquium of three universities held at Boston University. The tea, as is usual, preceded the colloquium. At one point an elderly gentleman, who looked more like a homeless man, appeared and asked me if I was Professor Kursunoglu, I said, "Yes sir, I am". He then said, "I drove thirty miles to hear this lecture. I hope you will make it interesting". I replied by thanking him for driving thirty miles for my lecture, I was greatly honored, but I could not assess the rating of my lecture before I had delivered it. I then was introduced and gave a sixty-minute talk.

The first question came from the elderly gentleman who I was told was George Wald, a professor emeritus from Harvard University, and was awarded a few years earlier Nobel prize for his work on vision; i.e. – on why and how we see. His question to me was "Why is the Universe expanding?" I was caught off-guard, since I did not think about it before hand. I said to him, "I do not know the answer, but I shall ask Professor Dirac when I go back to Coral Gables. Meanwhile, George Wald offered his own explanation about the expansion of the Universe; he proposed an imperceptively small difference between the unit of positive electric charge and the unit of negative electric charge producing net repulsive force everywhere. I countered by pointing out to him that an excess in one of the charges, say negative predominates over positive, would make it impossible for life to exist since all the molecules could have, however small, a net internal repulsive force and they could not bind long enough to originate and to propagate life.

When a few days later I returned to Coral Gables, I posed the question to Paul Dirac: "Why is the Universe expanding?" Dirac gave an elegant and simple answer by stating that "the Universe is expanding because that is the way it is observed" A more technical answer to the question of what is driving the expansion (measurable in units of r_o) of the universe is presented in what follows.

(b) Quintessential Universe.

The concept of quintessential universe was first introduced in my paper called "Exact solutions for confinement of electric charge via condensation of a spectrum of magnetic charges" which appeared in the 1995 proceedings of the Coral Gables Conference "Unified Symmetry in the Small and in the Large" (pp. 33-52, Plenum Publishing Corporation, New York, 1995, edited by Behram N. Kursunoglu et al.)

At the dawn of the Universe the material contents of the field could have consisted of the free positive and negative magnetic charge units g_n (n=0, 1, 2.....) the quintessential matter. This matter is being created from the vacuum continuously and is subject to monopole condensation which in turn confines electric charges to create quarks and then quarks by combining in the usual way generates elementary particles. In fact the quintessence is the source of repulsive gravitational field. While the ordinary matter itself provides attractive gravitation thus the expansion of the Universe is the result of a continuing competition between repulsive and attractive gravitational forces in which quintessential gravity is the winner. In fact there exists a quartet of gravitational forces. This can be seen in the equation of state of the unified field theory, which I had obtained while a student in Cambridge University in 1950, half a century old.

(c) Equation of State.

In the unified field theory the most general field variables are represented by a non-symmetric tensor

$$\hat{g}_{\mu\nu} = g_{\mu\nu} + q_o^{-1}\Phi_{\mu\nu}, \tag{24}$$

where the parameter q_o is a real or pure imaginary parameter and has the dimensions of an electric field. The compatibility of the field equations for the sixteen-component field variables $\hat{g}_{\mu\nu}$ requires the relationship

$$r_o^2 q_o^2 = \frac{c^4}{2G} \tag{25}$$

where the fundamental invariant intrinsic length parameter r_o represents a new cosmological variable and is not related to Einstein's cosmological constant but it performs all the functions required in the expansion of the universe and assumes values corresponding to the size of elementary particles as well as that of the universe.

In all these one important question should refer to how do we know that there is much more quintessential matter than the existing matter itself. In the first place matter is obtained as a result of continuous creation of quintessential substance from the vacuum. If all these results hold then we have provided a new cosmology where not only cosmic microwave background presence (CMBE) but in addition to which we have the role of the vacuum in the creation of quintessence. In the statement (16) if bothr_o and q_o are real then the gravitational constant G is positive and the corresponding gravitational force is attractive. The same situation can be seen in the case of choosing both parameters purely imaginary where again gravitational interaction is attraction. However if one of the two parameters r_o and q_o are pure imaginary numbers then corresponding gravitational forces are repulsive thus we have a quartet of gravitational forces. In this classification of the gravitational behavior we have two sets of forces 1) one short-range attractive gravitational force 2) long range repulsive gravitational force. This same state of gravitation appears once more where the short range one is repulsive and the long range one is attractive. This is the way things in gravity behave and I believe this theory of expansion of the Universe is based on solid grounds and differs from currently favored theories or models.

QUINTESSENCE AND COSMIC MICROWAVE BACKGROUND

Paul H. Frampton
Department of Physics and Astronomy, University of North Carolina,
Chapel Hill, NC 27599-3255.
frampton@physics.unc.edu

Abstract Analytic formulas for the position of the first acoustic peak in the CMB are derived and discussed. They are generalized to the case of a time-dependent dark energy component and it is shown how the cosmic parameters Ω_M and Λ_Λ, extracted from observations, have an intrinsic uncertainty until one knows whether the dark energy density is, or is not, time dependent.

1. CBR Temperature Anisotropy.

Although the Cosmic Background Radiation (CBR) was first discovered over thirty years ago , the detection of its temperature anisotropy waited until 1992 when the Cosmic Background Explorer (COBE) satellite provided its impressive experimental support[2, 3] for the Big Bang model. In particular, the COBE results were consistent with a scale-invariant spectrum of primordial scalar density perturbations[4, 5, 6, 7] such as might be generated by quantum fluctuations during an inflationary period.[8, 9, 10]

This discovery of temperature anisotropy in the CBR has inspired many further experiments which will be sensitive to smaller angle anisotropies than the COBE satellite was (about 1°). NASA has approved the flight of a satellite mission, the Microwave Anisotropy Probe (MAP) in the year 2000 and ESA has agreed to a more accurate later experiment called the Planck Surveyor. The expected precision of these measurements implies that the angular dependence of the temperature anisotropy will be known sufficiently well that the location of the first acoustic (Doppler) peak, and possibly subsequent ones, will be resolved. Actually, the BOOMERANG[35] data have already provided a good measurement.

The Role of Neutrinos, Strings, Gravity, and Variable Cosmological Constant in Elementary Particle Physics, Edited by Kursunoglu *et al.*, Kluwer Academic/Plenum Publishers, New York, 2001

Although the hot big bang theory is supported by at least three major triumphs: the expansion of the universe, the cosmic background radiation and the nucleosynthesis calculations, it leaves unanswered several questions. The most important unanswered questions are the horizon and flatness issues.

When the CBR last scattered, the age of the universe was about 100,000 years compared to its present age of some 10 billion years. As we shall see, the horizon size at the recombination time subtends now an angle of about (1/208) of π radians. On the celestial sphere there are therefore approximately 40,000 causally disconnected regions. Nevertheless, these different regions have a uniform CBR temperature to an accuracy of better than one part in 10^5. This is the horizon problem.

The flatness problem may be understood from the cosmological equation

$$\frac{k}{R^2} = (\Omega - 1)\frac{\dot{R}^2}{R^2} \tag{1}$$

Evaluating Eq.(1) at an arbitrary time t and dividing by the same relation at the present time $t = t_0$ and using $R \sim \sqrt{t} \sim T^{-1}$ gives

$$(\Omega - 1) = 4H_0^2 t^2 \frac{T^2}{T_0^2}(\Omega_0 - 1) \tag{2}$$

For high densities we write

$$\frac{\dot{R}^2}{R^2} = \frac{8\pi G\rho}{3} = \frac{8\pi G g a T^4}{6} \tag{3}$$

where a is the radiation constant and g is the effective number of degrees of freedom. This leads to the relation between time and temperature, after substituting the numerical values [$a = 7.56\times10^{-9} erg m^{-3} K^{-4}; G/c^2 = 0.742 \times 10^{-30} m/g; H_0 = 100 h_0 km/s/Mpc = 3.25 \times 10^{-18} h_0 s^{-1}$]

$$t(seconds) = (2.42 \times 10^{-6}) g^{-1/2} T_{GeV}^{-2} \tag{4}$$

Combining Eq.(2) with Eq.(4) leads to

$$(\Omega - 1) = 3.64 \times 10^{-21} h_0^2 g^{-1} T_{GeV}^{-2} (\Omega_0 - 1) \tag{5}$$

Given the proximity of Ω_0 to unity, we then deduce that Ω at, for example, $T = 1MeV$ ($t \sim$ 1second) must be equal to one within one part in 10^{14}! Otherwise the resultant cosmology will be incompatible with the present situation of our universe. This extraordinary fine-tuning is the flatness problem.

The goal[11, 12, 13, 14, 15, 16, 17] of the CBR experiments is to measure the temperature autocorrelation function. The fractional temperature perturbation as a function of the direction $\hat{\mathbf{n}}$ is expanded in spherical harmonics

$$\frac{\Delta T(\hat{\mathbf{n}})}{T} = \sum_{lm} a_{(lm)} Y_{lm}(\hat{\mathbf{n}}) \tag{6}$$

and the statistical isotropy and homogeneity of the universe imply that the coefficients have expectation values

$$< (a_{(lm)})^* a_{(l'm')} >= C_l \delta_{ll'} \delta_{mm'} \tag{7}$$

The plot of C_l versus l is expected to reflect oscillations in the baryon-photon fluid at the surface of last scatter. In particular, the first Doppler peak should be at the postion $l_1 = \pi/\Delta\theta$ where $\Delta\theta$ is the angle now subtended by the horizon at the time of the last scattering, namely the recombination time corresponding to a red shift $z_t \sim 1,100$.

The horizon and flatness problems described above can both be solved by the inflation scenario which has the further prediction that $\Omega_0 = 1$ if the cosmological constant vanishes or more generally that $\Omega_0 + \Omega_\Lambda = 1$ if the cosmological constant does not vanish.

The question we address here is restricted to the question of how much the value of l_1 alone - likely to be accurately determined in the next few years - will tell us about the values of the cosmic parameters Ω_0 and Ω_Λ?

In Section 2, the case $\Lambda = 0$ is discussed. In Section 3, there is the more general case; in Section 4 there is discussion of the Figures derived; finally, in Section 5, there is an amplification of the cosmological constant and its generalization to quintessence.

2. The Special Case $\Lambda = 0$, $0 < \Omega_0 < 1$

When the cosmological constant vanishes, the Einstein-Friedmann cosmological equations can be solved analytically (not the case, in general, when $\Lambda \neq 0$). So we shall begin by doing this special case explicitly. It gives rise to the well-known result that the position of the first Doppler peak (partial wave l_1) expected in the partial-wave analysis depending on the present matter-energy density Ω_0 (for $\Lambda = 0$) according to $l_1 \sim 1/\sqrt{\Omega_0}$ [13, 17]. We shall show in the next section how in the general case with $\Lambda \neq 0$ there is a rather serious "comic confusion" in disentangling the value of Ω_0 from the position l_1 of the first Doppler peak. Let us use the metric:

$$ds^2 = dt^2 - R^2[d\Psi^2 + sinh^2\Psi d\theta^2 + sinh^2\Psi sin^2\theta d\phi^2] \quad (8)$$

For a geodesic $ds^2 = 0$ and, in particular,

$$\frac{d\Psi}{dR} = \frac{1}{R} \quad (9)$$

Einstein's equation reads

$$\left(\frac{\dot{R}}{R}\right)^2 = \frac{8\pi}{3}G\rho + \frac{1}{R^2} \quad (10)$$

where we take curvature $k = -1$. Let us define:

$$\Omega_0 = \frac{8\pi G\rho_0}{3H_0^2}; \rho = \rho_0\left(\frac{R_0}{R}\right)^3; a = \Omega_0 H_0^2 R_0^3 \quad (11)$$

Then from Eq.(10) we find that

$$\dot{R}^2 R^2 = R^2 + aR \quad (12)$$

and so it follows that

$$\frac{d\Psi}{dR} = \frac{d\Psi}{dt}\left(\frac{dR}{dt}\right)^{-1} = \frac{1}{\dot{R}R} = \frac{1}{\sqrt{R^2 + aR}} \quad (13)$$

Since $\Psi_0 = 0$, the value at time t can be computed from the integral

$$\Psi_t = \int_{R_t}^{R_0} \frac{dR}{\sqrt{(R + a/2)^2 - (a/2)^2}} \quad (14)$$

This can be performed easily with the substitution $R = \frac{1}{2}a(coshV - 1)$ to give the result:

$$\Psi_t = cosh^{-1}\left(\frac{2R_0}{a} - 1\right) - cosh^{-1}\left(\frac{2R_t}{a} - 1\right) \quad (15)$$

From Eq.(10) evaluated at $t = t_0$ we see that

$$\frac{1}{a} = \frac{1 - \Omega_0}{R_0\Omega_0} \quad (16)$$

and so, using $sinh(cosh^{-1}x) = \sqrt{x^2 - 1}$ in Eq.(15) gives now

$$sinh\Psi_t = \sqrt{\left(\frac{2(1 - \Omega_0)}{\Omega_0} + 1\right)^2 - 1} - \sqrt{\left(\frac{2(1 - \Omega_0)R_t}{\Omega_0 R_0} + 1\right)^2 - 1} \quad (17)$$

The position of the first Doppler peak depends on the angle subtended by the horizon size at the time t equal to the recombination time. This corresponds to the distance $(H_t)^{-1}$. According to the metric of Eq.(8) the angle subtended is

$$\Delta\theta = \frac{1}{H_t R_t sinh \Psi_t} \tag{18}$$

and the position of the first Doppler peak corresponds to the partial wave l_1 given by

$$l_1 = \frac{\pi}{\Delta\theta} = \pi H_t R_t sinh \Psi_t \tag{19}$$

Now the red-shift at recombination is about $z_t = 1,100 \simeq (R_0/R_t) \gg 1$ so we may approximate in Eq.(17) to find

$$sinh \Psi_t \simeq \frac{2\sqrt{1-\Omega_0}}{\Omega_0} \tag{20}$$

Using $H_t^2 = 8\pi G\rho/3 + 1/R^2 \simeq \Omega_0 h_0^2 (R_0/R)^3$ gives

$$l_1(\Lambda = 0) = \frac{2\pi}{\sqrt{\Omega_0}} z_t^{1/2} \tag{21}$$

In particular, if $\Omega_0 = 1$ and $\Lambda = 0$, one has $l_1 \simeq 208.4$. If l_1 does have this value empirically it will favor this simplest choice, although as we shall see in the following subsection even here the conclusion has ambiguities.

In Fig. 1 of [22] is plotted l_1 versus Ω_0 for the particular case of $\Omega_\Lambda = 0$.

3. The General Case: $0 \leq \Omega_0 < 2; 0 \leq \Omega_\Lambda < 1$

For the general case of $0 \leq \Omega_\Lambda < 2; 0 < \Omega_0 < 1$ we use the more general Einstein cosmological equation:

$$\dot{R}^2 R^2 = -kR^2 + aR + \Lambda R^4/3 \tag{22}$$

where $a = \Omega_0 H_0^2 R_0^3$. We define

$$\Omega_0 = \frac{8\pi G\rho_0}{3H_0^2}; \Omega_\Lambda = \frac{\Lambda}{3H_0^2}; \Omega_C = \frac{-k}{H_0^2 R_0^2} \tag{23}$$

Substituting $R = R_0 r$ and $w = 1/r$ now gives rise to the integral[18] for Ψ_t

$$\Psi_t = \sqrt{\Omega_C} \int_1^\infty \frac{dw}{\sqrt{\Omega_\Lambda + \Omega_C w^2 + \Omega_0 w^4}} \tag{24}$$

in which $\Omega_\Lambda + \Omega_C + \Omega_0 = 1$.

Consider first the case of an open universe $\Omega_C > 0$. Then

$$l_1 = \pi H_t R_t sinh \Psi_t \tag{25}$$

We know that

$$\begin{aligned} H_t^2 &= \left(\frac{\dot{R}_t}{R_t}\right)^2 = \frac{8\pi G\rho}{3} + \frac{\Lambda}{3} + \frac{1}{R_t^2} \\ &= H_0^2 \left[\Omega \left(\frac{R_0}{R_t}\right)^3 + H_0^2 \Omega_\Lambda + \left(\frac{R_0}{R_t}\right)^2 \Omega_C\right] \end{aligned} \tag{26}$$

Since $R_0 \gg R_t$ we may approximate:

$$H_t \simeq \left(\frac{R_0}{R_t}\right)^{3/2} H_0 \sqrt{\Omega_0} \tag{27}$$

and hence

$$H_t R_t = \left(\frac{R_0}{R_t}\right)^{1/2} \sqrt{\frac{\Omega_0}{\Omega_C}} \tag{28}$$

It follows that for this case $\Omega_C > 0$ that

$$l_1 = \pi \sqrt{\frac{\Omega_0}{\Omega_C}} \left(\frac{R_0}{R_t}\right)^{1/2} sinh \left(\sqrt{\Omega_C} \int_1^\infty \frac{dw}{\sqrt{\Omega_\Lambda + \Omega_C w^2 + \Omega_0 w^3}}\right) \tag{29}$$

For the case $\Omega_C < 0 (k = +1)$ we simply replace *sinh* by *sin* in Eq. (29). Finally, for the special case $\Omega_C = 0$, the generalized flat case favored by inflationary cosmologies, Eq.(29) simplifies to:

$$l_1 = \pi \sqrt{\Omega_0} \left(\frac{R_0}{R_t}\right)^{1/2} \int_1^\infty \frac{dw}{\sqrt{\Omega_\Lambda + \Omega_0 w^3}} \tag{30}$$

In Fig. 2 of [22] is plotted the value of l_1 versus Ω_0 for the case $\Omega_C = 0$ (flat spacetime). The contrast with Fig 1 of [22] is clear: whereas l_1 increases with decreasing Ω_0 when $\Omega_\Lambda = 0$ (Fig. 1 of [22]) the opposite behaviour occurs when we constrain $\Omega_\Lambda = 1 - \Omega_0$ (fig.2 of [22]).

With Ω_0 and Ω_Λ unrestricted there are more general results. In Fig. 3 of [22] are displayed iso-l lines on a $\Omega_0 - \Omega_\Lambda$ plot. The iso-l lines are (from right to left) for the values $l_1 = 150, 160, 170, 180, 190, 200, 210, 220,$ $230, 240, 250, 260, 270$ respectively. One can see that from the *position*(l_1) only of the first Doppler peak there remains a serious ambiguity of interpretation without further information.

In Fig, 4 of [22], there is a three dimensional rendition of the value of l_1 versus the two variables Ω_0 and Ω_Λ.

4. Discussion of Cosmic Parameter Ambiguities.

Let us now turn to an interpretation of our Figures, from the point of view of determining the cosmic parameters.

In the case where $\Lambda = \Omega_\Lambda = 0$, Fig.1 of [22] is sufficient. In this case, there is the well-known dependence[13, 17] $l_1 = (208.4)/\sqrt{\Omega_0}$ illustrated in Fig.1. It would be straightforward to determine Ω_0 with an accuracy of a few percent from the upcoming measurements.

Of course there is a strong theoretical prejudice towards $\Lambda = 0$. But no underlying symmetry principle is yet known. If $\Omega_\Lambda \neq 0$, one knows that it is not bigger than order one; this is very many orders of magnitude smaller than expected[19] from the vacuum energy arising in spontaneous breaking of symmetries such as the electroweak group $SU(2) \times U(1)$.

Nevertheless, recent observations of high redshift Type 1a supernovae have led to the suggestion of an *increasing* Hubble parameter [20, 21]. An interpretation of this is that the cosmological constant is non-zero, possibly $\Omega_\Lambda \simeq 0.7$ but is still consistent with $\Omega_0 = 1 - \Omega_\Lambda$. These results are certainly enough to motivate a full consideration of non-zero values of Ω_Λ.

Thus we come to Fig. 2 of [22] which depicts the Ω_0 dependence of l_1 when $\Omega_0 + \Omega_\Lambda = 1$ is held fixed as in a generalized flat cosmology that could arise from inflation. We notice that here l_1 *decreases* as Ω_0 decreases from $\Omega_0 = 1$, the opposite behaviour to Fig. 1 of [22]. Thus even the shift of l_1 from $l_1 = 208.4$ depends on the size of Λ.

It is therefore of interest to find what are the contours of constant l_1 in the $\Omega_0 - \Omega_\Lambda$ plane. These iso-l lines are shown in Fig. 3 of [22]for $l_1 = 150, ..., 270$ in increments $\Delta l_1 = 10$. If we focus on the $l_1 = 210$ contour (the seventh contour from the left in Fig. 3 of [22]) as an example, we see that while this passes close to the $\Omega_0 = 1, \Lambda = 0$ point it also tracks out a line naturally between those shown in Figs. 1 and 2 (actually somewhat closer to the latter than the former).

Fig. 4 of [22] gives a three-dimensional rendition which includes the Figures 1-3 of [22] as special cases and provides a visualisation of the full functional dependence of $l_1(\Omega_0, \Omega_\Lambda)$.

Our main conclusion is that the position l_1 of the first Doppler peak will define the correct contour in our iso-l plot, Fig. 3 of [22]. More information will be necessary to determine Ω_0 and the validity of inflation.

5. The Cosmological Constant Reconsidered.

Our knowledge of the universe has changed dramatically even in the last few years. Not long ago the best guess, inspired partially by inflation, for the makeup of the present cosmological energy density was $\Omega_m = 1$ and $\Omega_\Lambda = 0$. However, the recent experimental data on the cosmic background radiation and the high - Z (Z = red shift) supernovae strongly suggest that both guesses were wrong. Firstly $\Omega_m \simeq 0.3 \pm 0.1$. Second, and more surprisingly, $\Omega_\Lambda \simeq 0.7 \pm 0.2$. The value of Ω_Λ is especially unexpected for two reasons: it is non-zero and it is ≥ 120 orders of magnitude below its "natural" value.

The fact that the present values of Ω_m and Ω_Λ are of comparable order of magnitude is a "cosmic coincidence" if Λ in the Einstein equation

$$R_{\mu\nu} - \frac{1}{2}g_{\mu\nu}R = 8\pi G_N T_{\mu\nu} + \Lambda g_{\mu\nu}$$

is constant. Extrapolate the present values of Ω_m and Ω_Λ back, say, to redshift $Z = 100$. Suppose for simplicity that the universe is flat $\Omega_C = 0$ and that the present cosmic parameter values are $\Omega_m = 0.300...$ exactly and $\Omega_\Lambda = 0.700...$ exactly. Then since $\rho_m \propto R(t)^{-3}$ (we can safely neglect radiation), we find that $\Omega_m \simeq 0.9999..$ and $\Omega_\Lambda \simeq 0.0000..$ at $Z = 100$. At earlier times the ratio Ω_Λ/Ω_m becomes infinitesimal. There is nothing to exclude these values but it does introduce a second "flatness" problem because, although we can argue for $\Omega_m + \Omega_\Lambda = 1$ from inflation, the comparability of the present values of Ω_m and Ω_Λ cries out for explanation.

In the present Section 5 we shall consider a specific model of quintessence. In its context we shall investigate the position of the first Doppler peak in the Cosmic Microwave Background (CMB) analysis using results published by two of us with Rohm earlier[22]. Other works on the study of CMB include[23, 24, 25, 26]. We shall explain some subtleties of the derivation given in [22] that have been raised since its publication mainly because the formula works far better than its expected order-of-magnitude accuracy. Data on the CMB have been provided recently in [27, 28, 29, 30, 31, 32, 33, 34] and especially in [35].

The combination of the information about the first Doppler peak and the complementary analysis of the deceleration parameter derived from observations of the high-red-shift supernovae[36, 37] leads to fairly precise values for the cosmic parameters Ω_m and Ω_Λ. We shall therefore also investigate the effect of quintessence on the values of these parameters.

In [22], by studying the geodesics in the post-recombination period a formula was arrived at for the position of the first Doppler peak, l_1. For example, in the case of a flat universe with $\Omega_C = 0$ and $\Omega_M + \Omega_\Lambda = 1$ and for a conventional cosmological constant:

$$l_1 = \pi \left(\frac{R_t}{R_0}\right) \left[\Omega_M \left(\frac{R_0}{R_t}\right)^3 + \Omega_\Lambda\right]^{1/2} \int_1^{\frac{R_0}{R_t}} \frac{dw}{\sqrt{\Omega_M w^3 + \Omega_\Lambda}} \tag{31}$$

If $\Omega_C < 0$ the formula becomes

$$\begin{aligned} l_1 &= \frac{\pi}{\sqrt{-\Omega_C}} \left(\frac{R_t}{R_0}\right) \left[\Omega_M \left(\frac{R_0}{R_t}\right)^3 + \Omega_\Lambda + \Omega_C \left(\frac{R_0}{R_t}\right)^2\right]^{1/2} \times \\ &\times \sin\left(\sqrt{-\Omega_C} \int_1^{\frac{R_0}{R_t}} \frac{dw}{\sqrt{\Omega_M w^3 + \Omega_\Lambda}}\right) \end{aligned} \tag{32}$$

For the third possibility of a closed universe with $\Omega_C > 0$ the formula is:

$$\begin{aligned} l_1 &= \frac{\pi}{\sqrt{\Omega_C}} \left(\frac{R_t}{R_0}\right) \left[\Omega_M \left(\frac{R_0}{R_t}\right)^3 + \Omega_\Lambda + 5\Omega_C \left(\frac{R_0}{R_t}\right)^2\right]^{1/2} \times \\ &\times \sinh\left(\sqrt{\Omega_C} \int_1^{\frac{R_0}{R_t}} \frac{dw}{\sqrt{\Omega_M w^3 + \Omega_\Lambda}}\right) \end{aligned} \tag{33}$$

The use of these formulas gives iso-l_1 lines on a $\Omega_M - \Omega_\Lambda$ plot in $25 \sim 50\%$ agreement with the corresponding results found from computer code. On the insensitivity of l_1 to other variables, see[38, 39]. The derivation of these formulas was given in [22]. Here we add some more details.

The formula for l_1 was derived from the relation $l_1 = \pi/\Delta\theta$ where $\Delta\theta$ is the angle subtended by the horizon at the end of the recombination transition. Let us consider the Legendre integral transform which has as integrand a product of two factors, one is the temperature autocorrelation function of the cosmic background radiation and the other factor is a Legendre polynomial of degree l. The issue is what is the lowest integer l for which the two factors reinforce to create the doppler peak? For small l there is no reinforcement because the horizon at recombination subtends a small angle about one degree and the CBR fluctuations average to zero in the integral of the Legendre transform. At large l the Legendre polynomial itself fluctuates with almost equispaced nodes and antinodes. The node-antinode spacing over which the Legendre polynomial varies from zero to a local maximum in magnitude is, in terms of angle, on average π divided by l. When this angle coincides with the angle subtended by the last-scattering horizon, the fluctuations of the two integrand factors are, for the first time with increasing l, synchronized and reinforce (constructive interference) and the corresponding partial wave coefficient is larger than for slightly smaller or slightly larger l. This explains the occurrence of π in the equation for the l_1 value of the first doppler peak written as $l_1 = \pi/\Delta\theta$.

Another detail concerns whether to use the photon or acoustic horizon, where the former is $\sqrt{3}$ larger than the latter? If we examine the evolution of the recombination transition given in [40] the degree of ionization is 99% at $5,000^0$K (redshift $Z = 1,850$) falling to 1% at $3,000^0$K ($Z = 1,100$). One can see qualitatively that during the recombination transition the fluctuation can grow. The agreement of the formula for l_1, using the photon horizon, with experiment shows phenomenologcally that the fluctuation does grow during the recombination transition and that is why there is no full factor of $\sqrt{3}$, as would arise using the acoustic horizon, in its numerator. When we look at the CMBFAST code below, we shall find a factor in l_1 of ~ 1.22, intermediate between 1 (optical) and $\sqrt{3}$ (acoustic).

To introduce our quintessence model as a time-dependent cosmological term, we start from the Einstein equation:

$$R_{\mu\nu} - \frac{1}{2}Rg_{\mu\nu} = \Lambda(t)g_{\mu\nu} + 8\pi G T_{\mu\nu} = 8\pi G \mathcal{T}_{\mu\nu} \tag{34}$$

where $\Lambda(t)$ depends on time as will be specified later and $T^\mu_\nu = \text{diag}(\rho, -p, -p,$ Using the Robertson-Walker metric, the '00' component of Eq.(34) is

$$\left(\frac{\dot{R}}{R}\right)^2 + \frac{k}{R^2} = \frac{8\pi G\rho}{3} + \frac{1}{3}\Lambda \tag{35}$$

while the 'ii' component is

$$2\frac{\ddot{R}}{R} + \frac{\dot{R}^2}{R^2} + \frac{k}{R^2} = -8\pi G p + \Lambda \tag{36}$$

Energy-momentum conservation follows from Eqs.(35,36) because of the Bianchi identity $D^\mu(R_{\mu\nu} - \frac{1}{2}g_{\mu\nu}) = D^\mu(\Lambda g_{\mu\nu} + 8\pi G T_{\mu\nu}) = D^\mu \mathcal{T}_{\mu\nu} = 0$.

Note that the separation of $\mathcal{T}_{\mu\nu}$ into two terms, one involving $\Lambda(t)$, as in Eq(10), is not meaningful except in a phenomenological sense because of energy conservation.

In the present cosmic era, denoted by the subscript '0', Eqs.(35,36) become respectively:

$$\frac{8\pi G}{3}\rho_0 = H_0^2 + \frac{k}{R_0^2} - \frac{1}{3}\Lambda_0 \tag{37}$$

$$-8\pi G p_0 = -2q_0 H_0^2 + H_0^2 + \frac{k}{R_0^2} - \Lambda_0 \tag{38}$$

where we have used $q_0 = -\frac{\ddot{R}_0}{R_0 H_0^2}$ and $H_0 = \frac{\dot{R}_0}{R_0}$.

For the present era, $p_0 \ll \rho_0$ for cold matter and then Eq.(38) becomes:

$$q_0 = \frac{1}{2}\Omega_M - \Omega_\Lambda \tag{39}$$

where $\Omega_M = \frac{8\pi G \rho_0}{3H_0^2}$ and $\Omega_\Lambda = \frac{\Lambda_0}{3H_0^2}$.

Now we can introduce the form of $\Lambda(t)$ we shall assume by writing

$$\Lambda(t) = bR(t)^{-P} \tag{40}$$

where b is a constant and the exponent P we shall study for the range $0 \leq P < 3$. This motivates the introduction of the new variables

$$\tilde{\Omega}_M = \Omega_M - \frac{P}{3-P}\Omega_\Lambda, \quad \tilde{\Omega}_\Lambda = \frac{3}{3-P}\Omega_\Lambda \tag{41}$$

It is unnecessary to redefine Ω_C because $\tilde{\Omega}_M + \tilde{\Omega}_\Lambda = \Omega_M + \Omega_\Lambda$. The case $P = 2$ was proposed, at least for late cosmological epochs, in [41].

The equations for the first Doppler peak incorporating the possibility of non-zero P are found to be the following modifications of Eqs.(31,32,33). For $\Omega_C = 0$

$$l_1 = \pi\left(\frac{R_t}{R_0}\right)\left[\tilde{\Omega}_M\left(\frac{R_0}{R_t}\right)^3 + \tilde{\Omega}_\Lambda\left(\frac{R_0}{R_t}\right)^P\right]^{1/2}\int_1^{\frac{R_0}{R_t}}\frac{dw}{\sqrt{\tilde{\Omega}_M w^3 + \tilde{\Omega}_\Lambda w^P}} \tag{42}$$

If $\Omega_C < 0$ the formula becomes

$$\begin{aligned} l_1 &= \frac{\pi}{\sqrt{-\Omega_C}}\left(\frac{R_t}{R_0}\right)\left[\tilde{\Omega}_M\left(\frac{R_0}{R_t}\right)^3 + \tilde{\Omega}_\Lambda\left(\frac{R_0}{R_t}\right)^P + \Omega_C\left(\frac{R_0}{R_t}\right)^2\right]^{1/2} \times \\ &\quad \times \sin\left(\sqrt{-\Omega_C}\int_1^{\frac{R_0}{R_t}}\frac{dw}{\sqrt{\tilde{\Omega}_M w^3 + \tilde{\Omega}_\Lambda w^P + \Omega_C w^2}}\right) \end{aligned} \tag{43}$$

For the third possibility of a closed universe with $\Omega_C > 0$ the formula is:

$$\begin{aligned} l_1 &= \frac{\pi}{\sqrt{\Omega_C}}\left(\frac{R_t}{R_0}\right)\left[\tilde{\Omega}_M\left(\frac{R_0}{R_t}\right)^3 + \tilde{\Omega}_\Lambda\left(\frac{R_0}{R_t}\right)^P + \Omega_C\left(\frac{R_0}{R_t}\right)^2\right]^{1/2} \times \\ &\quad \times \sinh\left(\sqrt{\Omega_C}\int_1^{\frac{R_0}{R_t}}\frac{dw}{\sqrt{\tilde{\Omega}_M w^3 + \tilde{\Omega}_\Lambda w^P + \Omega_C w^2}}\right) \end{aligned} \tag{44}$$

The dependence of l_1 on P is illustrated for constant $\Omega_M = 0.3$ in Fig. 1(a), and for the flat case $\Omega_C = 0$ in Fig. 1(b). (These figures are in [43]) For illustration we have varied $0 \leq P < 3$ but as will become clear later in the paper (see Fig 3 of [43] discussed below) only the much more restricted range $0 \leq P < 0.2$ is possible for a fully consistent cosmology when one considers evolution since the nucleosynthesis era.

We have introduced P as a parameter which is real and with $0 \leq P < 3$. For $P \to 0$ we regain the standard cosmological model. But now we must investigate other restrictions already necessary for P before precision cosmological measurements restrict its range even further.

Only for certain P is it possible to extrapolate the cosmology consistently for all $0 < w = (R_0/R) < \infty$. For example, in the flat case $\Omega_C = 0$ which our universe seems to approximate[35], the formula for the expansion rate is

$$\frac{1}{H_0^2}\left(\frac{\dot{R}}{R}\right)^2 = \tilde{\Omega}_M w^3 + \tilde{\Omega}_\Lambda w^P \tag{45}$$

This is consistent as a cosmology only if the right-hand side has no zero for a real positive $w = \hat{w}$. The root $\hat{w}$ is

$$\hat{w} = \left(\frac{3(1-\Omega_M)}{P - 3\Omega_M}\right)^{\frac{1}{3-P}} \tag{46}$$

If $0 < \Omega_M < 1$, consistency requires that $P < 3\Omega_M$.

In the more general case of $\Omega_C \neq 0$ the allowed regions of the $\Omega_M - \Omega_\Lambda$ plot for $P = 0, 1, 2$ are displayed in Fig. 2 of [43].

We see from Eq.(46) that if we do violate $P < 3\Omega_M$ for the flat case then there is a $\hat{w} > 0$ where the cosmology undergoes a bounce, with $\dot{R} = 0$ and $\dot{R}$ changing sign. This necessarily arises because of the imposition of $D^\mu T_{\mu\nu} = 0$ for energy conservation. For this example it occurs in the past for $\hat{w} > 1$. The consistency of big bang cosmology back to the time of nucleosynthesis implies that our universe has not bounced for any $1 < \hat{w} < 10^9$. It is also possible to construct cosmologies where the bounce occurs in the future! Rewriting Eq.(46) in terms of Ω_Λ:

$$\hat{w} = \left(\frac{3\Omega_\Lambda}{3\Omega_\Lambda - (3-P)}\right)^{\frac{1}{3-P}} \tag{47}$$

If $P < 3$, then any $\Omega_\Lambda < 0$ will lead to a solution with $0 < \hat{w} < 1$ corresponding to a bounce in the future. If $P > 3$ the condition for a future bounce is $\Omega_\Lambda < -\left(\frac{P-3}{3}\right)$. What this means is that for the flat case $\Omega_C = 0$ with quintessence $P > 0$ it is possible for the future cosmology to be qualitatively similar to a non-quintessence closed universe where $\dot{R} = 0$ at a finite future time with a subsequent big crunch.

Another constraint on the cosmological model is provided by nucleosynthesis which requires that the rate of expansion for very large w does not differ too much from that of the standard model.

The expansion rate for $P = 0$ coincides for large w with that of the standard model so it is sufficient to study the ratio:

$$(\dot{R}/R)^2_P/(\dot{R}/R)^2_{P=0} \overset{w\to\infty}{\to} (3\Omega_M - P)/((3-P)\Omega_M) \tag{48}$$

$$\overset{w\to\infty}{\to} (4\Omega_R - P)/((4-P)\Omega_R) \tag{49}$$

where the first limit is for matter-domination and the second is for radiation-domination (the subscript R refers to radiation).

The overall change in the expansion rate at the BBN era is therefore

$$(\dot{R}/R)^2_P/(\dot{R}/R)^2_{P=0} \overset{w\to\infty}{\to} (3\Omega_M - P)/((3-P)\Omega_M) \times \times (4\Omega_R^{trans} - P)/((4-P)\Omega_R^{trans}) \tag{50}$$

where the superscript "trans" refers to the transition from radiation domination to matter domination. Putting in the values $\Omega_M = 0.3$ and $\Omega_R^{trans} = 0.5$ leads to $P < 0.2$ in order that the acceleration rate at BBN be within 15% of its value in the standard model, equivalent to the contribution to the expansion rate at BBN of one chiral neutrino flavor.

Thus the constraints of avoiding a bounce ($\dot{R} = 0$) in the past, and then requiring consistency with BBN leads to $0 < P < 0.2$.

We may now ask how this restricted range of P can effect the extraction of cosmic parameters from observations. This demands an accuracy which has fortunately begun to be attained with the Boomerang data [35]. If we choose $l_1 = 197$ and vary P as $P = 0, 0.05, 0.10, 0.15, 0.20$ we find in the enlarged view of Fig 3 in [43] that the variation in the parameters Ω_M and Ω_Λ can be as large as $\pm 3\%$. To guide the eye we have added the line for deceleration parameter $q_0 = -0.5$ as suggested by [36, 37]. In the next decade, inspired by the success of Boomerang (the first paper of true precision cosmology) surely the sum $(\Omega_M + \Omega_\Lambda)$ will be examined at much better than $\pm 1\%$ accuracy, and so variation of the exponent of P will provide a useful parametrization of the quintessence alternative to the standard cosmological model with constant Λ.

Clearly, from the point of view of inflationary cosmology, the precise vanishing of $\Omega_C = 0$ is a crucial test and its confirmation will be facilitated by comparison models such as the present one.

We have also studied the use of the public code CMBFAST[42] and how its normalization compares to that in [22]. For example, with P = 0 and $\Omega_\Lambda = 0.3, h_{100} = 0.65$ we find using CMBFAST that

$\Omega_\Lambda = 0.5, l_1 = 284$ ($l_1 = 233$ from [22])
$\Omega_\Lambda = 0.6, l_1 = 254$ ($l_1 = 208$ from [22])
$\Omega_\Lambda = 0.7, l_1 = 222$ ($l_1 = 182$ from [22])
$\Omega_\Lambda = 0.8, l_1 = 191$ ($l_1 = 155$ from [22])

The CMBFAST l_1 values are consistently ~ 1.22 times the l_1 values from [22]. As mentioned earlier, this normalization is intermediate between that for the acoustic horizon ($\sqrt{3}$) and the photon horizon (1).

Finally, we remark that the quintessence model considered here is in the right direction to ameliorate the "age problem" of the universe. Taking the age as 14.5Gy for $\Omega_M = 0.3$, $\Omega_C = 0$ and $h_{100} = 0.65$ the age increases monotonically with P. It reaches slightly over 15 Gy at the highest-allowed value $P = 0.2$. This behavior is illustrated in Fig 4 of [43] which assumes $\Omega_M = 0.3$ and flatness as P is varied.

Acknowledgments

This work was supported in part by the US Department of Energy under Grant No. DE-FG02-97ER-41036.

References

[1] A.A. Penzias and R.W. Wilson, Ap. J. **142,** 419 (1965).

[2] G.F. Smoot *et al.*, Ap. J. Lett. **396,** L1 (1992).

[3] K. Ganga *et al.*, Ap.J. **410,** L57 (1993).

[4] J.M. Bardeen, P.J. Steinhardt and M.S. Turner, Phys. Rev. **D28,** 679 (1983).

[5] A.A. Starobinsky, Phys. Lett. **B117,** 175 (1982).

[6] A.H. Guth and S.-Y. Pi, Phys. Rev. Let. **49,** 1110 (1982).

[7] S.W. Hawking, Phys. Lett. **B115,** 295 (1982).

[8] A.H. Guth, Phys. Rev. **D28,** 347 (1981).

[9] A.D. Linde, Phys. Lett. **B108,** 389 (1982).

[10] A. Albrecht and P.J. Steinhardt, Phys. Rev. Lett. **48,** 1220 (1982).

[11] R.L. Davis, H.M. Hodges, G.F. Smoot, P.J. Steinhardt and M.S. Turner, Phys. Rev. Lett. **69**, 1856 (1992).

[12] J.R. Bond, R. Crittenden, R.L. Davis, G. Efstathiou and P.J. Steinhardt, Phys. Rev. Lett. **72**, 13 (1994).

[13] P.J. Steinhardt, Int. J. Mod. Phys. **A10**, 1091 (1995).

[14] A. Kosowsky, M. Kamionkowski, G. Jungman and D.N. Spergel, *Determining Cosmological Parameters from the Microwave Background.* Talk at 2nd Symposium on Critique of the Sources of Dark Matter in the Universe, Santa Monica, CA, Feb 14-16, 1996; Nucl. Phys. Proc. Suppl. **51B**, 49 (1996).

[15] M. Kamionkowski and A. Loeb, Phys. Rev. **D56**, 4511 (1997).

[16] M. Kamionkowski, *Cosmic Microwave Background Tests of Inflation.* To be published in Proceedings of the 5th International Workshop on Topics in Astroparticle and Underground Physi cs (TAUP97), Gran Sasso, Italy, September 7-11, 1997.

[17] M. Kamionkowski, *Cosmological-Parameter Determination with Cosmic Microwave Background Temperature Anisotropies and Polarizat ion.* To be published in the Proceedings of the 33rd rencontres de Moriond: Fundamental Parameters in Cosmology, Les Arcs, France, January 15-24, 1998. *astro-ph/9803168.*

[18] We note the similarity to the formula for the age of the universe:

$$t_0 = \frac{1}{H_0} \int_1^\infty \frac{dw}{w\sqrt{\Omega_\Lambda + \Omega_C w^2 + \Omega_0 w^3}} \tag{51}$$

which is also an elliptic integral that cannot be done analytically.

[19] For a review of the cosmological constant problem see, for example, S. Weinberg, Rev. Mod. Phys. **61**, 1 (1989); Y.J. Ng, Int. J. Mod. Phys. **D1**, 145 (1992).

[20] S.J. Perlmutter *et al.*,(The Supernova Cosmology Project) *Discovery of a Supernova Explosion at Half the Age of the Universe and Its Cosmological Implications. astro-ph/9712212.*

[21] S.J. Perlmutter *et al.*,(The Supernova Cosmology Project) *astro-ph/9608192.*

[22] P.H. Frampton, Y.J. Ng and R.M Rohm, Mod. Phys. Lett **A13**, 2541 (1998). **astro-ph/9806118.**

[23] M. Kamionkowski and A. Kosowsky, Ann. Rev. Nucl. Part. Sci. **49**, 77 (1999). M. Kamionkowski, Science **280**, 1397 (1998).

[24] J.R. Bond, in *Cosmology and Large Scale Structure.* Editors: R.Schaeffer *et al.* Elsevier Science, Amsterdam (1996).page 469.

[25] C.L. Bennett, M. Turner and M. White, Physics Today, November 1997. page 32

[26] C.R. Lawrence, D. Scott and M. White, PASP **111**, 525 (1999).

[27] C.H. Lineweaver, Science **284**, 1503 (1999).

[28] S. Dodelson and L. Knox, Phys. Rev. Lett. **84**, 3523 (2000).

[29] A. Melchiorri, *et al.* Ap.J. (in press). **astro-ph/9911444.**

[30] E. Pierpaoli, D. Scott and M. White, Science **287,** 2171 (2000).

[31] G. Efstathiou, to appear in Proceedings of NATO ASI: Structure Formation in the Universe, Editors: N. Turok, R . Crittenden. astro-ph/0002249.

[32] M. Tegmark and M. Zaldarriaga. astro-ph/0002091.

[33] O. Lahav *et al.* astro-ph/9912105.

[34] M. Le Dour *et al.* astro-ph/0004282.

[35] P. Bernardis, *et al.* (Boomerang experiment). Nature **404,** 955 (2000).

[36] S. Perlmutter *et al* (Supernova Cosmology Project). Nature **391,** 51 (1998).

[37] A.G. Reiss *et al.* Astron. J. **116,** 1009 (1998).
C.J. Hogan, R.P. Kirshner and N.B. Suntzeff, Sci. Am. **280,** 28 (1999).

[38] W. Hu and M. White, Phys. Rev. Lett. **77,** 1687 (1996).

[39] W. Hu and M. White, Ap. J. **471,** 30 (1996).

[40] P.J.E. Peebles, Ap.J. **153,** 1 (1968).

[41] R.D. Sorkin, Int. J. Th. Phys. **36,** 2759 (1997);
W. Chen and Y.S. Wu, Phys. Rev. **D41,** 695 (1990).

[42] See e.g. http://www.sns.ias.edu/~
matiasz/CMBFAST/cmbfast.html

[43] J.L. Crooks, J.O Dunn, P.H. Frampton, Y.J. Ng and R.M. Rohm, Mod. Phys. Lett. **A** (2001, in press). astro-ph/0010404

COSMIC ACCELERATION AND A NATURAL SOLUTION TO THE COSMOLOGICAL CONSTANT PROBLEM

Philip D. Mannheim*

We trace the origin of the cosmological constant problem to the assumption that Newton's constant G sets the scale for cosmology. And then we show that once this assumption is relaxed (so that the local G as measured in a local Cavendish experiment is no longer to be associated with global cosmology), the very same cosmic acceleration which has served to make the cosmological constant problem so very severe instead then serves to provide us with its potential resolution. In addition, we present an alternate cosmology, one based on conformal gravity (a theory which explicitly possesses no fundamental G), and show that once given only that the sign of the vacuum energy density Λ is explicitly the negative one associated with the spontaneous breakdown of the scale invariance of the conformal gravity theory (this actually being the choice of sign for Λ which precisely leads to cosmic acceleration in the conformal theory), then that alone, no matter how big Λ might actually be in magnitude, is sufficient to not only make its measurable contribution to current era cosmology naturally be of order one today, but to even do so in a way which is completely compatible with the recent high z supernovae cosmology data. Cosmology can thus live with either a fundamental G or with the large (and even potentially negative) Λ associated with elementary particle physics phase transitions but not with both. In an appendix we distinguish between the thermodynamic free energy and the thermodynamic internal energy, with it being the former which determines cosmological phase transitions and the latter which is the source of the gravitational field. Then we show that, even if we make the standard ad hoc assumption that Λ actually is quenched in standard gravity, vacuum energy density is nonetheless still found to dominate standard cosmology at the time of the phase transition which produced it. However, within conformal gravity no difficulty of this type is encountered.

*Department of Physics, University of Connecticut, Storrs, CT 06269; Email address: mannheim@uconnvm.uconn.edu. Report No. gr-qc/9903005 v2, February 15, 2001.

The Role of Neutrinos, Strings, Gravity, and Variable Cosmological Constant in Elementary Particle Physics, Edited by Kursunoglu *et al.*, Kluwer Academic/Plenum Publishers, New York, 2001

1 ORIGIN OF THE COSMOLOGICAL CONSTANT PROBLEM

The recent discovery [1, 2] of a cosmic acceleration has made the already extremely disturbing cosmological constant problem even more vexing than before. Specifically, a phenomenological fitting to the new high z supernovae Hubble plot data using the standard Einstein-Friedmann cosmological evolution equations

$$\dot{R}^2(t) + kc^2 = \dot{R}^2(t)(\Omega_M(t) + \Omega_\Lambda(t)) \quad (1)$$

$$\Omega_M(t) + \Omega_\Lambda(t) + \Omega_k(t) = 1 \quad (2)$$

$$q(t) = (n/2 - 1)\Omega_M(t) - \Omega_\Lambda(t) = (n/2 - 1)(1 + kc^2/\dot{R}^2(t)) - n\Omega_\Lambda(t)/2 \quad (3)$$

where $\Omega_M(t) = 8\pi G\rho_M(t)/3c^2H^2(t)$ is due to ordinary matter (viz. matter for which $\rho_M(t) = A/R^n(t)$ where $A > 0$ and $3 \leq n \leq 4$), where $\Omega_\Lambda(t) = 8\pi G\Lambda/3cH^2(t)$ is due to a cosmological constant Λ and where $\Omega_k(t) = -kc^2/\dot{R}^2(t)$ is due to the spatial 3-curvature k, has revealed that not only must the current era $\Omega_\Lambda(t_0)$ actually be non-zero today, it is even explicitly required to be of order one. Typically, the allowed parameter space compatible with the available data is found to be centered on the line $\Omega_\Lambda(t_0) = \Omega_M(t_0) + 1/2$ or so, with $\Omega_\Lambda(t_0)$ being found to be limited to the range $(1/2, 3/2)$ and (the presumed positive) $\Omega_M(t_0)$ to $(0, 1)$, with the current ($n = 3$) era deceleration parameter $q(t_0) = (n/2-1)\Omega_M(t_0) - \Omega_\Lambda(t_0)$ thus having to approximately lie within the $(-1/2, -1)$ interval. Thus, not only do we find that the universe is currently accelerating, but additionally we see that with there being no allowed $\Omega_\Lambda(t_0) = 0$ solution at all (unless $\Omega_M(t_0)$ could somehow be allowed to go negative), the longstanding problem (see e.g. [3, 4] for some recent reviews) of trying to find some way by which $\Omega_\Lambda(t_0)$ could be quenched by many orders of magnitude from both its quantum gravity and particle physics expectations (perhaps by making it vanish altogether) has now been replaced by the need to find a specific such mechanism which in practice (rather than just in principle) would explicitly put $\Omega_\Lambda(t_0)$ into this very narrow $(1/2, 3/2)$ box.† Further, even independent of any quantum considerations, the new high z data pose a problem for Eq. (1) even when considered purely from the viewpoint of classical physics. Specifically, since the $\Omega_\Lambda(t)/\Omega_M(t)$ ratio evolves as $R^n(t) \sim T^{-n}(t)$, its current closeness to one entails that in the early universe this same ratio would have had to have been fantastically small, with the universe only being able to evolve into its current state if this ratio had been extremely fine tuned in the early universe. Moreover, this particular fine tuning would have to be above and beyond that imposed by the flat inflationary universe model [5] since inflation only constrains the sum of $\Omega_M(t)$ and $\Omega_\Lambda(t)$ to be one and does not fix their ratio. Thus at the present time neither inflationary nor quantum cosmology can readily accommodate the new high z data at all.

In order to try to diagnose the nature of the problem in as general a way as possible, we note that in any cosmology with a big bang, the early universe $\dot{R}(t = 0)$ would have to be divergent (or at least be extremely large), with Eq. (1) then requiring the quantity $(\Omega_M(t = 0) + \Omega_\Lambda(t = 0))$ to be equal to one no matter what the value of the spatial curvature k. Thus, given the radically different

†In an appendix we show how contrived the current situation actually needs to be.

temporal behaviors of $\Omega_M(t)$ and $\Omega_\Lambda(t)$, in standard gravity no cosmology, flat or non-flat, could ever evolve into one in which $\Omega_\Lambda(t_0) \simeq \Omega_M(t_0) \simeq O(1)$ today without extreme fine tuning, to thus bring standard cosmology to a rather severe impasse which challenges its viability. Further, if $\dot{R}(t=0)$ does start off divergent, it must diminish as the universe evolves, with the early universe thus decelerating. Since the current universe now appears to be accelerating, the fine tuning problem can be viewed as the need to adjust parameters in such a way that the cosmology can exhibit diametrically opposite deceleration and acceleration behaviors in differing epochs. Since the big bang singularity itself derives from the fact that standard gravity is always attractive (since G controls gravity on all distance scales including those much larger than the solar system one on which standard gravity was first established), while acceleration is more naturally associated with repulsion (cf. the fits to the supernovae data in which $\Omega_M(t_0)$ actually is taken to be negative), it is thus suggestive that we might be able to more readily balance the early and current universes if there were no initial singularity at all, and if cosmological gravity in fact got to be repulsive in all epochs, with the universe then expanding from some initially hot state characterized by the non-singular $\dot{R}(t=0)=0$ instead. To achieve this would thus appear to require the removal of G from the fundamental gravitational action. Additionally, if $\dot{R}(t=0)$ were indeed to vanish, the initial value of $\Omega_\Lambda(t)$ would then be infinite no matter what, and would thus never require early universe fine tuning. Moreover, continuing in this same vein, we note that for an accelerating universe which actually accelerates indefinitely, then, no matter what may or may not have actually occurred in the early universe, at late times $\dot{R}(t)$ will eventually become arbitrarily large, with Eq. (1) then requiring the quantity $(\Omega_M(t)+\Omega_\Lambda(t))$ to asymptote to one at late times, again independent of the value of k. However, because of their differing time behaviors, we see that in the late universe it would precisely be Ω_Λ which would then have to tend to one no matter what its early universe value. Thus at late enough times the cosmological constant problem would actually get solved, and in fact would get solved by cosmology itself (i.e. no matter how big Λ might actually be, in permanently accelerating universes there will eventually come a time in which the measurable consequence of Ω_Λ will be that it will make a contribution to the expansion of the universe which will be of order one). Thus even while the discovery of cosmic acceleration makes the cosmological constant problem more acute, nonetheless, its very existence also suggests a possible resolution of the issue.

In order to see how we might be able to take advantage of this possibility, it is very instructive to analyze [6] de Sitter geometry in a purely kinematic way which requires no commitment to any particular dynamical equation of motion. Specifically, suppose we know only that a given geometry is de Sitter, i.e. that its Riemann tensor is given by

$$R^{\lambda\rho\sigma\nu} = \alpha(g^{\sigma\rho}g^{\lambda\nu} - g^{\nu\rho}g^{\lambda\sigma}). \quad (4)$$

For such a geometry contraction then yields the kinematic relation

$$R^{\mu\nu} - g^{\mu\nu}R^{\sigma}{}_{\sigma}/2 = 3\alpha g^{\mu\nu}, \quad (5)$$

a relation which reduces to

$$\dot{R}^2(t) + kc^2 = \alpha c^2 R^2(t) \quad (6)$$

when evaluated in the Robertson-Walker coordinate system. On defining $\Omega_\Lambda(t) = \alpha c^2 R^2(t)/\dot{R}^2(t)$ we obtain $-q(t) = \Omega_\Lambda(t) = 1 + kc^2/\dot{R}^2(t)$, with $R(t)$, $q(t)$ and $\Omega_\Lambda(t)$ being found [6] to given by

$$\begin{aligned}
R(t, \alpha < 0, k < 0) &= (k/\alpha)^{1/2} sin((-\alpha)^{1/2} ct), \\
R(t, \alpha = 0, k < 0) &= (-k)^{1/2} ct, \\
R(t, \alpha > 0, k < 0) &= (-k/\alpha)^{1/2} sinh(\alpha^{1/2} ct), \\
R(t, \alpha > 0, k = 0) &= R(t = 0) exp(\alpha^{1/2} ct), \\
R(t, \alpha > 0, k > 0) &= (k/\alpha)^{1/2} cosh(\alpha^{1/2} ct),
\end{aligned} \quad (7)$$

$$\begin{aligned}
\Omega_\Lambda(t, \alpha < 0, k < 0) &= -q(t, \alpha < 0, k < 0) = -tan^2((-\alpha)^{1/2} ct), \\
\Omega_\Lambda(t, \alpha = 0, k < 0) &= -q(t, \alpha = 0, k < 0) = 0, \\
\Omega_\Lambda(t, \alpha > 0, k < 0) &= -q(t, \alpha > 0, k < 0) = tanh^2(\alpha^{1/2} ct), \\
\Omega_\Lambda(t, \alpha > 0, k = 0) &= -q(t, \alpha > 0, k = 0) = 1, \\
\Omega_\Lambda(t, \alpha > 0, k > 0) &= -q(t, \alpha > 0, k > 0) = coth^2(\alpha^{1/2} ct)
\end{aligned} \quad (8)$$

in all of the various allowable cases. As we thus see, when the parameter α is positive, each associated solution corresponds to a permanently accelerating universe, and that in each such universe $\Omega_\Lambda(t)$ will eventually reach one no matter how big the parameter α might be, and independent in fact of whether or not G even appears in the cosmological evolution equations at all. Moreover, while $\Omega_\Lambda(t, \alpha > 0, k > 0)$ will reach one at late times, quite remarkably, $\Omega_\Lambda(t, \alpha > 0, k < 0)$ will be bounded between zero and one at all times, no matter how large α might be. Thus unlike the unbounded $\alpha < 0$ case, we see that when α is greater or equal to zero, $\Omega_\Lambda(t)$ is either bounded at all times or approaches a bound at late times. Late time $\alpha \geq 0$ de Sitter cosmologies will thus always quench the contribution of a cosmological constant to cosmology no matter how large it may be, and thus the key task is to find a cosmology in which the current era is already sufficiently (but not too) late. Since $\Omega_M(t_0)$ is not zero today, the standard cosmology would not immediately appear to be an appropriate candidate, but, as we shall now see, this bounding mechanism will precisely be found to occur in conformal gravity, a theory which has recently been advanced as an alternative to standard gravity and its standard dark matter paradigm, a theory in which G does not in fact set the scale for cosmology.

2 SOLUTION TO THE COSMOLOGICAL CONSTANT PROBLEM

Conformal gravity (viz. gravity based on the locally conformal invariant Weyl action

$$I_W = -\alpha_g \int d^4x(-g)^{1/2} C_{\lambda\mu\nu\kappa} C^{\lambda\mu\nu\kappa} \tag{9}$$

where $C^{\lambda\mu\nu\kappa}$ is the conformal Weyl tensor and where α_g is a purely dimensionless gravitational coupling constant) has recently been advanced as a candidate gravitational theory because it has been found capable of addressing so many of the problems (such as dark matter) which currently afflict standard gravity (see e.g. [6, 7, 8] ‡). Its original motivation was the desire to give gravity a dimensionless coupling constant just like those associated with the three other fundamental interactions. And indeed [9], the local conformal symmetry invoked to do this then not only excludes the existence of any fundamental mass scales such as a fundamental cosmological constant, even after mass scales are induced by spontaneous breakdown of the conformal symmetry, the (still) traceless energy-momentum tensor then constrains any induced cosmological constant term to be of the same order of magnitude as all the other terms in $T^{\mu\nu}$, neither smaller nor larger. Thus, unlike standard gravity, precisely because of its additional symmetry, conformal gravity has a great deal of control over the cosmological constant. Essentially, with all mass scales - of gravity and particle physics both - being jointly generated by spontaneous breakdown of the scale symmetry, conformal gravity knows exactly where the zero of energy is. And, moreover, with the theory possessing no mass scale at all in the absence of symmetry breaking, the very lowering of the vacuum energy needed to actually spontaneously break the scale symmetry in the first place necessarily yields an induced cosmological constant which is expressly negative. Conformal gravity thus has control not merely of the magnitude of the cosmological constant, it even has control of its sign, and it is the purpose of this paper to show that it is this very control which will provide for a completely natural accounting of the new high z supernovae data.

The cosmology associated with conformal gravity was first presented in [10] where it was shown to possess no flatness problem, to thus release conformal cosmology from the need for the copious amounts of cosmological dark matter required of the standard theory. Subsequently [11, 6], the cosmology was shown to also possess no horizon problem, no universe age problem, and, through negative spatial curvature, to naturally lead to cosmic repulsion.§ To discuss conformal cosmology

‡These references also give references to some of the large general literature on conformal gravity as well as provide further details of the work described in this paper.

§In the same way as a standard $k = 0$ spatially flat inflationary cosmology with $\Lambda = 0$ leads to $q(t_0) = 1/2$, the analogous $\Lambda = 0$ conformal cosmology presented in [10] is a necessarily spatially open negatively curved $k < 0$ one with a current era deceleration parameter which obeys $q(t_0) = 0$ [11]. Thus, as had actually been noted well in advance of the recent supernovae data, even without a cosmological constant conformal cosmology already possesses a repulsion not present in the standard theory.

it is convenient to consider the conformal matter action

$$I_M = -\hbar \int d^4x(-g)^{1/2}[S^\mu S_\mu/2 + \lambda S^4 - S^2 R^\mu{}_\mu/12 + i\bar{\psi}\gamma^\mu(x)(\partial_\mu + \Gamma_\mu(x))\psi - gS\bar{\psi}\psi] \tag{10}$$

for generic massless scalar and fermionic fields. For such an action, when the scalar field acquires a non-zero expectation value S_0, the entire energy-momentum tensor of the theory is found (for a perfect matter fluid $T^{\mu\nu}_{kin}$ of fermions) to take the form

$$T^{\mu\nu} = T^{\mu\nu}_{kin} - \hbar S_0^2(R^{\mu\nu} - g^{\mu\nu}R^\alpha{}_\alpha/2)/6 - g^{\mu\nu}\hbar\lambda S_0^4; \tag{11}$$

with the complete solution to the scalar field, fermionic field, and gravitational field equations of motion in a background Robertson-Walker geometry (viz. a geometry in which the Weyl tensor vanishes) then reducing to just one relevant equation, viz.

$$T^{\mu\nu} = 0, \tag{12}$$

a remarkably simple condition which immediately fixes the zero of energy. We thus see that the evolution equation of conformal cosmology looks identical to that of standard gravity save only that the quantity $-\hbar S_0^2/12$ has replaced the familiar $c^3/16\pi G$. This change in sign compared with standard gravity leads to a cosmology in which gravity is globally repulsive rather than attractive, even while local solar system gravity remains attractive (and completely standard [12]) in the conformal theory.[¶] Because of this change in sign, conformal cosmology thus has no initial singularity (i.e. it expands from a finite minimum radius), and is thus precisely released from the standard big bang model constraints described earlier. Similarly, because of this change in sign the contribution of $\rho_M(t)$ to the expansion of the universe is now effectively repulsive, to (heuristically) mesh with the phenomenological high z data fits in which $\Omega_M(t)$ was allowed to go negative. Apart from a change in sign, we see that through S_0 there is also a change in the strength of gravity compared to the standard theory. It is this feature which will now enable us to provide a complete accounting of the high z data.

Given the equation of motion $T^{\mu\nu} = 0$, the ensuing conformal cosmology evolution equation is then found to take the form (on setting $\Lambda = \hbar\lambda S_0^4$)

$$\dot{R}^2(t) + kc^2 = -3\dot{R}^2(t)(\Omega_M(t) + \Omega_\Lambda(t))/4\pi S_0^2 L_{PL}^2 \equiv \dot{R}^2(t)(\bar{\Omega}_M(t) + \bar{\Omega}_\Lambda(t)),$$
$$\bar{\Omega}_M(t) + \bar{\Omega}_\Lambda(t) + \Omega_k(t) = 1, \tag{13}$$

with the deceleration parameter now being given as

$$q(t) = (n/2 - 1)\bar{\Omega}_M(t) - \bar{\Omega}_\Lambda(t) = (n/2 - 1)(1 + kc^2/\dot{R}^2(t)) - n\bar{\Omega}_\Lambda(t)/2. \tag{14}$$

As we see, Eq. (13) is remarkably similar in form to Eq. (1), with conformal cosmology thus only containing familiar ingredients. As an alternate cosmology then,

---

[¶]The sign of the gravity generated by local gravitational inhomogeneities is fixed [12] by the sign of the coupling constant α_g in the Weyl action I_W of Eq. (9), a quantity which simply makes no contribution at all in highly symmetric cosmologically relevant geometries where the Weyl tensor vanishes, with the signs of local and global gravity thus being totally decoupled in the conformal theory, to thereby allow inhomogeneous locally attractive Cavendish gravity and homogeneous globally repulsive cosmological gravity to coexist.

conformal gravity thus gets about as close to standard gravity as it would appear possible for an alternative to get while nonetheless still being different. Moreover, even though that had not been its intent, because of this similarity, we see that phenomenological fits in which $\Omega_M(t)$ and $\Omega_\Lambda(t)$ are allowed to vary freely in Eq. (1) are thus also in fact phenomenological fits to Eq. (13), with the various $\Omega(t)$ simply being replaced by their barred counterparts.

In order to see whether conformal gravity can thus fit into the relevant $\bar{\Omega}_\Lambda(t_0) = \bar{\Omega}_M(t_0)+1/2$ window (a window which we note includes $\bar{\Omega}_M(t_0) = 0$, $\bar{\Omega}_\Lambda(t_0) = 1/2$), it is thus necessary to analyze the solutions to Eq. (13). Before doing this, however, we note that unlike the situation in standard gravity where the sign of $\Omega_\Lambda(t)$ is not a priori known (its sign is only ascertained via the supernovae fitting itself), in conformal gravity we see that since, as noted earlier, the sign of Λ is necessarily negative in the conformal theory, the cosmic repulsion associated with the negative effective gravitational coupling $-\hbar S_0^2/12$ then entails that $\bar{\Omega}_\Lambda(t)$ itself is uniquely positive. And, indeed, not only is this sign unambiguously known in the conformal theory, according to Eq. (14) it immediately leads to none other than cosmic acceleration. Moreover, in conformal gravity not only is the sign of $\bar{\Omega}_\Lambda(t)$ known a priori, so also is the sign of k, with it having been shown [8] that at temperatures above all elementary particle physics phase transitions, the condition $T^{\mu\nu}_{kin} = 0$ (which is what Eqs. (11) and (12) reduce to when $S_0 = 0$) is only satisfiable non-trivially when k is negative, with the positive energy density of ordinary matter then being identically canceled by the negative gravitational energy density associated with negative spatial curvature. Beyond this theoretical argument, actual observational support for k being negative in conformal gravity has even been independently obtained from a recent study of galactic rotation curves [13], where an effect due to a global $k < 0$ cosmology on local galactic dynamics was identified, one which turns out to provide for a complete accounting of rotation curve systematics without the need to ever introduce any of the galactic dark matter required in standard gravity. Thus in the following we shall explicitly take k to be negative in the conformal theory while noting immediately that according to Eq. (14) such negative k not only also leads to cosmic acceleration, but with its associated $\Omega_k(t)$ contribution being positive, Eq. (13) then even enables the necessarily positive $\bar{\Omega}_\Lambda(t)$ to be less than one. Given such structure in the conformal theory we shall thus now turn to the explicit solutions to the conformal theory to see whether the theory is capable of not only giving some possible cosmic acceleration but whether it can even give just the amount detected.

Explicit solutions to the conformal theory are readily obtained [6], and they can be classified according to the signs of λ and k. (For completeness and comparison purposes in the following we explore the behavior of the theory for all possible allowable signs of λ and k rather than for just our preferred negative ones.) In the simpler to treat high temperature era where $\rho_M(t) = A/R^4 = \sigma T^4$ the complete family of possible solutions is given as

$$R^2(t, \alpha < 0, k < 0) = k(1-\beta)/2\alpha + k\beta sin^2(ct\sqrt{-\alpha})/\alpha,$$
$$R^2(t, \alpha = 0, k < 0) = -2A/k\hbar c S_0^2 - kc^2t^2,$$
$$R^2(t, \alpha > 0, k < 0) = -k(\beta-1)/2\alpha - k\beta sinh^2(\alpha^{1/2}ct)/\alpha,$$

$$R^2(t, \alpha > 0, k = 0) = (-A/\hbar\lambda c S_0^4)^{1/2} cosh(2\alpha^{1/2}ct),$$
$$R^2(t, \alpha > 0, k > 0) = k(1+\beta)/2\alpha + k\beta sinh^2(\alpha^{1/2}ct)/\alpha, \tag{15}$$

where we have introduced the parameters $\alpha = -2\lambda S_0^2$, and $\beta = (1 - 16A\lambda/k^2\hbar c)^{1/2}$. Similarly the associated deceleration parameters take the form

$$q(\alpha < 0, k < 0) = tan^2(ct\sqrt{-\alpha}) - 2(1-\beta)cos(2ct\sqrt{-\alpha})/\beta sin^2(2ct\sqrt{-\alpha}),$$
$$q(\alpha = 0, k < 0) = -2A/k^2\hbar c^3 S_0^2 t^2,$$
$$q(\alpha > 0, k < 0) = -tanh^2(\alpha^{1/2}ct) + 2(1-\beta)cosh(2\alpha^{1/2}ct)/\beta sinh^2(2\alpha^{1/2}ct),$$
$$q(\alpha > 0, k = 0) = -1 - 2/sinh^2(2\alpha^{1/2}ct),$$
$$q(\alpha > 0, k > 0) = -coth^2(\alpha^{1/2}ct) - 2(1-\beta)cosh(2\alpha^{1/2}ct)/\beta sinh^2(2\alpha^{1/2}ct). \tag{16}$$

Now while Eq. (15) yields a variety of temporal behaviors for $R(t)$, it is of great interest to note that every single one of them begins with $\dot{R}(t=0)$ being zero (rather than infinite) just as desired above; and that when λ is precisely our preferred negative value (viz. $\alpha > 0$) each associated such solution corresponds to a universe which permanently expands (only the $\lambda > 0$ solution recollapses, with conformal cosmology thus correlating the long time behavior of $R(t)$ with the sign of λ rather than with that of k). We thus need to determine the degree to which the permanently expanding universes have by now already become permanently accelerating.

To this end we note first from Eq. (16) that with β being greater than one when λ is negative, both the $\alpha > 0$, $k < 0$ and the $\alpha > 0$, $k = 0$ cosmologies are in fact permanently accelerating ones no matter what the values of their parameters. To explore the degree to which they have by now already become asymptotic, as well as to determine the acceleration properties of the $\alpha > 0$, $k > 0$ cosmology, we note that since each of the solutions given in Eq. (15) has a non-zero minimum radius, each associated $\alpha > 0$ cosmology has some very large but finite maximum temperature T_{max} given by

$$T^2_{max}(\alpha > 0, k < 0)/T^2(t, \alpha > 0, k < 0) = 1 + 2\beta sinh^2(\alpha^{1/2}ct)/(\beta - 1),$$
$$T^2_{max}(\alpha > 0, k = 0)/T^2(t, \alpha > 0, k = 0) = cosh(2\alpha^{1/2}ct),$$
$$T^2_{max}(\alpha > 0, k > 0)/T^2(t, \alpha > 0, k > 0) = 1 + 2\beta sinh^2(\alpha^{1/2}ct)/(\beta + 1), \tag{17}$$

with all the permanently expanding ones thus necessarily being below their maximum temperatures today, and actually being way below once given enough time. To obtain further insight into these solutions it is convenient to introduce an effective temperature according to $-c\hbar\lambda S_0^4 = \sigma T_V^4$. In terms of this T_V we then find that in all the $\lambda < 0$ cosmologies the energy density terms take the form

$$\bar{\Omega}_\Lambda(t) = (1 - T^2/T^2_{max})^{-1}(1 + T^2T^2_{max}/T_V^4)^{-1},$$
$$\bar{\Omega}_M(t) = -(T^4/T_V^4)\bar{\Omega}_\Lambda(t), \tag{18}$$

where $(\beta - 1)/(\beta + 1) = T_V^4/T^4_{max}$ for the $k < 0$ case, and where $(\beta - 1)/(\beta + 1) = T^4_{max}/T_V^4$ for the $k > 0$ case. With β being greater than one, we find that for the $k > 0$ case T_V is greater than T_{max}, for $k = 0$ T_V is equal to T_{max}, and for $k < 0$ T_V is less than T_{max}, with the energy in curvature (viz. the energy in the gravitational field itself) thus making a direct contribution to the maximum temperature of the universe. Hence, simply because T_{max} is overwhelmingly larger than the current temperature $T(t_0)$ (i.e. simply because the universe has been expanding and cooling for such a long time now), we see, that without any fine tuning at all, in both the $k > 0$ and $k = 0$ cases (i.e. cases where $T_V \geq T_{max} \gg T(t_0)$) the quantity $\bar{\Omega}_\Lambda(t_0)$ must already be at its asymptotic limit of one today, that $\bar{\Omega}_M(t_0)$ must be completely suppressed, and that $q(t_0)$ must be equal to minus one.

For the $k < 0$ case however (the only $\alpha > 0$ case where T_V is less than T_{max}, with a large T_V thus entailing an even larger T_{max}) an even more interesting and relevant outcome is possible. Specifically, since in this case the quantity $(1 + T^2 T_{max}^2 / T_V^4)^{-1}$ is always bounded between zero and one no matter what the relative magnitudes of T_V, T_{max} and $T(t)$, the simple fact that $T_{max} \gg T(t_0)$ (something which must anyway be the case if $T_V \gg T(t_0)$) entails that rather than having already reached its asymptotic value of one, $\bar{\Omega}_\Lambda(t_0)$ instead need only be bounded by it. With the temporal evolution of the $\alpha > 0$, $k < 0$ cosmology in Eq. (17) being given by

$$tanh^2(\alpha^{1/2} ct) = (1 - T^2/T_{max}^2)/(T_{max}^2 T^2/T_V^4 + 1), \tag{19}$$

we see, and again without any fine tuning at all, that simply since because T_{max} is so much greater than $T(t_0)$, the current value of $\bar{\Omega}_\Lambda(t_0)$ is given by the nicely bounded $tanh^2(\alpha^{1/2} ct_0)$, that $\bar{\Omega}_M(t_0)$ is again completely suppressed in the current era, and that the curvature contribution is given by the non-negative $\Omega_k(t_0) = 1 - \bar{\Omega}_\Lambda(t_0) = sech^2(\alpha^{1/2} ct_0)$ ($\Omega_k(t_0)$ is positive for negative k), with negative spatial curvature now being able to explicitly contribute to current era cosmological evolution. Moreover, in the $\alpha > 0$, $k < 0$ case, the bigger the spatial curvature contribution gets to be the further $\bar{\Omega}_\Lambda(t_0)$ will lie below one, with it taking a value close to one half the closer to T_V^2/T_{max} the current temperature gets to be.

Thus we see that in all three of the $\alpha > 0$ cases the single simple requirement that $T_{max} \gg T(t_0)$ ensures that $\bar{\Omega}_M(t_0)$ is completely negligible at current temperatures (it can thus only be relevant in the early universe), with the current era Eq. (13) then reducing to

$$\dot{R}^2(t) + kc^2 = \dot{R}^2(t)\bar{\Omega}_\Lambda(t) = -\dot{R}^2(t)q(t), \tag{20}$$

to thus not only yield as a current era conformal cosmology what in the standard theory could only possibly occur as a very late one, but to also yield one which enjoys all the nice purely kinematic properties of a de Sitter geometry which we identified above. Since studies of galaxy counts indicate that the purely visible matter contribution to $\Omega_M(t_0)$ is of order one (actually of order 10^{-3} or so in theories in which dark matter is not considered), it follows from Eq. (13) that current era suppression of $\bar{\Omega}_M(t_0)$ will in fact be achieved if the conformal cosmology scale parameter S_0 is many orders of magnitude larger than L_{PL}^{-1}, a condition which is actually compatible with a large rather than a small T_V. Comparison with Eq. (1) shows that current era $\lambda < 0$ conformal cosmology looks exactly like a low mass standard model cosmology, except that instead of $\Omega_M(t_0)$ being negligibly small (something difficult to understand in the standard theory) it is $\bar{\Omega}_M(t_0) = -3\Omega_M(t_0)/4\pi S_0^2 L_{PL}^2$ which is negligibly small instead ($\Omega_M(t_0)$ itself need not actually be negligible in conformal gravity - rather, it is only the contribution of $\rho_M(t)$ to the evolution of the current universe which needs be small). Thus, to conclude we see that when λ is negative, conformal cosmology automatically leads us to $\bar{\Omega}_M(t_0) = 0$ and to $0 \leq \bar{\Omega}_\Lambda(t_0) \leq 1$, with $\bar{\Omega}_\Lambda(t_0)$ being smaller the more important the contribution $\Omega_k(t_0)$ of the spatial curvature of the universe to its expansion gets to be. Moreover, with it being the case that, as had been noted earlier, k actually is negative in the conformal theory, we see that conformal gravity thus leads us directly to $\bar{\Omega}_M(t_0) = 0$, $\bar{\Omega}_\Lambda(t_0) = tanh^2(\alpha^{1/2}ct_0)$, i.e. precisely right into the phenomenological region favored by the new high z data.

As regards our treatment of the cosmological constant, it is important to stress that there is a big distinction between trying to make Λ itself small (the standard way to try to address the cosmological constant problem) and trying to make its current contribution ($\Omega_\Lambda(t_0)$ or $\bar{\Omega}_\Lambda(t_0)$) to observational cosmology be small, with this latter possibility being all that is required by actual observational information. Moreover, independent of whether superstring quantum gravity is or is not capable of quenching a Planck density cosmological constant, spontaneous breakdown effects such as those associated with a Goldstone boson pion or with massive intermediate vector bosons are clearly very much in evidence in current era particle physics experiments, and thus not quenched apparently. Hence all the evidence of particle physics is that its contribution to Λ should in fact be large rather than small today, with the essence of our work here being that even in such a case $\bar{\Omega}_\Lambda(t_0)$ can nonetheless still be small today. Indeed, as our model independent study of de Sitter geometry showed, cosmic acceleration will cause the standard gravity $\Omega_\Lambda(t)$ and the conformal gravity $\bar{\Omega}_\Lambda(t)$ to both become small at late enough times no matter how large Λ might be. Moreover, the G independent ratio $\Omega_M(t)/\Omega_\Lambda(t) = \bar{\Omega}_M(t)/\bar{\Omega}_\Lambda(t) = -T^4/T_V^4$ will also become very small at late times no matter what value of G might be measured in a local Cavendish experiment. With the overall normalization of the contribution of $\rho_M(t)$ to cosmology being the only place where G could possibly play any role cosmologically, we see that the standard gravity fine tuning problem associated with having $\Omega_M(t_0) \simeq \Omega_\Lambda(t_0)$ today can be viewed as being not so much one of trying to understand why it is $\Omega_\Lambda(t_0)$ which is of order one after 15 or so billion years, but rather of trying to explain why the matter density contribution to cosmology should be of order one after that much time rather than a factor T^4/T_V^4 smaller. Since this latter problem is readily resolved if G does not in fact control cosmology, but if cosmology is instead controlled by some altogether smaller scale such as $-1/S_0^2$ (indeed the essence of conformal cosmology is that the larger S_0, i.e. the larger rather than the smaller Λ, the faster $\bar{\Omega}_M(t)$ decouples from cosmology), we see that the origin of the entire cosmological constant problem can directly be traced to the assumption that gravity is controlled by Newton's constant G on each and every distance scale. The author wishes to thank Drs. G. V. Dunne, B. R. Holstein, V. A. Miransky and M. Sher for helpful comments. This work has been supported in part by the Department of Energy under grant No. DE-FG02-92ER40716.00.

3 APPENDIX - DIFFICULTIES IN QUENCHING Λ

Beyond the fact that a quenching of the cosmological constant down from its quantum gravity and particle physics expectations has yet to be achieved for standard gravity, it is important to note how contrived such a quenching would anyway need to be even it were to be achieved. Specifically, suppose that some explicit quenching does take place in the very early Planck temperature dominated standard model quantum gravity universe (some candidate mechanisms which might be able to do this are discussed in [3, 4]), so that at the end of that era (or at the end of a subsequent but still early inflationary universe era) Λ then takes some specific and particular value (this value might even be due to a fundamental or anthropically induced cosmological constant which simply appears in the fundamental gravitational action as an a priori fundamental constant). As the ensuing universe then expands and cools it will potentially go through a whole sequence of elementary particle physics phase transitions, in each one of which the vacuum energy would be lowered (this being the definition of a phase transition). The residual Λ from the early universe quenching would then have to be such that it would just almost (but not quite completely, now given the new supernovae data) cancel the net drop in vacuum energy due to all the particle physics phase transitions (transitions that would occur only after the early universe quenching had already taken place - unless each such phase transition is to be accompanied by its own quenching that is), so that just today, i.e. conveniently just for our own particular epoch, $\Omega_\Lambda(t_0)$ would then be of order one. Difficult as this is to even conceive of let alone demonstrate in an explicit dynamical model, we note, however, that even if such a delicate current era balancing were to actually take place, the resulting universe would then be one in which there might not necessarily have been such delicate cancellations in epochs prior to the current one. Thus in earlier epochs there could well have been substantial cosmological constant contributions, to thus potentially give the universe a history and cosmology very different from the one conventionally considered.

To see how an undesired scenario such as this might actually occur, it is instructive to consider the standard elementary particle physics electroweak symmetry breaking phase transition, a transition which purely for illustrative purposes we initially model by a Ginzburg-Landau theory with free energy effective potential

$$V(\phi,T) = \lambda\phi^4/24 - \mu^2(T)\phi^2 \tag{21}$$

where ϕ is the relevant order parameter and where $\mu^2(T)$ is given [14] by a typical form such as $\mu^2(T) = \lambda(T_V^2 - T^2)/12$. When the temperature T is less than the transition temperature T_V, the potential $V(\phi,T)$ possesses a non-trivial minimum at $\phi^2 = T_V^2 - T^2$ in which it takes the temperature dependent value

$$V_{min}(T < T_V) = -\lambda(T_V^2 - T^2)^2/24, \tag{22}$$

a value which is expressly negative (with respect to the zero value which $V_{min}(T > T_V)$ takes above the critical temperature), with the zero temperature $V_{min}(T = 0)$ taking the value $-\lambda T_V^4/24$. Now while the phase diagram of the phase transition is described by the free energy $V(\phi,T)$, gravity itself couples to the internal energy density [15], i.e. to

$$U(T) = V(\phi, T) - TdV(\phi, T)/dT, \tag{23}$$

an energy density which evaluates to

$$U_{min}(T < T_V) = -\lambda(T_V^4 + 2T_V^2T^2 - 3T^4)/24 \tag{24}$$

at the minimum of the free energy.

To this energy density we must now add on a very specifically and uniquely chosen additional residual vacuum energy density from the early universe, viz.

$$U_{res} = \lambda(T_V^4 + 2T_V^2T^2(t_0) - 3T^4(t_0))/24 + \sigma T^4(t_0) \tag{25}$$

so as to thus give us a total energy density

$$U_{tot}(T < T_V) = -\lambda(T^2 - T^2(t_0))(2T_V^2 - 3T^2 - 3T^2(t_0))/24 + \sigma T^4(t_0) \tag{26}$$

and a current era $U_{tot}(T(t_0))$ which would then precisely be of order the energy density in order matter (generically given as $\sigma T^4(t_0)$) today.[‖] However, as we immediately see, for somewhat earlier epochs where $T_V > T \gg T(t_0)$, the total vacuum energy density $U_{tot}(T < T_V)$ would then be of order $-\lambda T_V^2 T^2$ and (with λ being related to the standard Higgs and vector boson masses as $\lambda \simeq e^2 M_H^2/M_W^2$) would thus be substantially larger than the black body energy density at the same temperature T. Since the temperature dependence of a standard $k = 0$ radiation era cosmology with non-zero Λ is given by $\dot{T}^2/T^2 = (8\pi G/3c^2)(\sigma T^4 + c\Lambda)$, we see that our very ability to track current era cosmology all the way back to the early universe is contingent on demonstrating that vacuum terms such as $\lambda T_V^2 T^2$ are never of relevance. It is thus not sufficient to simply find a mechanism which quenches the cosmological constant once (in some particular chosen epoch). Rather one needs renewed, temperature dependent, quenching each and every time there is phase transition or new contribution to the vacuum energy.[**]

Now, as regards loop correction modifications to tree approximation based wisdom, we recall (see [16] or the recent review of [17]) that the occupied energy mode one loop correction to the tree approximation $V_{tree}(\phi) = m^2\phi^2/2 + \lambda\phi^4/24$ potential yields a net effective potential

$$V(\phi) = m^2\phi^2/2 + \lambda\phi^4/24 + [M^4 ln(M^4/m^4) - 3(M^2 - 2m^2/3)^2]/128\pi^2 + V_T(\phi) \tag{27}$$

where $M^2 = m^2 + \lambda\phi^2/2$ and where

$$V_T(\phi) = \frac{k^4T^4}{2\pi^2}\int_0^\infty dx x^2 ln(1 - e^{-(x^2 + M^2/k^2T^2)^{1/2}}), \tag{28}$$

with the associated energy density then being given by

$$U(T) = V(\phi) - TdV_T(\phi)/dT \tag{29}$$

‖ Why exactly the residual early universe U_{res} should so explicitly and so specifically depend on the current temperature $T(t_0)$ so that just our particular epoch of observers would see only the quenched energy density $\sigma T^4(t_0)$ remains of course to be explained.

** While U_{res} can cancel the leading term in $U_{min}(T < T_V)$, because it is temperature independent U_{res} leaves the non-leading terms untouched.

in this case.[††] Since the temperature dependent $V_T(\phi)$ is exponentially suppressed and thus insignificant at low temperatures, this time we need to add on to U_{min} an early universe residual

$$U_{res} = -V(\phi_0) + \sigma T^4(t_0) \tag{30}$$

where

$$V(\phi_0) = m^2\phi_0^2/2 + \lambda\phi_0^4/24 + [M_0^4 ln(M_0^4/m^4) - 3(M_0^2 - 2m^2/3)^2]/128\pi^2 \tag{31}$$

is the value of $V(\phi)$ at its low temperature minimum ϕ_0 (here $M_0^2 = m^2 + \lambda\phi_0^2/2$), to give us a total energy density

$$U_{tot}(T) = V(\phi) - TdV_T(\phi)/dT + U_{res}, \tag{32}$$

an energy density whose current era value at $\phi = \phi_0$ is then the requisite $\sigma T^4(t_0)$. However, now extrapolating $U_{tot}(T)$ back to the critical region where $V_T(\phi)$ no longer is suppressed, and noting that the one and higher multi loop corrections to the effective potential will collectively generate an entire power series of $(T_V^2 - T^2)^n$ effective Ginzburg-Landau type terms with both the complete all order $V(\phi)$ and the complete all order $dV(\phi)dT = (dV(\phi)/d\phi)(d\phi/dT)$ then vanishing at the critical point, we see that U_{res} will remain uncanceled there and thus dominate the expansion rate of the universe in the critical region. Since this U_{res} will dominate over black body radiation until temperatures above the critical point, we see that its presence could potentially yield a substantial modification to the cosmological history of the universe in such critical regions. And while its presence would not affect standard big bang nucleosynthesis (which occurs at temperatures way below the last phase transition, temperatures at which $V_T(\phi)$ is suppressed), nonetheless our ability to extrapolate back beyond the electroweak scale would still be severely impaired.

While we thus see that any (still to be found) current era cosmological constant cancellation mechanism will still create problems for cosmology in other epochs, we conclude this paper by noting that the above described extrapolation problem is not in fact encountered in the conformal cosmology theory which we have presented in this work, since in the conformal case no residual U_{res} is ever needed. Rather, the complete all order energy density $U_{min}(T < T_V) = [V(\phi) - TdV_T(\phi)/dT]|_{\phi=\phi_0}$ itself is the entire contribution associated with vacuum breaking, an energy density which then vanishes identically at the critical point, one which, even while it takes the very large unquenched value $V(\phi_0)$ at low temperatures, still leads to a nicely quenched current era $\bar{\Omega}_\Lambda(t_0)$ of order one.

[††] With the leading term in $V_T(\phi)$ being $-\pi^2k^4T^4/90$ at high temperatures, Eq. (29) then yields $U = \pi^2k^4T^4/30$, the standard spinless particle black body energy density.

4 REFERENCES

1. A. G. Riess et. al., Observational evidence from supernovae for an accelerating universe and a cosmological constant, Astronom. J. **116**, 1009 (1998).
2. S. Perlmutter et. al., Measurements of Ω and Λ from 42 high-redshift supernovae, Astrophys. J. **517**, 565 (1999).
3. S. Weinberg, The cosmological constant problem, Rev. Mod. Phys. **61**, 1 (1989).
4. Y. J. Ng, The cosmological constant problem, Int. J. Mod. Phys. D **1**, 145 (1992).
5. A. H. Guth, Inflationary universe: A possible solution to the horizon and flatness problems, Phys. Rev. D **23**, 347 (1981).
6. P. D. Mannheim, Implications of cosmic repulsion for gravitational theory, Phys. Rev. D **58**, 103511 (1998).
7. P. D. Mannheim, Cosmic acceleration as the solution to the cosmological constant problem, astro-ph/9910093.
8. P. D. Mannheim, Attractive and repulsive gravity, Founds. Phys. **30**, 709 (2000).
9. P. D. Mannheim, Conformal cosmology with no cosmological constant, Gen. Relativ. Gravit. **22**, 289 (1990).
10. P. D. Mannheim, Conformal gravity and the flatness problem, Astrophys. J. **391**, 429 (1992).
11. P. D. Mannheim, Conformal cosmology and the age of the universe, astro-ph/9601071.
12. P. D. Mannheim and D. Kazanas, Newtonian limit of conformal gravity and the lack of necessity of the second order Poisson equation, Gen. Relativ. Gravit. **26**, 337 (1994).
13. P. D. Mannheim, Are galactic rotation curves really flat?, Astrophys. J. **479**, 659 (1997).
14. S. Weinberg, Gauge and global symmetries at high temperature, Phys. Rev. D **9**, 3357 (1974).
15. J. F. Donoghue, B. R. Holstein and R. W. Robinett, Gravitational coupling at finite temperature, Phys. Rev. D **34**, 1208 (1986).
16. L. Dolan and R. Jackiw, Symmetry behavior at finite temperature, Phys. Rev. D **9**, 3320 (1974).
17. M. Sher, Electroweak Higgs potential and vacuum stability, Phys. Rept. **179**, 273 (1989).

BEYOND THE STANDARD MODEL

SPONTANEOUS VIOLATION OF LORENTZ AND CPT SYMMETRY

Don Colladay *

1. INTRODUCTION

The standard model as well as many of its modern day extensions preserves Lorentz and CPT symmetry. In fact, symmetry under the Lorentz group is a fundamental assumption of virtually any fundamental theory used to describe elementary particle physics. Under very mild assumptions, the postulates of a point particle theory that preserves Lorentz invariance imply that CPT is also preserved [1].

In this proceedings, I will discuss the construction of quantum field theories that break Lorentz and CPT symmetry. There are both experimental and theoretical motivations to develop such theories.

Many sensitive experimental tests of Lorentz and CPT symmetry have been performed. For example, high precision tests involving atomic systems [2, 3], clock comparisons [4, 5], and neutral meson oscillations [6, 7] provide stringent tests of Lorentz and CPT symmetry. Recently, a pendulum with a net macroscopic spin angular momentum has been constructed and used to investigate spin-dependent Lorentz and CPT violation [8]. In the past, such experiments have placed bounds on phenomenological parameters that lack any clear connection with the microscopic physics of the standard model. One motivation of constructing a theory in the context of the standard model that allows for Lorentz and CPT violation is the desire to have a single theory within the context of conventional quantum field theory that could relate various experiments and be used to motivate future investigations.

This begs the question as to how such effects might arise naturally within the current framework of quantum field theory. The main idea is that miniscule low-energy remnant effects that violate fundamental symmetries may arise in theories underlying the standard model. One example is string theory in which nontrivial structure of the vacuum solutions may induce observable Lorentz and CPT violations [9, 10, 11].

Rather than attempting a construction based directly on a specific underlying model, such as string theory, we proceed using the generic mechanism of spontaneous symmetry violation to implement the breaking. Terms involving standard model fields that violate Lorentz and CPT symmetry are assumed to arise from a

*New College of the University of South Florida, Sarasota, FL, 34243.

The Role of Neutrinos, Strings, Gravity, and Variable Cosmological Constant in Elementary Particle Physics, Edited by Kursunoglu *et al.*, Kluwer Academic/Plenum Publishers, New York, 2001

general spontaneous symmetry breaking mechanism in which vacuum expectation values for tensor fields are generated in the underlying theory [12]. The approach then is to construct all possible terms that can arise through spontaneous symmetry breaking that are consistent with the gauge invariance of the standard model and power-counting renormalizability. These conditions are imposed on the model to limit the deviation from the conventional standard model by preserving gauge symmetries and renormalizability. It is somewhat analogous to imposing R-parity in supersymmetry to eliminate pesky lepton number violating interactions.

The resulting terms lead to modified field equations that can be analyzed within the context of conventional quantum field theory. In this proceedings, I will develop the modified Feynman rules for a model theory and will explore some possible consequences for quantum electrodynamics. Other topics in Lorentz and CPT violation including a detailed analysis of causality and stability issues [13] and an investigation of effects on neutrino oscillations [14] are also being discussed at this meeting.

2. SPONTANEOUS BREAKING OF LORENTZ SYMMETRY

In this section, a general spontaneous symmetry breaking mechanism is applied to the fermion sector to generate an example of the types of interactions that arise. The conventional mechanism of this type occurs in theories that contain scalar field potentials with nontrivial minima, such as the conventional Higgs mechanism of the standard model in which Yukawa couplings generate the fermion masses after spontaneous symmetry breaking of the scalar Higgs field. In conventional theories of this sort, internal symmetries of the original Lagrangian such as gauge invariance may be violated, but Lorentz symmetry is always maintained.

The key element in preserving Lorentz invariance when symmetry is broken spontaneously is the fact that a Lorentz scalar obtains an expectation value. Spontaneous Lorentz breaking may occur in a fundamental theory containing a potential for a tensor field that has nontrivial minima. For example, consider a lagrangian describing a fermion ψ and a tensor T of the form

$$\mathcal{L} = \mathcal{L}_0 - \mathcal{L}' \quad , \tag{1}$$

where

$$\mathcal{L}' \supset \frac{\lambda}{M^k} T \cdot \overline{\psi} \Gamma (i\partial)^k \psi + \text{h.c.} + \text{V(T)} \quad . \tag{2}$$

In this expression, λ is a dimensionless coupling constant, M is some heavy mass scale of the underlying theory, Γ denotes a general gamma matrix element in the Dirac algebra, and $V(T)$ is a potential for the tensor field. (indices are suppressed for notational simplicity) The potential $V(T)$ is assumed to arise in a more fundamental theory underlying the standard model. Note that terms contributing to $V(T)$ are precluded from conventional renormalizable four-dimensional field theories, making this type of violation impossible. However, these terms are naturally generated in the low-energy limit of more general theories such as string theory [9, 10].

If the function $V(T)$ has nontrivial minima, a nonzero expectation value of T will be generated in the vacuum. The lagrangian will then contain a term of the form

$$\mathcal{L}' \supset \frac{\lambda}{M^k} \langle T \rangle \cdot \overline{\psi} \Gamma (i\partial)^k \psi + \text{h.c.} \quad , \tag{3}$$

that is bilinear in the fermion fields and can violate Lorentz invariance and various discrete symmetries C, P, T, CP, and CPT.

3. RELATIVISTIC QUANTUM MECHANICS

To illustrate the techniques for treating terms of the form generated in Eq.(3), we examine a subset of all the possible terms. An example applicable to the standard model fermions is furnished by the choice $k = 0$ (no derivatives) and $\Gamma \sim \gamma^\mu$ or $\Gamma \sim \gamma^5\gamma^\mu$, the most general nonderivative terms that violate CPT symmetry. With these restrictions, the model lagrangian for a single fermion ψ becomes

$$\mathcal{L} = \frac{i}{2}\overline{\psi}\gamma^\mu \overleftrightarrow{\partial_\mu} \psi - a_\mu\overline{\psi}\gamma^\mu\psi - b_\mu\overline{\psi}\gamma_5\gamma^\mu\psi - m\overline{\psi}\psi \quad , \tag{4}$$

where the parameters a_μ and b_μ are real constant coefficients that denote the tensor expectation values and coupling constants that are present in (3).

Several features of this theory can be immediately deduced from the structure of the lagrangian. First, the lagrangian is hermitian and therefore preserves probability. This means that conventional quantum mechanics can be used to evolve the particle states in time. The model lagrangian is invariant under translations and U(1) gauge transformations which leads to conservation of energy, momentum, and charge. The resulting Dirac equation

$$(i\gamma^\mu\partial_\mu - a_\mu\gamma^\mu - b_\mu\gamma_5\gamma^\mu - m)\psi = 0 \quad . \tag{5}$$

obtained by minimizing the variation of the action with respect to the fermion field is linear in ψ. Equation (5) can be solved exactly using the plane-wave solutions

$$\psi(x) = e^{\pm i p_\mu x^\mu} w(\vec{p}) \quad , \tag{6}$$

where $p^0(\vec{p}) \equiv E(\vec{p})$ is the energy (defined as the magnitude of the eigenvalue of the hamiltonian acting on the state) determined by setting the determinant of the matrix acting on $w(\vec{p})$ equal to zero.

The general form of the resulting dispersion relation is complicated, so we content ourselves here by investigating the special case $\vec{b} = 0$. The exact solutions for the energies are found to be

$$E_+(\vec{p}) = \left[m^2 + (|\vec{p} - \vec{a}| \pm b_0)^2\right]^{1/2} + a_0 \quad , \tag{7}$$

$$E_-(\vec{p}) = \left[m^2 + (|\vec{p} + \vec{a}| \mp b_0)^2\right]^{1/2} - a_0 \quad . \tag{8}$$

Some interesting consequences of the breaking is apparent. Note that the conventional energy degeneracy of the fermion and antifermion states is broken by a_μ while b_0 splits the degeneracy of the helicities. These energy splittings are indicative of the effect of the general Lorentz violating terms in the standard model extension. The corresponding spinor solutions form an orthogonal basis of states as a result of the hermiticity of the hamiltonian.

An interesting feature of the above dispersion relations is the modified relationship that exists between the velocity of a wave packet and its corresponding momentum. For instance, a wave packet formed from a superposition of positive helicity fermions with a four-momentum $p^\mu = (E, \vec{p})$ has a corresponding expectation value for the velocity operator $\vec{v} = i[H, \vec{x}] = \gamma^0\vec{\gamma}$ of

$$\langle\vec{v}\rangle = \langle\frac{(|\vec{p} - \vec{a}| - b^0)}{(E - a^0)}\frac{(\vec{p} - \vec{a})}{|\vec{p} - \vec{a}|}\rangle \quad . \tag{9}$$

Note that the velocity of the packet and the conserved momentum are not in the same direction. Examination of the velocity using a general nonzero b_μ reveals that $|v_j| < 1$, and that the limiting velocity as $\vec{p} \to \infty$ is 1. This implies that effects due to the CPT violating terms are mild enough to preserve causality[1]. This will be verified independently from the perspective of quantum field theory that will now be discussed.

4. CONSTRUCTION OF THE FREE FIELD THEORY

The approach to quantization taken here is similar to the conventional one in which the quantization conditions on the fields are deduced from the requirement of positivity of the energy. The wave function ψ is expanded in terms of its four solutions as

$$\begin{aligned}\psi(x) &= \int \frac{d^3p}{(2\pi)^3} \sum_{\alpha=1}^{2} \left[\frac{m}{E_u^{(\alpha)}} b_{(\alpha)}(\vec{p}) e^{-ip_u^{(\alpha)}\cdot x} u^{(\alpha)}(\vec{p}) \right. \\ &\qquad \left. + \frac{m}{E_v^{(\alpha)}} d^*_{(\alpha)}(\vec{p}) e^{ip_v^{(\alpha)}\cdot x} v^{(\alpha)}(\vec{p}) \right] , \end{aligned} \tag{10}$$

and is promoted to an operator acting on a Hilbert space of basis states.

Translational invariance is used to define the conserved energy and momentum as

$$P_\mu = \int d^3x \Theta^0{}_\mu = \int d^3x \tfrac{1}{2} i \overline{\psi} \gamma^0 \overleftrightarrow{\partial}_\mu \psi \quad . \tag{11}$$

The time component P_0 is interpreted as the energy (after normal ordering of the operators) and is positive definite (for $|a^0| < m$) provided the following anticommutation relations are imposed:

$$\begin{aligned}\{b_{(\alpha)}(\vec{p}), b^\dagger_{(\alpha')}(\vec{p}\,')\} &= (2\pi)^3 \frac{E_u^{(\alpha)}}{m} \delta_{\alpha\alpha'} \delta^3(\vec{p} - \vec{p}\,') \quad , \\ \{d_{(\alpha)}(\vec{p}), d^\dagger_{(\alpha')}(\vec{p}\,')\} &= (2\pi)^3 \frac{E_v^{(\alpha)}}{m} \delta_{\alpha\alpha'} \delta^3(\vec{p} - \vec{p}\,') \quad . \end{aligned} \tag{12}$$

The resulting equal-time anticommutators of the fields are

$$\begin{aligned}\{\psi_\alpha(t,\vec{x}), \psi^\dagger_\beta(t,\vec{x}')\} &= \delta_{\alpha\beta}\delta^3(\vec{x} - \vec{x}') \quad , \\ \{\psi_\alpha(t,\vec{x}), \psi_\beta(t,\vec{x}')\} &= 0 \quad , \\ \{\psi^\dagger_\alpha(t,\vec{x}), \psi^\dagger_\beta(t,\vec{x}')\} &= 0 \quad . \end{aligned} \tag{13}$$

These relations show that conventional Fermi statistics remain unaltered by the CPT violation.

The conserved charge Q and conserved momentum P_μ are now computed explicitly as

$$Q = \int \frac{d^3p}{(2\pi)^3} \sum_{\alpha=1}^{2} \left[\frac{m}{E_u^{(\alpha)}} b^\dagger_{(\alpha)}(\vec{p}) b_{(\alpha)}(\vec{p}) - \frac{m}{E_v^{(\alpha)}} d^\dagger_{(\alpha)}(\vec{p}) d_{(\alpha)}(\vec{p}) \right] , \tag{14}$$

[1] Note that while causality is preserved, there can be problems with stability at energies nearing the heavy mass scale of the underlying theory. See Kostelecký, this proceedings.

$$P_\mu = \int \frac{d^3p}{(2\pi)^3} \sum_{\alpha=1}^{2} \left[\frac{m}{E_u^{(\alpha)}} p_{u\mu}^{(\alpha)} b_{(\alpha)}^\dagger(\vec{p}) b_{(\alpha)}(\vec{p}) \right.$$
$$\left. + \frac{m}{E_v^{(\alpha)}} p_{v\mu}^{(\alpha)} d_{(\alpha)}^\dagger(\vec{p}) d_{(\alpha)}(\vec{p}) \right] \quad . \tag{15}$$

From these expressions, it is observed that the charge of the fermion is unperturbed and the energy and momentum satisfy the same energy-momentum relations that were found using relativistic quantum mechanics.

To preserve causality, it is necessary that the anticommutation relations of the fermion fields at unequal times are zero for spacelike separations. Explicit integration for the special case of $\vec{b} = 0$ reveals that

$$\{\psi_\alpha(x), \overline{\psi}_\beta(x')\} = 0 \quad , \tag{16}$$

for spacelike separations $(x - x')^2 < 0$. This result indicates that physical observables separated by spacelike intervals will in fact commute. This agrees with our previous results regarding the velocity obtained using the relativistic quantum mechanics approach. An analysis of causality and stability issues for other Lorentz- and CPT-violating terms in the fermion sector of the standard model extension has recently been performed [15]. Some similar issues in the photon sector have also been addressed [16].

5. INTERACTING FIELD THEORY

Next, the issue of extending this free field theory to interacting theory is addressed. Much of the conventional formalism developed for perturbative calculations in conventional interacting field theory carries over essentially unchanged to the present case. The asymptotic in and out states are defined as in the usual case using the free field solutions. The LSZ reduction procedure is then used to express the transition-matrix elements in terms of Green's functions for the theory. Dyson's formalism is then used to express the time-ordered products of the interacting fields in terms of the asymptotic fields. Wick's theorem remains unaffected by the modifications.

A central result is that the usual Feynman rules apply provided that the Feynman propagator is modified to

$$S_F(p) = \frac{i}{p_\mu\gamma^\mu - a_\mu\gamma^\mu - b_\mu\gamma_5\gamma^\mu - m + i\epsilon} \quad , \tag{17}$$

and the exact spinor solutions of the modified free fermion theory are used on external legs. The main reason that conventional techniques apply seems to be due to the fact that the Lorentz violating modifications are linear in the fermion fields.

6. QED EXTENSION AND THE PHOTON

In this section, some implications of Lorentz breaking for photon propagation are investigated. The conventional QED lagrangian is

$$\mathcal{L}_{\text{electron}}^{\text{QED}} = \tfrac{1}{2} i\overline{\psi}\gamma^\mu \overleftrightarrow{D_\mu} \psi - m_e\overline{\psi}\psi - \tfrac{1}{4}F_{\mu\nu}F^{\mu\nu} \quad , \tag{18}$$

where ψ is the electron field, m_e is its mass, and $F^{\mu\nu}$ is the photon field strength tensor. When all possible Lorentz-violating contributions from spontaneous symmetry breaking consistent with gauge invariance and power-counting renormalizability are introduced into the standard model, the resulting modifications to the photon sector are [12]

$$\mathcal{L}_{\text{photon}}^{\text{CPT-even}} = -\tfrac{1}{4}(k_F)_{\kappa\lambda\mu\nu}F^{\kappa\lambda}F^{\mu\nu} \quad , \tag{19}$$

and

$$\mathcal{L}_{\text{photon}}^{\text{CPT-odd}} = +\tfrac{1}{2}(k_{AF})^{\kappa}\epsilon_{\kappa\lambda\mu\nu}A^{\lambda}F^{\mu\nu} \quad . \tag{20}$$

The parameters k_F and k_{AF} are fixed background fields related to vacuum expectation values of tensors coupled to the photon in the underlying theory. These two couplings are even and odd under CPT respectively. The CPT-odd terms have been treated in detail elsewhere [17]. Here the special case of $(k_{AF})^{\mu} = 0$ (no CPT-odd piece), and $(k_F)_{0j0k} = -\frac{1}{2}\beta_j\beta_k$ is examined for the sake of a specific example.

The resulting modifications to the Maxwell equations are linear just as the modified Dirac equation is in the fermion case. Plane waves can therefore be used to solve the modified equations of motion. A solution exists provided p_μ satisfies

$$(p_o)^2 = 0 \quad , \tag{21}$$

$$(p_e)^2 = -\frac{(\vec{\beta}\times\vec{p}_e)^2}{1+\vec{\beta}^2} \quad , \tag{22}$$

where p_o denotes an ordinary mode and p_e denotes an extraordinary mode of photon propagation. The ordinary mode propagates as a conventional photon, while the extraordinary mode has a modified dispersion relation.

For the special direction of propagation for which $\vec{\beta}\cdot\vec{p} = 0$, the ordinary mode is polarized with $\vec{A}_o$ along the direction of $\vec{p}\times\vec{\beta}$ while the extraordinary mode $\vec{A}_e$ is polarized along $\vec{\beta}$. Both polarizations are perpendicular to the momentum vector of the wave $\vec{p}$. The group velocities of wave packets $\vec{v}_g \equiv \vec{\nabla}_p p^0$ are calculated as

$$\vec{v}_{g,o} = \hat{p} \quad , \quad \vec{v}_{g,e} = \frac{1}{\sqrt{1+\vec{\beta}^2}}\,\hat{p} \quad . \tag{23}$$

The extraordinary mode is seen to travel with a modified velocity that is slightly less than the velocity of the ordinary mode. As a result, an initially plane polarized wave will in general become elliptically polarized after traveling a distance

$$r \simeq \frac{\pi}{2\left(\sqrt{1+\vec{\beta}^2}-1\right)p^0} \simeq \frac{\pi}{\vec{\beta}^2 p^0} \quad , \tag{24}$$

where the approximation holds for $\vec{\beta}^2 \propto k_F << 1$. The magnetic field behaves analogously. Terms of this form can also have implications for photon birefringence. In particular they can contribute to polarization rotation from distant quasars [18].

7. ACKNOWLEDGMENTS

This work was supported in part by the United States Department of Energy under grant number DE-FG02-91ER40661.

8. REFERENCES

1. See, for example, J. Schwinger, Phys. Rev. **82** (1951) 914.
2. See, for example, R.S. Van Dyck, Jr., P.B. Schwinberg, and H.G. Dehmelt, Phys. Rev. Lett. **59** (1987) 26; G. Gabrielse *et al.*, *ibid.*, **74** (1995) 3544.
3. R. Bluhm, V.A. Kostelecký and N. Russell, Phys. Rev. Lett. **79** (1997) 1432; Phys. Rev. D **57** (1998) 3932; IUHET 388 (1988).
4. See, for example, J.D. Prestage *et al.*, Phys. Rev. Lett. **54** (1985) 2387; S.K. Lamoreaux *et al.*, *ibid.*, **57** (1986) 3125; T.E. Chupp *et al.*, *ibid.*, **63** (1989) 1541.
5. V.A. Kostelecký and C.D. Lane, Phys. Rev. D **60** (1999) 116010; J. Math. Phys. **40** (1999) 6245.
6. See, for example B. Schwingenheuer *et al.*, Phys. Rev. Lett. **74** (1995) 4376; OPAL Collaboration, R. Ackerstaff *et al.*, Z. Phys. C **76** (1997) 401.
7. D. Colladay and V.A. Kostelecký, Phys. Lett. B **344** (1995) 259; Phys. Rev. D **52** (1995) 6224; V.A. Kostelecký and R. Van Kooten, Phys. Rev. D **54** (1996) 5585.
8. R. Bluhm and V.A. Kostelecký, Phys. Rev. Lett. **84** (2000) 1381, and references therein.
9. V.A. Kostelecký and R. Potting, Nucl. Phys. B **359** (1991) 545; Phys. Lett. B **381** (1996) 389.
10. V.A. Kostelecký, R. Potting, and S. Samuel, in S. Hegarty et al., eds., *Proceedings of the 1991 Joint International Lepton-Photon Symposium and Europhysics Conference on High Energy Physics,* World Scientific, Singapore, 1992;
    V.A. Kostelecký and R. Potting, in D.B. Cline, ed., *Gamma Ray–Neutrino Cosmology and Planck Scale Physics* (World Scientific, Singapore, 1993) (hep-th/9211116).
11. V.A. Kostelecký and R. Potting, Phys. Rev. D **51** (1995) 3923.
12. D. Colladay and V.A. Kostelecký, Phys. Rev. D **55** (1997) 6760; Phys. Rev. D **58** (1998) 116002.
13. V.A. Kostelecký, this proceedings.
14. V. Barger, S. Pakvasa, T. Weiler, and K. Whisnant, Phys. Rev. Lett. **85** (2000) 5055; T. Weiler, these proceedings.
15. V.A. Kostelecký and R. Lehnert, To be published in Phys. Rev. D **63** (2001) 065008.
16. C. Adam and F. Klinkhamer, hep-ph/0101087.
17. S.M. Carroll, G.B. Field, and R. Jackiw, Phys. Rev. D **41** (1990) 1231;
18. See P.P. Kronberg in V.A. Kostelecký, ed., *CPT and Lorentz Symmetry*, World Scientific, Singapore, 1999.

TOPICS IN LORENTZ AND CPT VIOLATION

V. Alan Kostelecký[1]

1. INTRODUCTION

Invariance under Lorentz and CPT transformations is a fundamental requirement of local relativistic quantum field theories, including the standard model of particle physics. This invariance also seems to be realized in nature, as no clear signals for violations have been observed despite numerous experimental tests. Nonetheless, several facts offer motivation for theoretical studies of possible Lorentz and CPT violation [1]. One is that quantitative statements about the degree to which nature exhibits Lorentz and CPT symmetry are best expressed within a consistent and general theoretical framework that allows for violations [2, 3, 4]. Another more subtle fact is that the exceptional sensitivity of present experimental tests [5] implies access to highly suppressed Lorentz and CPT violations that might arise at scales well beyond the standard model in the context of Planck-scale physics [6].

At earlier conferences in this series [7] I have discussed the possibility that Lorentz and CPT symmetry might be violated as a result of new physics in a theory underlying the standard model, perhaps including string theory [6]. I have also described the Lorentz- and CPT-violating standard-model extension that allows for the associated low-energy effects in a very general context [2, 3] and outlined some of the numerous existing and future experiments that test these ideas. These experiments include, for example, studies of neutral-meson oscillations [8, 2, 9, 10, 11, 12], comparative tests of QED in Penning traps [13, 14, 15, 16], spectroscopy of hydrogen and antihydrogen [17, 18], measurements of muon properties [19, 20], clock-comparison experiments [21, 22, 23, 24], observations of the behavior of a spin-polarized torsion pendulum [25, 26], measurements of cosmological birefringence [27, 3, 28, 29], and observations of the baryon asymmetry [30].

In the present talk I briefly describe some of our recent theoretical analyses, emphasizing in particular topics at the level of quantum field theory [2, 3] and the issues of causality and stability in Lorentz-violating theories [4]. Other studies of Lorentz and CPT violation in the context of the standard-model extension are being presented at this conference [33, 34]. The reader may also find of interest recent

[1]Physics Department, Indiana University, Bloomington, IN 47405, U.S.A.

The Role of Neutrinos, Strings, Gravity, and Variable Cosmological Constant in Elementary Particle Physics, Edited by Kursunoglu *et al.*, Kluwer Academic/Plenum Publishers, New York, 2001

efforts to create a classical analogue for CPT violation [31, 32], which lie outside the scope of this talk. A treatment of experimental results released in the year since the previous conference in this series is also outside the present scope. I mention here only the newly published results on Lorentz and CPT violation involving protons [18] and neutrons [24], and the preliminary announcement of results in the muon sector obtained from muonium hyperfine spectroscopy [20].

2. CONCEPTUAL BASICS

Over the past decade, a framework allowing for Lorentz and CPT violation within realistic field-theoretic models has been developed [6] that leads to a phenomenology for Lorentz and CPT violation at the level of the standard model and quantum electrodynamics (QED) [2]. The resulting general standard-model extension [3] can be chosen to preserve the usual SU(3)×SU(2)×U(1) gauge structure and to be power-counting renormalizable. Energy and momentum are conserved, and conventional canonical methods for quantization apply. In this part of the talk, I summarize some useful basic concepts for the theoretical results to follow.

Observer and particle Lorentz transformations. By construction, the standard-model extension is invariant under rotations or boosts of an observer's inertial frame, called observer Lorentz transformations. These must be contrasted with rotations or boosts of the localized fields in a fixed observer coordinate system, called particle Lorentz transformations, which can change the physics [3]. The observer Lorentz invariance of the standard-model extension, together with its generality, means that it is the low-energy limit of *any* realistic underlying theory in which the physics is coordinate independent in an inertial frame but in which Lorentz symmetry is broken. Lorentz- and CPT-violating effects could therefore provide a unique low-energy signature for qualitatively new physics from the Planck scale.

One attractive scenario for generating Lorentz- and CPT-violating terms while maintaining observer Lorentz invariance is to invoke spontaneous Lorentz breaking in an underlying fully Lorentz-covariant theory at the Planck scale, perhaps string theory [6]. Since observer Lorentz invariance constrains the physical behavior under suitable coordinate changes made by an external observer, an underlying theory incorporating this property cannot lose it through internal interactions such as those leading to spontaneous Lorentz violation. If instead the underlying theory were to break Lorentz invariance explicitly, then observer Lorentz invariance would appear unnatural and imposing it would involve an extra requirement.

Stability and causality. Two crucial features of acceptable physical quantum field theories are stability and causality. In relativistic field theories, these are closely linked to Lorentz invariance [35]. Among their implications are the requirements of energy positivity at arbitrary momenta and of commutativity of spacelike-separated observables, both of which must hold in all observer inertial frames. For the standard-model extension, which allows for (particle) Lorentz violation, it is natural to ask about the implications of these requirements both within the theory and in the larger context of the underlying Planck-scale physics [4].

In addressing many aspects of this question, the complications of the full standard-model extension can be avoided by limiting attention to the case of the quadratic fermion part of a general renormalizable lagrangian with explicit Lorentz-

and CPT-breaking terms. This is the single-fermion limit of the free-matter sector in the general standard-model extension, given by [3]:

$$\mathcal{L} = \tfrac{1}{2} i \overline{\psi} \Gamma^\nu \overleftrightarrow{\partial_\nu} \psi - \overline{\psi} M \psi, \tag{1}$$

where

$$\Gamma^\nu \equiv \gamma^\nu + c^{\mu\nu}\gamma_\mu + d^{\mu\nu}\gamma_5\gamma_\mu + e^\nu + i f^\nu \gamma_5 + \tfrac{1}{2} g^{\lambda\mu\nu}\sigma_{\lambda\mu} \tag{2}$$

and

$$M \equiv m + a_\mu \gamma^\mu + b_\mu \gamma_5 \gamma^\mu + \tfrac{1}{2} H^{\mu\nu}\sigma_{\mu\nu}. \tag{3}$$

The parameters a_μ, b_μ, $c_{\mu\nu}$, ..., $H_{\mu\nu}$ control the degree of Lorentz and CPT violation, and their properties are discussed in Ref. [3]. Note that observer Lorentz symmetry is manifest in this model because the lagrangian (1) is independent of the choice of coordinate system. However, particle Lorentz transformations modify the fields but leave invariant the coefficients a_μ, b_μ, ..., $H_{\mu\nu}$, thereby breaking Lorentz symmetry.

A satisfactory investigation of stability and causality for Eq. (1) or the standard-model extension must incorporate the implications of observer Lorentz invariance [4]. For example, if energy positivity is found to be violated for specified coefficients for Lorentz violation in some inertial frame, then a corresponding difficulty must be present in every other inertial frame. Conversely, if complete consistency can be established in any inertial frame, then observer Lorentz invariance ensures consistency in all other inertial frames.

For simplicity and definiteness in what follows, I take the mass m of the fermion in Eq. (1) to be nonzero. This is certainly appropriate for the non-neutrino fermionic sector of the standard-model extension, and also applies if neutrinos have mass with possible minor modifications for Majorana fermions. Many of the results obtained can also be applied to bosons and to the massless case, although some care would be required to handle correctly the additional complications arising from distinctions between finite- and zero-mass representations of the Lorentz group. This is particularly true for gauge bosons [4], for which a satisfactory analysis of causality and stability remains an open problem. Some results about causality restricted to the special case of a single Lorentz-violating term in the photon sector have recently been obtained [36].

Concordant frames. The coefficients for Lorentz violation a_μ, b_μ, $c_{\mu\nu}$, ..., $H_{\mu\nu}$ in Eq. (1) transform as nontrivial representations of the observer Lorentz group O(3,1). Since this group is noncompact, individual components of these coefficients can become arbitrarily large. Under certain circumstances, including both exact and perturbative calculations, it is therefore valuable to define a special class of inertial frames called concordant frames, in which the coefficients for Lorentz and CPT violation represent only a small perturbation relative to the ordinary Dirac case [4].

Since no Lorentz and CPT violation has been observed in nature, any effects are presumably minuscule in an Earth-based laboratory. Barring unexpected surprises such as fortuitous cancellations, this suggests that all the coefficients in Eq. (2) are well below 1 and those in Eq. (3) are well below m. It can then be regarded as an experimental fact that any inertial frame in which the Earth moves nonrelativistically is a concordant frame.

If small Lorentz- and CPT-violating effects in nature indeed arise from an underlying theory at some large scale M_P, such as the Planck scale, then the natural dimensionless suppression factor is some power of the ratio m/M_P [2]. The size of the coefficients $c_{\mu\nu}$, $d_{\mu\nu}$, e_μ, f_μ, $g_{\lambda\mu\nu}$ in Eq. (2) is therefore likely to be no larger than m/M_P, although smaller values are possible. Similarly, the size of the coefficients a_μ, b_μ, $H_{\mu\nu}$ in Eq. (3) is likely to be no larger than m^2/M_P.

High-energy physics. Within standard special relativity, the separation between high- and low-energy physics is frame independent. Thus, high-energy physics in one inertial frame corresponds to high-energy physics in another frame. However, this correspondence fails in the presence of Lorentz violation [4]. The point is that the coefficients for Lorentz and CPT violation determining the physics of a high-energy particle in one inertial frame can be very different from those determining the high-energy physics in another frame. The breaking of particle Lorentz invariance therefore implies that high-energy physics can change between inertial frames, despite the observer Lorentz invariance. In particular, statements concerning Lorentz-breaking effects restricted to high energies may be observer dependent.

Since the standard concept of high and low energy is ambiguous, a cleaner definition is useful [4]. A useful option is to take the separation between high and low energies relative to the scale of the underlying theory as being defined in a concordant frame. This definition is experimentally reasonable and compatible with intuition and common usage, since any laboratory frame moves nonrelativistically relative to a concordant frame, so high- and low-energy physics are similar in both.

3. QUANTUM MECHANICS AND QUANTUM FIELD THEORY

In a concordant frame, where the coefficients for Lorentz violation are small, the usual methods of relativistic quantum mechanics and quantum field theory can be adopted. The first step is the construction of the relativistic quantum hamiltonian H from the lagrangian $\mathcal{L}$ of Eq. (1). Care is required because $\mathcal{L}$ contains extra time-derivative terms. A spinor redefinition can be used to eliminate these couplings in a concordant frame [13]: define $\psi = A\chi$ and require the nonsingular spacetime-independent matrix A to obey $A^\dagger\gamma^0\Gamma^0 A = I$. The reader is warned that the explicit form of A depends on the chosen inertial frame. In any case, it follows that $\mathcal{L}[\chi]$ contains no time derivatives other than the usual one, $\frac{1}{2}i\overline{\chi}\gamma^0\overleftrightarrow{\partial_0}\chi$. Since the conversion of ψ to χ can be regarded as a change of basis in spinor space, the physics is unaffected.

It is known that A exists if and only if all the eigenvalues of $\gamma^0\Gamma^0$ are positive. Quantitatively, a parameter δ^0 can be defined as the upper bound on the size of certain coefficients for Lorentz and CPT violation such that A exists [4]. It can be proved that $\delta^0 < 1/480$, which represents a value far larger than the maximum size of δ^0 likely to be acceptable on experimental grounds. The spinor redefinition involving A therefore always exists in a concordant frame and is applicable to realistic situations in nature.

After implementing the spinor redefinition from ψ to χ, one can use the Euler-Lagrange equations to obtain a modified Dirac equation in terms of χ. This takes the form

$$(i\partial_0 - H)\chi = 0, \tag{4}$$

where the hamiltonian $H = A^\dagger \gamma^0 (i\Gamma^j \partial_j - M)A$ is hermitian. Various explicit forms for this hamiltonian can be found in Ref. [22].

The modified Dirac equation (4) is solved via a superposition of plane spinor waves, as usual: $\chi(x) = w(\lambda)\exp(-i\lambda_\mu x^\mu)$. The quantity λ_μ obeys the dispersion relation

$$\det(\Gamma^\mu \lambda_\mu - M) = 0. \tag{5}$$

This dispersion relation is displayed as an explicit polynomial in Ref. [4]. It can be regarded as a quartic equation for $\lambda^0(\vec{\lambda})$. For a particle with definite 3-momentum, the dispersion relation fixes the exact eigenenergies in the presence of Lorentz and CPT violation. All four roots of Eq. (5) are necessarily real, since H is hermitian in a concordant frame. They are also independent of the spinor redefinition. The dispersion relation is observer Lorentz invariant, so λ_μ must be an observer Lorentz 4-vector.

In the usual Dirac case, the roots of the dispersion relation exhibit a fourfold degeneracy. However, in the presence of Lorentz and CPT violation this degeneracy is typically lifted. Nonetheless, for sufficiently small Lorentz and CPT violation, the roots can still be separated into two positive ones and two negative ones. The criterion for the existence of this separation can be quantitatively expressed in terms of a parameter δ, defined in Ref. [4]. It is known that the bound $\delta < m/124$ is sufficient. The values of this bound is again much larger than experimental observations are likely to allow, so the existence of Lorentz and CPT violation in nature would have no effect on the separation between positive and negative roots. Note that this bound is independent of the spinor redefinition. The two bounds on δ^0 and δ provide criteria quantitatively constraining the definition of a concordant frame.

As usual, the eigenfunctions of the modified Dirac equation (4) corresponding to the two negative roots can be reinterpreted as positive-energy reversed-momentum wave functions. The four resulting spinors u and v are eigenvectors of the hermitian hamiltonian H. They span the spinor space and can be used to write a general solution of Eq. (4) as a Fourier superposition of plane wave solutions with complex weights in the standard way.

To convert from relativistic quantum mechanics to quantum field theory, the complex weights in the Fourier expansion are promoted to creation and annihilation operators on a Fock space, as usual. The spinors ψ and χ become quantum fields, related through the redefinition $\psi = A\chi$ as before. The standard nonvanishing anticommutation relations can be imposed on the creation and annihilation operators. The resulting equal-time anticommutators for the fields χ are conventional, while the nonvanishing equal-time anticommutators for the original fields ψ become [4]

$$\{\psi_j(t,\vec{x}), \overline{\psi}_l(t,\vec{x}\,')\Gamma^0_{lk}\} = \delta_{jk}\delta^3(\vec{x}-\vec{x}\,'), \tag{6}$$

where the spinor indices are explicitly shown. Note the generalization from the usual Dirac case of the canonical conjugate of ψ, which in the presence of Lorentz and CPT violation takes the form $\pi_\psi = \overline{\psi}\Gamma^0$, with Γ^0 given by Eq. (2).

In a concordant frame, the vacuum state $|0\rangle$ of the Hilbert space is defined as the state that vanishes when the annihilation operators are applied. The creation operators then act on $|0\rangle$ to yield states describing particles and antiparticles with 4-momenta appropriately determined by the dispersion relation (5). An important consequence is that the zero components of these 4-vectors are positive definite. This implies positivity of the energy for the Hilbert-space states in the concordant frame [4].

4. STABILITY AND CAUSALITY

The derivation of the quantum physics associated with the lagrangian (1) can be performed as outlined in the previous section provided the constraints on δ^0 and δ are satisfied. These constraints ensure that the Lorentz-violating time-derivative terms can be eliminated and that the usual separation holds between particles and antiparticles, and they quantify the notion of concordant frame. They involve specific components of the parameters for Lorentz and CPT violation and so are noninvariant under observer Lorentz transformations, as expected. A class of observers therefore exists for whom these bounds are violated and the derivation of the quantum physics fails. These observers are strongly boosted relative to a concordant frame. Nonetheless, when combined with the requirement of observer Lorentz invariance, their existence indicates some difficulty must occur even for the quantization scheme in a concordant frame. These are associated with the stability and causality of the theory, which are considered next.

Stability. In conventional relativistic field theory with Lorentz symmetry, energy positivity in a given frame implies that the vacuum is stable in any frame provided certain conditions are met. One is that for all one-particle states in the given frame the 4-momenta are timelike or lightlike with nonnegative zeroth components. Since the signs of these zeroth components are invariant under an observer Lorentz transformation, energy positivity is a Lorentz-invariant concept despite being a statement about a 4-vector component. Thus, for example, the usual free Dirac theory exhibits energy positivity in all observer frames.

In contrast, for a theory with Lorentz and CPT violation, results demonstrating energy positivity in one frame are insufficient to ensure energy positivity or stability in all inertial frames. This is true, for example, of the energy positivity discussed in the previous section for the theory (1) in a concordant frame. At least one of the usual assumptions fails: certain energy-momentum 4-vectors satisfying the dispersion relation (5) may in fact be spacelike in all observer frames. An example is provided by the dispersion relation for the special case of the theory (1) in which only the b_μ coefficient is nonzero. No matter how small b_μ is chosen, an observer frame can always be found in which spacelike 4-vectors λ^μ exist that satisfy the dispersion relation [4].

Observer Lorentz invariance ensures that the instabilities due to the existence of spacelike solutions exist in any frame, including concordant ones. However, they are perhaps most intuitively appreciated by considering an appropriate observer boost. With a boost velocity less than 1, it is always possible to convert a spacelike vector with a positive zeroth component to one with a negative zeroth component. This guarantees the existence of a class of observer frames for which a single root of the dispersion relation involves both positive and negative energies, and it implies the canonical quantization procedure fails.

It is of interest to determine the scale $\tilde{M}$ of the 3-momentum at which the 4-momentum turns spacelike. As an example, consider the special case of the theory (1) with only a nonzero timelike b_μ. In the observer frame with $b_\mu = (b_0, \vec{0})$, and taking $b_0 \sim O(m^2/M_P)$ according to the discussion of scales in section 2, it follows that $\tilde{M} \gtrsim O(M_P)$. This shows that instabilities arise only for Planck-scale 4-momenta in any of the concordant frames. The associated negative-energy problem in boosted frames emerges only for observers undergoing Planck-scale boosts. The concordant-frame quantization described in section 3 therefore maintains stability for all experimentally attainable physical momenta and in all experimentally attainable observer frames.

The existence of observer Lorentz invariance implies that the negative-energy instabilities in strongly boosted frames must have a counterpart in concordant frames, albeit restricted to particles with Planck-scale energies. This is indeed the case: single-particle states with Planck-scale energies in a concordant frame are unstable to decay. For example, a Planck-energy fermion can explicitly be shown to be unstable to the emission of a fermion-antifermion pair [4]. In conventional QED, this process is kinematically forbidden. However, the presence of spacelike momenta in the context of the Lorentz- and CPT-violating QED extension makes the emission proceed at Planck energies. Other conventionally forbidden processes are also likely to occur. A single-particle state describing a fermion of sufficiently large 3-momentum is therefore unstable even in a concordant frame.

Causality. Microcausality holds in a quantum field theory provided any two local observables with spacelike separation commute. In a theory of Dirac fermions, the local quantum observables are fermion bilinears. Microcausality therefore holds for the modified theory (1) if

$$iS(x - x') = \{\psi(x), \overline{\psi}(x')\} = 0, \quad (x - x')^2 < 0 \tag{7}$$

is satisfied. Note that the original field ψ is involved, rather than the redefined field χ, because the definition of the latter depends on the choice of inertial frame.

An integral representation for $S(x - x')$ provides a useful tool in studying the conditions for microcausality. In a concordant frame, an integral represention for the anticommutator function $S(x - x')$ can be obtained as [4]

$$S(z) = \text{cof}(\Gamma^\mu i\partial_\mu - M) \int_C \frac{d^4\lambda}{(2\pi)^4} \frac{e^{-i\lambda\cdot z}}{\det(\Gamma^\mu \lambda_\mu - M)}. \tag{8}$$

Provided $c_{\mu\nu} = d_{\mu\nu} = e_\mu = f_\mu = g_{\lambda\mu\nu} = 0$, the derivative couplings take the standard form with $\Gamma^\mu = \gamma^\mu$. In this case, a hermitian hamiltonian always exists and the four poles remain on the real axis. Explicit calculation of the contour integration in Eq. (8) shows that $S(z)$ vanishes outside the light cone in this case. It follows that the theory (1) restricted to nonzero a_μ, b_μ and $H_{\mu\nu}$ is microcausal. However, in the unrestricted theory (1), the poles of the integrand in Eq. (8) can lie away from the real λ^0 axis, in which case the contour C may fail to encircle them. This difficulty occurs when the bound on δ^0 discussed in section 3 is violated. The hamiltonian then cannot be made hermitian, and the roots of the dispersion relation may be complex.

In discussions of causality, it is advantageous to define the velocity of a particle at arbitrary 3-momentum. However, the definition of the quantum velocity operator is nontrivial even for the Lorentz- and CPT-invariant case and becomes involved when these symmetries are violated [3]. One useful concept is the group velocity, which can be defined for a monochromatic wave in terms of the dispersion relation by $\vec{v}_g = \partial E/\partial \vec{p}$, as usual. It can be shown that the flow velocities of the conserved momentum and the conserved charge for one-particle states agree with the group velocity. Also, explicit checks in special cases suggest that $\langle d\vec{x}/dt \rangle = \vec{v}_g$ in relativistic quantum mechanics, and that the maximal group velocity attainable has magnitude equal to the maximal signal speed obtained from $S(z)$.

The group velocity can be used to establish the scale $\tilde{M}$ of microcausality violation. Setting the group velocity to 1 and solving for the magnitude $|\vec{p}|$ of the 3-momentum determines the scale $\tilde{M}$ through $\tilde{M} = |\vec{p}|$. For example, consider the theory (1) with only the term e_μ nonzero. Suppose e_μ is timelike, and take $\vec{e} = 0$ in the chosen concordant frame. Then, assuming the maximal expected Lorentz violation $e_0 \sim \mathcal{O}(m/M_P)$ following the discussion in section 2, the scale $\tilde{M}$ of microcausality violation is obtained as $\tilde{M} \gtrsim \mathcal{O}(M_P)$. Thus, the e_μ model violates microcausality at the scale M_P.

Intuition about the connection between hermiticity of the hamiltonian H and microcausality can also be obtained from the theory involving only e_μ. In this case, the nonzero entries of the matrix $\gamma^0\Gamma^0$ in the Pauli-Dirac representation consist only of diagonal entries $1 \pm e_0$. For $|e_0| < 1$, the spectrum of $\gamma^0\Gamma^0$ is therefore positive, and both a suitable matrix A and a hermitian hamiltonian H exist. However, when $|e_0| > 1$ the spectrum of $\gamma^0\Gamma^0$ includes two negative eigenvalues, and neither A nor a hermitian H exists. In the dispersion relation, the associated difficulty is that for $|e_0| > 1$ it is always possible to find an observer frame in which the roots become complex.

Intermediate-scale physics. The above results indicate that stability and causality violations emerge at a scale $\mathcal{O}(M_P)$. However, this conclusion may fail for the special case of theories of the form (1) with a nonzero coefficient $c_{\mu\nu}$ [4]. The point is that field operators with closely related derivative and spinor structures are involved for both the usual Dirac kinetic term and the term with coefficient $c_{\mu\nu}$, which means the latter behaves in many respects as a first-order correction to a zeroth-order result. This feature is unique to the $c_{\mu\nu}$ term.

To illustrate this point with a definite example, consider the special case of the lagrangian (1) with only the coefficient c_{00} nonzero in a concordant frame, and suppose in accordance with the discussion in section 2 that $c_{00} \sim \mathcal{O}(m/M_P)$. The dispersion relation for this model in an arbitrary frame is

$$(\eta_{\alpha\mu} + c_{\alpha\mu})(\eta^{\alpha}{}_{\nu} + c^{\alpha}{}_{\nu})\lambda^\mu\lambda^\nu - m^2 = 0. \tag{9}$$

For the case $c_{00} > 0$ it can then be shown that spacelike 4-momenta occur at a scale $\tilde{M} \gtrsim \mathcal{O}(\sqrt{mM_P})$, and so instabilities occur at energies well below the scale M_P of the underlying theory in this case. If instead $c_{00} < 0$, then it can be shown that at the same scale microcausality violations arise instead: the integration in (8) can be performed analytically, revealing that the anticommutator function $S(z)$ can be nonzero outside the region defined by $z^0 < (1 + c_{00})|\vec{z}|$ and that signal propagation could therefore occur with maximal speed $1/(1 + c_{00}) > 1$.

These results concerning intermediate scales are of interest because energies comparable to the order of $\sqrt{mM_P}$ in the concordant frame are exhibited by certain physical phenomena. For instance, it has been suggested that effects from c_{00}-type terms might be responsible for the apparent excess of cosmic rays in the region of 10^{19} GeV [37, 38]. The above analysis suggests that these effects can be traced to stability or causality violations, which surely must be absent in a satisfactory underlying theory. It therefore seems plausible that the effective dispersion relation involving the coefficients $c_{\mu\nu}$ is modified already at these scales, perhaps along the lines described in the next section, but in any case in a way that preserves stability and causality. A corresponding modification of the predictions for cosmic rays and other phenomena at the scale $\sqrt{mM_P}$ would then be likely.

5. PLANCK-SCALE EFFECTS

The results described in section 4 indicate that problems with stability and causality in the theory (1) arise primarily for Planck-scale 4-momenta in a concordant frame or for observers undergoing a Planck boost relative to this frame. Although perhaps strictly unnecessary from the phenomenological viewpoint of current laboratory physics, it would be of interest to establish a framework for Lorentz and CPT violation in which both stability and microcausality hold.

A natural question is whether spontaneous Lorentz and CPT breaking suffices to avoid the issues with stability and causality. Since this type of breaking could occur in a theory with a Lorentz-invariant lagrangian and hence with Lorentz-covariant dynamics, it is likely to avoid at least some of the problems faced by theories with explicit Lorentz and CPT violation. For example, one important advantage of spontaneous Lorentz violation is the natural occurrence of observer Lorentz invariance, which eliminates the coordinate dependence and related problems faced by theories with explicit violation. However, spontaneous Lorentz violation manifests itself physically because the Fock-space states are constructed on a noninvariant vacuum, whereas stability and causality in a relativistic theory depend partly on the existence of a Lorentz-invariant vacuum. It is therefore to be expected that, despite the advantages of spontaneous Lorentz violation, some difficulties with stability and causality remain.

This expectation can be directly confirmed by studying the fermion sector of quantum field theories with spontaneous Lorentz violation [4]. The key point is that the lagrangian (1) includes by construction the most general terms quadratic in fermion fields that appear in a renormalizable theory. This means that implementing spontaneous Lorentz and CPT violation in any conventional fermion field theory must result in free-fermion Fock-space states with dispersion relations described by Eq. (5) or a restriction of it. If indeed all possible dispersion relations have either stability or causality violations at some large scale, then no fermion lagrangian with spontaneous Lorentz and CPT violation can have a completely satisfactory perturbative Hilbert space in conventional quantum field theory. Maintaining stability and causality would therefore require an additional ingredient beyond conventional quantum field theory, in accordance with the notion that Lorentz and CPT violation is a unique potential signal for Planck-scale physics.

The above discussion suggests that a theory with a quadratic lagrangian that maintains both stability and causality would need to include terms beyond the ones in Eq. (1). In the low-energy limit of an underlying realistic theory at the Planck scale with spontaneous Lorentz violation, any such terms would emerge as higher-dimensional nonrenormalizable operators that must be included in the standard-model extension at energies determined by the Planck scale. The structure of the dispersion relation would correspondingly change. In a concordant frame, it would need to remain of the form (5) for small 3-momenta but would avoid spacelike 4-momenta and group velocities exceeding 1 for large 3-momenta. In fact, it can be shown by explicit construction that appropriate modifications of the dispersion relation can satisfy both the stability and causality requirements [4]. It suffices to introduce a suitable factor suppressing the coefficients for Lorentz and CPT violation only at large 3-momenta. Since the notion of large 3-momenta is frame dependent, the suppression factor itself must be frame-dependent and hence must involve Lorentz- and CPT-violating coefficients.

As an example, consider the dispersion relation (5) in the special case with only c_{00} nonzero and negative in a concordant frame. In an arbitrary frame, this dispersion relation has the form (9). To minimize complications arising from the small size of the coefficients for Lorentz and CPT violation, let the mass and the value of the Lorentz-violating coefficients be of order 1 in appropriate units. If each factor of c_{00} is combined with an exponential factor $\exp(c_{00}\lambda_0^2)$, then in an arbitrary frame the dispersion relation (9) becomes

$$(\eta_{\alpha\mu} + c_{\alpha\mu} \exp(c_{\beta\gamma}\lambda^\beta\lambda^\gamma))(\eta^\alpha{}_\nu + c^\alpha{}_\nu \exp(c_{\beta\gamma}\lambda^\beta\lambda^\gamma))\lambda^\mu\lambda^\nu - m^2 = 0. \qquad (10)$$

The presence of the exponential factors eliminates the violations of microcausality that occurred at large λ_μ in Eq. (9). The group velocity now lies below 1 for all $\vec{\lambda}$. It can also be shown that no stability problems are introduced [4].

It would evidently be interesting to identify theories in which dispersion relations of this type naturally arise. If transcendental functions of the momenta are indeed necessary to overcome the polynomial Lorentz-violating behavior in Eq. (5), a suitable lagrangian must incorporate derivative couplings of arbitrary order. It then follows that spontaneous Lorentz and CPT violation in a nonlocal theory can naturally provide the required structure for stability and causality at all scales.

The emergence of nonlocality as an important ingredient is noteworthy in part because string theories are nonlocal objects. Moreover, calculations with string field theory provided the original motivation for identifying spontaneous Lorentz and CPT violation as a potential Planck-scale signal [6] and for the development of the standard-model extension as the relevant low-energy limit [2, 3]. In fact, it can be shown that the structure of the field theory for the open bosonic string is compatible with dispersion relations of the desired type [4]. Thus, suppose in this string theory there exist nonzero Lorentz- and CPT-violating expectation values of tensor fields resulting from spontaneous symmetry breaking. Then, for example, the dispersion relation for the scalar tachyon mode contains a piece closely related to Eq. (9), but it includes also nonlocal terms of the type needed to maintain stability and causality. Although the bosonic string is merely a toy model in this context, the structural features of interest are generic to other string theories, including ones with fermions. This provides support for the existence of a stable and causal realistic fundamental theory exhibiting spontaneous Lorentz violation.

6. ACKNOWLEDGMENTS

This work is supported in part by the United States Department of Energy under grant number DE-FG02-91ER40661.

7. REFERENCES

1. For overviews of various theoretical ideas, see, for example, V.A. Kostelecký, ed., *CPT and Lorentz Symmetry,* World Scientific, Singapore, 1999.
2. V.A. Kostelecký and R. Potting, Phys. Rev. D **51** (1995) 3923.
3. D. Colladay and V.A. Kostelecký, Phys. Rev. D **55** (1997) 6760; *ibid.* **58**, 116002 (1998).
4. V.A. Kostelecký and R. Lehnert, Phys. Rev. D **63** (2001) 065008.
5. A list of particle-physics CPT tests can be found in, for example, *Review of Particle Physics,* Eur. Phys. J. C **15** (2000) 1. See also the specific experimental references below.
6. V.A. Kostelecký and S. Samuel, Phys. Rev. D **39** (1989) 683; *ibid.* **40** (1989) 1886; Phys. Rev. Lett. **63** (1989) 224; *ibid.* **66** (1991) 1811; V.A. Kostelecký and R. Potting, Nucl. Phys. B **359** (1991) 545; Phys. Lett. B **381** (1996) 89; Phys. Rev. D **63** (2001) 046007; V.A. Kostelecký, M. Perry, and R. Potting, Phys. Rev. Lett. **84** (2000) 4541.
7. V.A. Kostelecký, in B.N. Kursunoglu, S.L. Mintz, and A. Perlmutter, eds., *High-Energy Physics and Cosmology,* Plenum, New York, 1997 (hep-ph/9704264); *Physics of Mass,* Plenum, New York, 1999 (hep-ph/9810239); *Elementary Particles and Gravitation,* Plenum, New York, 1999 (hep-ph/hep-ph/0005280).
8. E731 Collaboration, L.K. Gibbons et al., Phys. Rev. D **55** (1997) 6625.
9. D. Colladay and V.A. Kostelecký, Phys. Lett. B **344** (1995) 259; Phys. Rev. D **52** (1995) 6224; V.A. Kostelecký and R. Van Kooten, Phys. Rev. D **54** (1996) 5585.
10. OPAL Collaboration, R. Ackerstaff *et al.*, Z. Phys. C **76** (1997) 401; DELPHI Collaboration, M. Feindt *et al.*, preprint DELPHI 97-98 CONF 80 (July 1997).
11. V.A. Kostelecký, Phys. Rev. Lett. **80** (1998) 1818; Phys. Rev. D **61** (2000) 16002.
12. KTeV Collaboration, presented by Y.B. Hsiung at the KAON 99 conference, Chicago, June 1999.
13. R. Bluhm, V.A. Kostelecký and N. Russell, Phys. Rev. Lett. **79** (1997) 1432; Phys. Rev. D **57** (1998) 3932.
14. G. Gabrielse *et al.*, in Ref. [1]; Phys. Rev. Lett. **82** (1999) 3198.
15. H. Dehmelt *et al.*, Phys. Rev. Lett. **83** (1999) 4694.
16. R. Mittleman, I. Ioannou, and H. Dehmelt, in Ref. [1]; R. Mittleman *et al.*, Phys. Rev. Lett. **83** (1999) 2116.
17. R. Bluhm, V.A. Kostelecký and N. Russell, Phys. Rev. Lett. **82** (1999) 2254.
18. D. Phillips *et al.*, Phys. Rev. D, Rapid Communications, in press (physics/0008230).
19. R. Bluhm, V.A. Kostelecký and C.D. Lane, Phys. Rev. Lett. **84** (2000) 1098.
20. V.W. Hughes *et al.*, presented at the Hydrogen II Conference, Tuscany, Italy, June 2000.

21. V.W. Hughes, H.G. Robinson, and V. Beltran-Lopez, Phys. Rev. Lett. **4** (1960) 342; R.W.P. Drever, Philos. Mag. **6** (1961) 683; J.D. Prestage *et al.*, Phys. Rev. Lett. **54** (1985) 2387; S.K. Lamoreaux *et al.*, Phys. Rev. Lett. **57** (1986) 3125; Phys. Rev. A **39** (1989) 1082; T.E. Chupp *et al.*, Phys. Rev. Lett. **63** (1989) 1541; C.J. Berglund *et al.*, Phys. Rev. Lett. **75** (1995) 1879.
22. V.A. Kostelecký and C.D. Lane, Phys. Rev. D **60** (1999) 116010; J. Math. Phys. **40** (1999) 6245.
23. L.R. Hunter *et al.*, in Ref. [1] and references therein.
24. D. Bear *et al.*, Phys. Rev. Lett. **85** (2000) 5038.
25. R. Bluhm and V.A. Kostelecký, Phys. Rev. Lett. **84** (2000) 1381.
26. B. Heckel *et al.*, in B.N. Kursunoglu, S.L. Mintz, and A. Perlmutter, eds., ***Elementary Particles and Gravitation***, Plenum, New York, 1999.
27. S.M. Carroll, G.B. Field, and R. Jackiw, Phys. Rev. D **41** (1990) 1231.
28. R. Jackiw and V.A. Kostelecký, Phys. Rev. Lett. **82** (1999) 3572.
29. M. Pérez-Victoria, Phys. Rev. Lett. **83** (1999) 2518; J.M. Chung, Phys. Lett. B **461** (1999) 138.
30. O. Bertolami *et al.*, Phys. Lett. B **395** (1997) 178.
31. J. Rosner, Am. J. Phys. **64** (1996) 982; J. Rosner and S.A. Slezak, Am. J. Phys. **69** (1996) 44.
32. V.A. Kostelecký and A. Roberts, Phys. Rev. D, in press (hep-ph/0012381).
33. V. Barger, S. Pakvasa, T. Weiler, and K. Whisnant, Phys. Rev. Lett. **85** (2000) 5055; T. Weiler, these proceedings.
34. D. Colladay, these proceedings.
35. W. Pauli, Phys. Rev. **58** (1940) 716.
36. C. Adam and F. Klinkhamer, hep-ph/0101087.
37. S. Coleman and S. Glashow, Phys. Rev. D **59** (1999) 116008.
38. O. Bertolami and C.S. Carvalho, gr-qc/9912117.

DARK MATTER CAUSTICS

P. Sikivie, W. Kinney*

The late infall of cold dark matter onto an isolated galaxy such as our own produces flows with definite velocity vectors at any physical point in the galactic halo. It also produces caustics which are places where the dark matter density is very large. The outer caustics are topological spheres whereas the inner caustics are rings. The self-similar model of galactic halo formation predicts that the caustic ring radii a_n follow the approximate law $a_n \sim 1/n$. In a recent study of 32 extended and well-measured galactic rotation curves, we found evidence for this law.

1 INTRODUCTION

Before the onset of galaxy formation but after the time t_{eq} of equality between matter and radiation, the velocity dispersion of the cold dark matter candidates is very small, of order $\delta v_a(t) \sim 3 \cdot 10^{-17} \left(\frac{10^{-5} eV}{m_a}\right) \left(\frac{t_0}{t}\right)^{2/3}$ for axions and $\delta v_W(t) \sim 10^{-11} \left(\frac{GeV}{m_W}\right)^{1/2} \left(\frac{t_0}{t}\right)^{2/3}$ for WIMPs, where t_0 is the present age of the universe and m_a and m_W are respectively the masses of the axion and the WIMP. In the context of galaxy formation, such small velocity dispersions are entirely negligible. Massive neutrinos, on the other hand, have primordial velocity dispersion $\delta v_\nu(t) \simeq 5.3 \; 10^{-4} \left(\frac{eV}{m_\nu}\right) \left(\frac{t_0}{t}\right)^{2/3}$ which is comparable to the virial velocity in galaxies and therefore non-negligible in the context of galaxy formation[1]. This is the reason why massive neutrinos are called 'hot dark matter'.

Collisionless dark matter particles lie on a thin 3-dimensional (3D) sheet in 6D phase-space. The thickness of this sheet is the primordial velocity dispersion δv. If each of the aforementioned species of collisionless particles is present, the

*Department of Physics, University of Florida, Gainesville, FL 32611, USA

The Role of Neutrinos, Strings, Gravity, and Variable Cosmological Constant in Elementary Particle Physics, Edited by Kursunoglu *et al.*, Kluwer Academic/Plenum Publishers, New York, 2001

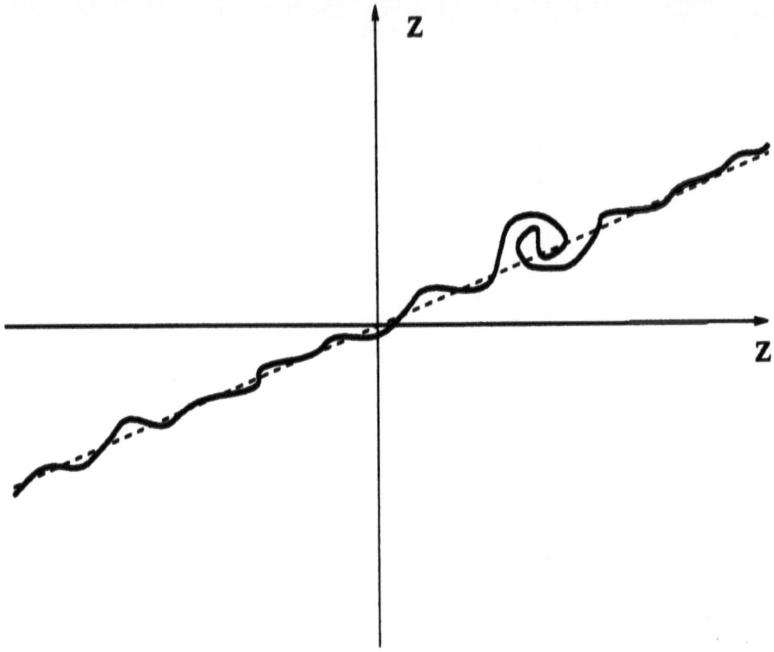

Figure 1: The wiggly line represents the intersection of the $(z, \dot{z})$ plane with the 3D sheet on which the collisionless dark matter particles lie in phase-space. The thickness of the line is the primordial velocity dispersion. The amplitude of the wiggles in the $\dot{z}$ direction is the velocity dispersion associated with density perturbations. Where an overdensity grows in the non-linear regime, the line winds up in clockwise fashion. One such overdensity is shown.

phase-space sheet has three layers, a very thin layer of axions, a medium layer of WIMPs and a thick layer of neutrinos. The phase-space sheet is located on the 3D hypersurface of points $(\vec{r}, \vec{v})$: $\vec{v} = H(t)\vec{r} + \Delta\vec{v}(\vec{r}, t)$ where $H(t) = \frac{2}{3t}$ is the Hubble expansion rate and $\Delta\vec{v}(\vec{r}, t)$ is the peculiar velocity field. Fig. 1 shows a 2D section of 6D phase-space along the $(z, \dot{z})$ plane. The wiggly line is the intersection of the 3D sheet on which the particles lie in phase-space with the plane of the figure. The thickness of the line is the velocity dispersion δv, whereas the amplitude of the wiggles in the line is the peculiar velocity Δv. If there were no peculiar velocities, the line would be straight since $\dot{z} = H(t)z$ in that case.

The peculiar velocities are associated with density perturbations and grow by gravitational instability as $\Delta v \sim t^{2/3}$. On the other hand the primordial velocity dispersion decreases on average as $\delta v \sim t^{-2/3}$, consistently with Liouville's theorem. When an overdensity enters the non-linear regime, the particles in its vicinity fall back onto it. This implies that the phase-space sheet 'winds up' there in clockwise fashion. One such overdensity is shown in Fig. 1. In the linear regime, there is only one value of velocity, i.e. one single flow, at a typical location in physical space, because the phase-space sheet covers physical space only once. On the other hand, inside an overdensity in the non-linear regime, the phase-space sheet covers physical space multiple times implying that there are several (but always an odd number of) flows at such locations.

At the boundary surface between two regions one of which has n flows and the other $n + 2$ flows, the physical space density is very large because the phase-space sheet has a fold there. At the fold, the phase-space sheet is tangent to velocity space and hence, in the limit of zero velocity dispersion ($\delta v = 0$), the density diverges since it is the integral of the phase-space density over velocity space. The structure associated with such a phase-space fold is called a 'caustic'. It is a surface in physical space. It is easy to show that, in the limit of zero velocity dispersion, the density diverges as $d \sim \frac{1}{\sqrt{\sigma}}$ when the caustic is approached from the side with $n+2$ flows, where σ is the distance to the caustic. Velocity dispersion cuts off the divergence.

As mentioned above, the process of galactic halo formation involves the local winding up of the phase-space sheet of collisionless dark matter particles. If the galactic center is approached from an arbitrary direction at a given time, the local number of flows increases. First, there is one flow, then three flows, then five, seven... The number of flows at our location in the Milky Way galaxy today has been estimated[2] to be of order 100. The boundary between the region with one (three, five, ...) and the region with three (five, seven, ...) flows is the location of a caustic which is topologically a sphere surrounding the galaxy. When these caustic spheres are approached from the inside the density diverges as $d \sim \frac{1}{\sqrt{\sigma}}$ in the zero velocity dispersion limit. These spheres are the outer caustics in the phase-space structure of galactic halos. In addition there are inner caustics.

It is a little more difficult to see why there must be inner caustics, and to derive their structure. See refs. [3,4] for details. The inner caustics are rings. They are located near where the particles with the most angular momentum in a given in and out flow are at their distance of closest approach to the galactic center. A ring is a closed tube whose cross-section is a D_{-4} catastophe[5]. The cross-section is shown in Fig. 2 in the limit of axial and reflection symmetry, and where the transverse dimensions, p and q, are much smaller than the ring radius a. In the absence of any symmetry, the cross-section of the tube does not have the exact shape shown in Fig. 2 but it still has that shape qualitatively, i.e. it is still a closed line with three cusps one of which points away from the galactic center.

The existence of caustic rings of dark matter follows from only two assumptions:

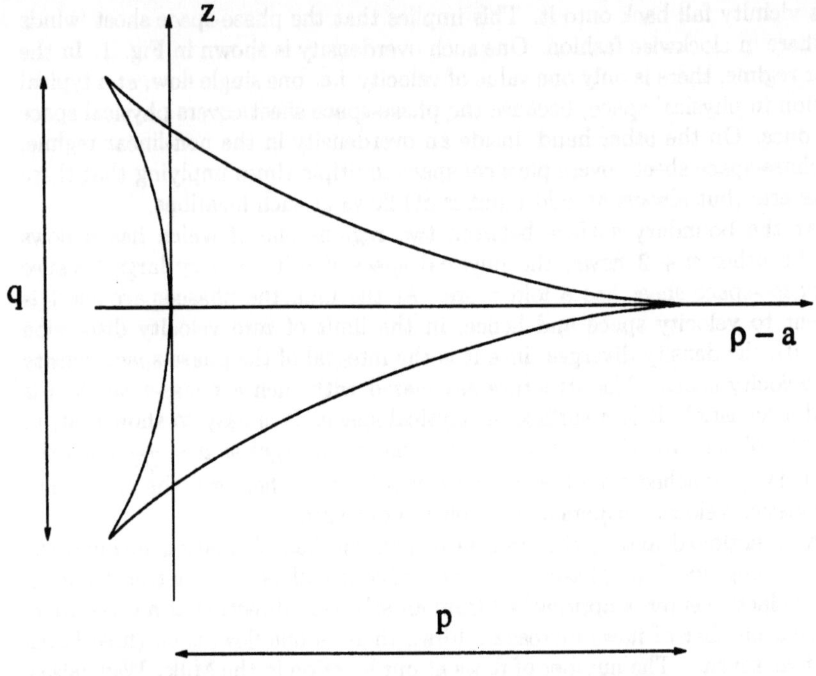

Figure 2: Cross-section of a caustic ring in the case of axial and reflection symmetry. The galactic center is to the left of the figure. The z-direction is orthogonal to the galactic plane. The ρ-direction is radial. a is the caustic ring radius. The closed line with three cusps is the location of a caustic surface. The density diverges when the surface is approached from the inside as $\sigma^{-1/2}$ where σ is the distance to the surface.

1. the existence of collisionless dark matter

2. that the velocity dispersion of the infalling dark matter is much less, by a factor ten say, than the rotation velocity of the galaxy.

Only the second assumption requires elaboration. Velocity dispersion has the effect of smoothing out caustics. The question is when is the velocity dispersion so large as to smooth caustic rings over distance scales of order the ring radius a, thus making the notion of caustic ring meaningless. In ref. [3] this critical velocity dispersion was estimated to be 30 km/s $= 10^{-4}$ for the caustic rings in our own galaxy, whose rotation velocity is 220 km/s. 10^{-4} is much less than the *primordial* velocity dispersion δv of the cold dark matter candidates. However the velocity dispersion Δv associated with density perturbations also smoothes caustics in coarse grained observations. So the question is whether the velocity dispersion Δv of cold dark matter particles associated with density perturbations falling onto our galaxy is less than 30 km/s. The answer is yes with high probability since the infalling dark matter particles are not associated with any observed inhomogeneities. 30 km/s is of order the velocity dispersion of the Magellanic Clouds. For the velocity dispersion of the dark matter particles presently falling onto our galaxy to be as large 30 km/s, these particles would have to be part of clumps whose mass/size ratio is of order that of the Magellanic Clouds. But if that were the case, why did these clumps fail to become luminous?

One might ask whether caustic rings can be seen in N-body simulations of galaxy formation. The generic surface caustics associated with simple folds of the phase-space sheet have been seen[6]. However, caustic rings would require far greater resolution than presently available, at least in a 3D simulation of our own halo. Indeed, the largest ring in our galaxy has been estimated[3] to have radius of order 40 kpc. It is part of an in and out flow that extends to the Galaxy's current turnaround radius, of order 2 Mpc. To resolve this first ring, the spatial resolution would have to be considerably smaller than 10 kpc. Hence a minimum of $2 \cdot \frac{1}{(10\text{kpc})^3} \frac{4\pi}{3} (2\text{Mpc})^3 \simeq 7 \cdot 10^7$ particles would be required to see the caustic ring in a simulation of this one flow. However, the number of flows at 40 kpc in our halo[8] is of order 10. So it appears that 10^9 particles is a strict minimum in a 3D simulation of our halo. Even so, this addresses only the kinematic requirement of resolving the halo in phase-space, assuming moreover that the particles are approximately uniformly distributed on the phase-space sheets. There is a further dynamical requirement that 2-body collisions do not artificially 'fuzz up' the phase-space sheets. Indeed 2-body collisions are entirely negligible in the flow of cold dark matter particles such as axions or WIMPs. On the other hand, 2-body collisions are present, and hence the velocity dispersion is artificially increased, in the simulations. This may occur to such an extent that the caustics are washed away even if 10^9 particles are used.

In the self-similar infall model[7,8] of galactic halo formation the caustic ring radii a_n are predicted[3]:

$$\{a_n : n = 1, 2, ...\} \simeq (39,\ 19.5,\ 13,\ 10,\ 8, ...)\text{kpc} \left(\frac{j_{\max}}{0.25}\right) \left(\frac{0.7}{h}\right) \left(\frac{v_{rot}}{220\frac{km}{s}}\right) \tag{1}$$

where h is the present Hubble rate in units of 100 km/s.Mpc, v_{rot} is the rotation velocity of the galaxy and $j_{\max}$ is the maximum of its dimensionless angular momentum distribution as defined in ref. [8]. In Eq. (1) we assume that the parameter[7,8] $\epsilon = 0.3$.

Eq. (1) predicts the caustic ring radii of a galaxy in terms of its first ring radius a_1. If the caustic rings lie close to the galactic plane they cause bumps in the rotation curve, at the caustic ring radii. As a possible example of this effect, consider[3] the rotation curve of NGC3198, one of the best measured. It has three faint bumps at radii: 28, 13.5 and 9 kpc, assuming $h = 0.75$. The ratios happen to be consistent with Eq. (1) assuming the bumps are caused by the first three ($n = 1, 2, 3$) ring caustics of NGC3198. Moreover, since $v_{rot} = 150$ km/s, j_{max} is determined to equal 0.28. The uncertainty in h is a systematic effect that can be corrected for when determining j_{max} because the bump radii scale like $1/h'$ where h' is the Hubble rate assumed by the observer in constructing the rotation curve, and the caustic ring radii scale as $1/h$. Rises in the inner rotation curve of the Milky Way were also interpreted[3] as due to caustics $n = 6, 7, 8, 9, 10, 11, 12$ and 13. This determined the value of j_{max} of our own galaxy to be 0.263. The first five caustic ring radii in our galaxy are then predicted to be: 41, 20, 13.3, 10, 8 kpc.

2 EVIDENCE FOR UNIVERSAL STRUCTURE IN GALACTIC HALOS

Motivated by these findings, we analyzed[9] a set of 32 extended well-measured galactic rotation curves which had been previously selected[10] under the criteria that each is an accurate tracer of the galactic radial force law, and that it extends far beyond the edge of the luminous disk.

According to the self-similar caustic ring model, each galaxy has its own value of j_{max}. Over the set of 32 galaxies selected in ref. [10], j_{max} has some unknown distribution. However, the fact that the values of j_{max} of NGC3198 and of the Milky Way happen to be close to one another, within 7%, suggests that the j_{max} distribution may be peaked near a value of 0.27 . Our strategy is to rescale each rotation curve according to

$$r \to \tilde{r} = r\,(\frac{220\,\text{km/s}}{v_{rot}}) \tag{2}$$

and to add them in some way. Since Eq. (1) predicts the nth caustic radius a_n to be distributed like j_{max} for all n, and it fixes the ratios $a_n/a_1 \simeq 1/n$, the sum of rotation curves should show the j_{max} distribution, once for $n = 1$, then at about half the $n = 1$ radii for $n = 2$, then at about 1/3 the $n = 1$ radii for $n = 3$,and so on. If the j_{max} distribution is broad, the sum of rotation curves is unlikely to show any feature. However, if it is peaked, then the sum should show a peak for $n = 1$ at some radius, then again at 1/2 that radius for $n = 2$, at 1/3 the radius for $n = 3$, and so on. If the j_{max} distribution is peaked at 0.263 (the value for the Milky Way) the peaks in the sum of rotation curves should appear at 41 kpc, 20 kpc, 13.3 kpc ...

The procedure followed to add the 32 rotation curves is described in detail in ref. [9]. Briefly, we proceeded as follows. For each rotation curve, all data points with rescaled radii $\tilde{r} < 10$ kpc were deleted to remove the effect of the luminous disk. The remaining points were then fitted to a line. The rotation

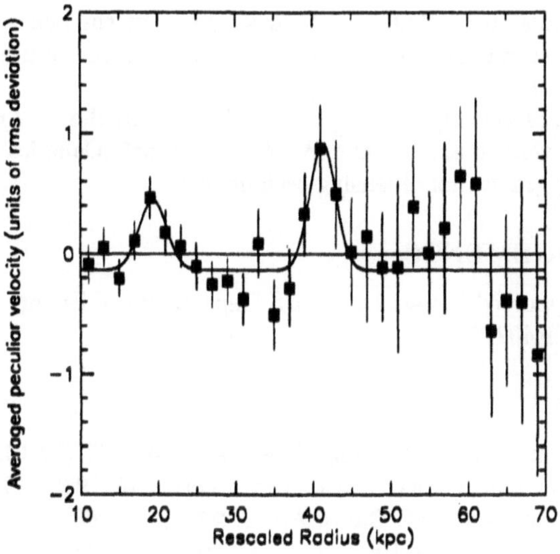

Figure 3: Binned data for 32 galaxy sample, with peaks fit to Gaussians.

velocity v_{rot} used to rescale the radii in Eq. (2) is the average of that line. The rms deviation $\sqrt{\langle \delta v^2 \rangle}$ from the linear fit was determined for each galaxy. This was taken to be the error on the residuals δv_i, i.e. the differences between the measured velocities in a rotation curve and the linear fit. Finally the sample of 32 galaxies was averaged in 2 kpc radial bins:

$$b_i \equiv \frac{1}{N_i} \sum_{j=1}^{N_i} \delta \tilde{v}_j, \tag{3}$$

where N_i is the number of data points in the bin. The assigned error on each b_i is then simply $1/\sqrt{N_i}$. Fig. 3 shows the result.

There are two features evident at roughly 20 and 40 kpc. A fit to two Gaussians plus a constant indicates features at 19.4 ± 0.7 kpc and 41.3 ± 0.8 kpc, with overall significance of 2.4σ and 2.6σ, respectively. Fig. 3 shows the fitted curve. When the same fit is applied to the same data in 1 kpc bins, the significance of the two peaks is 2.6 and 3.0 σ respectively. The locations of the features agrees with the predictions of the self-similar caustic ring model with the j_{max}

distribution peaked at 0.27. The use of Gaussians to fit the peaks in the combined rotation curve was an arbitrary choice in the absence of information on the j_{max} distribution.

The existence of velocity peaks and caustic rings in the cold dark matter distribution is relevant to axion[11] and WIMP searches[12]. Caustics may also be investigated using gravitational lensing techniques[13].

ACKNOWLEDGEMENTS:

This work was supported in part by the US Department of Energy under grant No. DEFG05-86ER40272.

REFERENCES

1. S. Tremaine and J.E. Gunn, Phys. Rev. Lett. **42** 407 (1979); J.R. Bond, G. Efstathiou and J. Silk, Phys. Rev. Lett. **45** 1980 (1980); S.D.M. White, C.S. Frenk and M. Davis, Ap. J. **274** L1 (1983).
2. J.R. Ipser and P. Sikivie, Phys. Lett. **B291** 288 (1992).
3. P. Sikivie, Phys. Lett. B. **432** 139 (1998).
4. P. Sikivie, Phys. Rev. **D60** 063501 (1999).
5. R. Gilmore, *Catastrophy Theory for Scientists and Engineers*, Wiley, 1981.
6. A.G. Doroshkevich et al., M.N.R.A.S. **192** 321 (1980); A.A. Klypin and S.F. Shandarin, M.N.R.A.S. **204** 891 (1983); J.M. Centrella and A.L. Melott, Nature **305** 196 (1983); A.L. Melott and S.F. Shandarin, Nature **346** 633 (1990).
7. J. A. Filmore and P. Goldreich, Ap. J. **281** 1 (1984); E. Bertschinger, Ap. J. Suppl. **58** 39 (1985).
8. P. Sikivie, I.I. Tkachev and Y. Wang, Phys. Rev. Lett. **75** 2911 (1995) and Phys. Rev. **D56** 1863 (1997).
9. W. Kinney and P. Sikivie, Phys. Rev. **D**, 61:087305 (2000).
10. K. G. Begeman, A. H. Broeils and R. H. Sanders, MNRAS **249** 523 (1991); R. H. Sanders, Ap. J. **473** 117 (1996).
11. P. Sikivie, astro-ph/9810286, published in *The Identification of Dark Matter*, edited by N. Spooner and V. Kudryavtsev, World Scientific 1999, pp. 58-65,
12. C. Copi, J. Heo and L. Krauss, Phys. Lett. **B461** 43 (1999).
13. C. Hogan, Ap. J. **527** 42 (1999).

MANIFEST SUPERSYMMETRY AND BACKGROUND RAMOND FLUX

L. Dolan

Department of Physics, University of North Carolina
Chapel Hill, North Carolina 27599-3255, USA

1. INTRODUCTION

Ramond-Ramond p-form fields appear still to contain some secrets in string theory. Recent advances in their quantization using new methods of K-theory[1] have emphasized their special role. Even background Ramond flux has been difficult to incorporate in superstring theory. It couples in the worldsheet action to Ramond vertex operators. In the conventional Ramond-Neveu-Schwarz description, these are given by spin fields which have non-meromorphic OPE's with the supercurrents and therefore would break superconformal symmetry. Unlike Neveu-Schwarz vertex operators which are given by ordinary worldsheet conformal fields, the Ramond states traditionally have been been less well understood.

The Role of Neutrinos, Strings, Gravity, and Variable Cosmological Constant in Elementary Particle Physics, Edited by Kursunoglu *et al.*, Kluwer Academic/Plenum Publishers, New York, 2001

In view of the exciting conjectures relating strings on anti-de Sitter (AdS) space to non-perturbative Yang-Mills gauge theories, and the fact that most of these examples require non-zero Ramond flux to accompany the AdS metric, it has become essential to find a less complicated worldsheet treatment of target space fermions in string theory. Furthermore, it is difficult to quantize the superstring on backgrounds with Ramond flux even using the light-cone gauge Green-Schwarz formalism[2].

Noticeably lacking is a manifestly supersymmetric covariant quantization of the superstring. Either super-Poincare, or other supergroup manifest invariance (relevant for curved background metrics) appears to lead to modified worldsheet variables, so that the original Ramond states now have ordinary conformal field vertex operators instead of spin fields[4−10].

In six dimensions, an $N = (4,4)$ topological description[8] provides worldsheet fields without using spin fields, although only 8 of the 16 supersymmetries are actually manifest.

In ten dimensions, a proposed manifestly supersymmetric covariant formulation[4] involves the superspace variables on the worldsheet x^m, θ^α ($0 \le m \le 9$ and $1 \le \alpha \le 16$), as well as the worldsheet boson field with a spacetime spinor index $\lambda^\alpha(z,\bar{z})$ with the condition $\lambda\gamma^m\lambda = 0$. In contrast to the six-dimensional case, where an $N = (4,4)$ worldsheet conformal field theory is constructed, here the topological description involves only the following structures given in flat space by:

$$T(z) = -\frac{1}{2}\partial x^m \partial x_m - p_\alpha \partial\theta^\alpha + \dots$$
$$Q(z) = \lambda^\alpha d_\alpha$$
$$j_{\text{ghost}}(z) = \gamma\beta$$

where $d_\alpha = p_\alpha + \gamma^m_{\alpha\beta}\partial x_m\theta^\beta + \frac{1}{2}\gamma^m_{\alpha\beta}\gamma_{m\,\gamma\delta}\theta^\beta\theta^\gamma\partial\theta^\delta$, and β, γ are fields conjugate to the spinor[4]. The non-vanishing OPE's are $x^m(z,\bar{z})x^n(\zeta,\bar{\zeta}) = -\eta^{mn}\ln|z-\zeta|$, $p_\alpha(z)\theta^\beta(\zeta) = (z-\zeta)^{-1}\delta^\beta_\alpha$. There are corresponding anti-holomorphic currents denoted by barred generators. In terms of these variables the thirty-two supersymmetry generators take the simple form

$$q^\alpha = \frac{1}{2\pi i}\oint dz\, p^\alpha\,; \qquad \bar{q}^{\bar{\alpha}} = \frac{1}{2\pi i}\oint d\bar{z}\, \bar{p}^{\bar{\alpha}}$$

and all the supersymmetry is *manifest.*

2. SUPERSTRING AS TOPOLOGICAL STRING THEORY

An $N=2$ topological string is characterized by a twisted $N=2$ superconformal algebra:

$$\begin{aligned}
\tilde{T}(z)\tilde{T}(\zeta) &= (z-\zeta)^{-2}2\tilde{T}(\zeta) + (z-\zeta)^{-1}\partial\tilde{T}(\zeta)\,,\\
\tilde{T}(z)G^+(\zeta) &= (z-\zeta)^{-2}G^+(\zeta) + (z-\zeta)^{-1}\partial G^+(\zeta)\,,\\
\tilde{T}(z)G^-(\zeta) &= (z-\zeta)^{-2}2G^-(\zeta) + (z-\zeta)^{-1}\partial G^-(\zeta)\,,\\
G^+(z)G^-(\zeta) &= (z-\zeta)^{-3}\frac{c}{3} + (z-\zeta)^{-2}J(\zeta) + (z-\zeta)^{-1}\tilde{T}(\zeta)\,,\\
G^+(z)G^+(\zeta) &= 0\,, && *\\
G^-(z)G^-(\zeta) &= 0\,,\\
\tilde{T}(z)J(\zeta) &= (z-\zeta)^{-3}(-\frac{c}{3}) + (z-\zeta)^{-2}J(\zeta) + (z-\zeta)^{-1}\partial J(\zeta)\,,\\
J(z)J(\zeta) &= (z-\zeta)^{-2}\frac{c}{3}\,, && *\\
J(z)G^+(\zeta) &= (z-\zeta)^{-1}G^+(\zeta)\,, && *\\
J(z)G^-(\zeta) &= -(z-\zeta)^{-1}G^-(\zeta)\,. && (1.1)
\end{aligned}$$

They are related to the generators of the $N = 2$ superconformal algebra:

$$\begin{aligned}
L(z)L(\zeta) &= (z-\zeta)^{-4}\frac{c}{2} + (z-\zeta)^{-2}2L(\zeta) + (z-\zeta)^{-1}\partial L(\zeta)\,, \\
L(z)G^{\pm}(\zeta) &= (z-\zeta)^{-2}\tfrac{3}{2}G^{\pm}(\zeta) + (z-\zeta)^{-1}\partial G^{\pm}(\zeta)\,, \\
G^{+}(z)G^{-}(\zeta) &= (z-\zeta)^{-3}\frac{c}{3} + (z-\zeta)^{-2}J(\zeta) + (z-\zeta)^{-1}(L(\zeta) + \tfrac{1}{2}\partial J(\zeta))\,, \\
G^{+}(z)G^{+}(\zeta) &= 0\,, \quad G^{-}(z)G^{-}(\zeta) = 0\,, \\
L(z)J(\zeta) &= (z-\zeta)^{-2}J(\zeta) + (z-\zeta)^{-1}\partial J(\zeta)\,, \\
J(z)J(\zeta) &= (z-\zeta)^{-2}\frac{c}{3}\,, \\
J(z)G^{\pm}(\zeta) &= \pm(z-\zeta)^{-1}G^{\pm}(\zeta)\,,
\end{aligned} \tag{1.2}$$

where the generators differ only by the twisted Virasoro generator $\tilde{T}(z) \equiv L(z) + \frac{1}{2}\partial J(z)$. Since the OPE's (1.1) resemble somewhat those of the bosonic string (1.3), we can define physical fields relative to Q_0 cohomology where $G^{+}(z) = \sum_n Q_n z^{-n-1}$. Physical fields will be chiral primaries $\Phi^{+}(z)$ with ghost charge $+1$ and dimension 0, so that $\{Q_0, \Phi^{+}(\zeta)\} = 0$. (For $c = 9$, the starred expressions in (1.1) differ from the corresponding equations in (1.3), nonetheless both sets satisfy Jacobi relations. In the six-dimensional topological version of the superstring[4], $c = 6$ instead of 9, which forces the introduction of an $N = (4,4)$ symmetry.)

Bosonic string theory is described (in BRST language) by a two-dimensional conformal field theory together with the following currents (we concentrate on left-movers and will denote right-movers with barred notation):

$$
\begin{aligned}
T_{\rm tot}(z)T_{\rm tot}(\zeta) &= (z-\zeta)^{-2}\,2T_{\rm tot}(\zeta) + (z-\zeta)^{-1}\partial T_{\rm tot}(\zeta)\,, \\
T_{\rm tot}(z)j_{\rm BRST}(\zeta) &= (z-\zeta)^{-2} j_{\rm BRST}(\zeta) + (z-\zeta)^{-1}\partial j_{\rm BRST}(\zeta)\,, \\
T_{\rm tot}(z)b(\zeta) &= (z-\zeta)^{-2}2b(\zeta) + (z-\zeta)^{-1}\partial b(\zeta)\,, \\
j_{\rm BRST}(z)b(\zeta) &= (z-\zeta)^{-3}3 + (z-\zeta)^{-2} j_{\rm ghost}(\zeta) + (z-\zeta)^{-1}T_{\rm tot}(\zeta)\,, \\
j_{\rm BRST}(z)j_{\rm BRST}(\zeta) &= -j_{\rm BRST}(\zeta)j_{\rm BRST}(z) \\
&= 2\frac{\partial}{\partial\zeta}[(z-\zeta)^{-2} \,{}^{\times}_{\times}\, \partial c(\zeta)\, c(\zeta) \,{}^{\times}_{\times}] \neq 0 \\
b(z)b(\zeta) &= 0\,, \\
T_{\rm tot}(z)j_{\rm ghost}(\zeta) &= (z-\zeta)^{-3}(-3) + (z-\zeta)^{-2} j_{\rm ghost}(\zeta) \\
&\quad + (z-\zeta)^{-1}\partial j_{\rm ghost}(\zeta)\,, \\
j_{\rm ghost}(z)j_{\rm ghost}(\zeta) &= (z-\zeta)^{-2}\,, \\
j_{\rm ghost}(z)j_{\rm BRST}(\zeta) &= (z-\zeta)^{-3}4c(\zeta) + (z-\zeta)^{-2}2\partial c(\zeta) \\
&\quad + (z-\zeta)^{-1} j_{\rm BRST}(\zeta)\,, \\
j_{\rm ghost}(z)b(\zeta) &= -(z-\zeta)^{-1}b(\zeta)\,,
\end{aligned}
\tag{1.3}
$$

where the generators are

$$
\begin{aligned}
T_{\rm tot} &= T_m^{N=0} + T_g^{N=0} = T_m^{N=0} - 2\,{}^{\times}_{\times}\, b\partial c \,{}^{\times}_{\times} - {}^{\times}_{\times}\, \partial b\, c \,{}^{\times}_{\times} \\
j_{\rm BRST}(z) &= cT_m + \tfrac{1}{2}\,{}^{\times}_{\times}\, cT_g \,{}^{\times}_{\times} + \tfrac{3}{2}\partial^2 c = cT_m - {}^{\times}_{\times}\, cb\partial c \,{}^{\times}_{\times} + \tfrac{3}{2}\partial^2 c \\
& b(z) \\
j_{\rm ghost}(z) &= {}^{\times}_{\times}\, cb \,{}^{\times}_{\times}
\end{aligned}
\tag{1.4}
$$

and $c(z)b(\zeta) = -b(\zeta)c(z) = (z-\zeta)^{-1} + {}^{\times}_{\times}\, c(z)b(\zeta) \,{}^{\times}_{\times}$. The normal ordering has been defined putting the annihilation operators to the right of the creation operators, where $b_n|0\rangle^{bc} = 0$ for $n \geq -1$; $c_n|0\rangle^{bc} = 0$ for $n \geq 2$. The BRST charge is $Q \equiv \frac{1}{2\pi i}\oint dz j_{BRST}(z) = \sum_m c_m L_m^X - \frac{1}{2}\sum_{m,n}(m-n) :c_{-m}c_{-n}b_{n+m}:$. Then $Q\left(|0\rangle^{bc} \otimes |0\rangle_X\right) = 0$. The ghost charge is $J_0 \equiv \sum_n :c_n b_{-n}:$. Then $(c_1|0\rangle^{bc} \otimes |\phi\rangle_X)$ has ghost charge eigenvalue, *i.e.* ghost number, equal to one. The physical state conditions in the "old-covariant" formalism of the bosonic string are $(L_0^X - 1)|\phi\rangle_X = 0$, $L_n^X|\phi\rangle_X = 0$ for $n > 0$. Since $Q|\psi\rangle = 0$ implies $\left(c_0(L_0^X - 1) + \sum_n c_{-n}L_n^X\right)|\psi\rangle = 0$ when $|\psi\rangle = c_1|0\rangle^{bc} \otimes |\phi\rangle_X$, then in the BRST formalism the physical vertex operators are defined by ghost number one fields $\Phi^+(z)$ that obey $\{Q, \Phi^+(z)\} = 0$, *i.e.* the OPE of $j_{BRST}(z)\Phi^+(\zeta)$ has no single pole. Every physical field has a representative that is of the form $\Phi^+(z) = c(z)\phi_X(z)$.

As we shall see, the topological description streamlines both the old-covariant and BRST formalisms in that it incorporates physical state conditions as a result of the cohomology, and introduces different worldsheet variables which intermingle, *i.e.* do not divide into matter + ghost parts. These in turn allow for more of the supersymmetry to be manifest.

The $N = 4$ topological string version[8] of the superstring on flat space has generators:

$$
\begin{aligned}
\tilde{T}(z) &= T_m^{N=1} + T_g^{N=1} \\
&= T_m^{N=1} - 2:b\partial c: - :\partial b c: - \tfrac{3}{2}:\beta\partial\gamma: - \tfrac{1}{2}:\partial\beta\gamma: \\
G^+(z) &= \gamma G_m + c(T_m - \tfrac{3}{2}\beta\partial\gamma - \tfrac{1}{2}\partial\beta\gamma - b\partial c) - \gamma^2 b + \partial^2 c + \partial(c\xi\eta) \\
G^-(z) &= b \\
J(z) &= cb + \eta\xi \\
J^+(z) &= c\eta
\end{aligned}
$$

$$
\begin{aligned}
J^-(z) &= b\xi \\
\tilde{G}^+(z) &= \eta \\
\tilde{G}^-(z) &= b(ie^{\phi}G_m + \eta e^{2\phi}\partial b - c\partial\xi) \\
&\quad + \xi(T_m - \tfrac{3}{2}\beta\partial\gamma - \tfrac{1}{2}\gamma\partial\beta - 2b\partial c + c\partial b) + \partial^2\xi\,,
\end{aligned}
\tag{1.5}
$$

with $c(z)b(\zeta) = -b(\zeta)c(z) = (z-\zeta)^{-1} + {}^{\times}_{\times} c(z)b(\zeta) {}^{\times}_{\times}$. Also the super-reparametrization ghosts with $\gamma(z)\beta(\zeta) = \beta(\zeta)\gamma(z) = (z-\zeta)^{-1} + {}^{\times}_{\times} \gamma(z)\beta(\zeta) {}^{\times}_{\times}$ have been bosonized as ($\beta = ie^{-\phi}\partial\xi, \gamma = -i\eta e^{\phi}$) with $\xi(z)\eta(\zeta) = -\eta(\zeta)\xi(z) = (z-\zeta)^{-1} + {}^{\times}_{\times} \xi(z)\eta(\zeta) {}^{\times}_{\times}$, and $\phi(z)\phi(\zeta) = -\ln(z-\zeta) + {}^{\times}_{\times} \phi(z)\phi(\zeta) {}^{\times}_{\times}$ so that $e^{-\phi(z)}e^{\phi(\zeta)} =: e^{-\phi(z)+\phi(\zeta)} : (z-\zeta)$. These satisfy the ($c = 6$) twisted $N = 4$ superconformal algebra. For the IIB superstring we have both a holomorphic $N = 4$ superconformal algebra and another anti-holomorphic one. In terms of new worldsheet variables (that display manifest 6D spacetime supersymmetry) the holomorphic ones specialized for IIB compactified to 6D (1.5) are given by:

$$
\begin{aligned}
\tilde{T}(z) &= -\tfrac{1}{2}\partial x^m\partial x_m - p_a\partial\theta^a - \tfrac{1}{2}\partial\rho\partial\rho - \tfrac{1}{2}\partial\sigma\partial\sigma + \tfrac{3}{2}\partial^2(\rho+i\sigma) + \tilde{T}_C \\
G^+(z) &= -e^{-2\rho-i\sigma}(p)^4 + \tfrac{i}{2}e^{-\rho}p_a p_b\partial x^{ab} + e^{i\sigma}(-\tfrac{1}{2}\partial x^m\partial x_m \\
&\quad - p_a\partial\theta^a - \tfrac{1}{2}\partial(\rho+i\sigma)\partial(\rho+i\sigma) + \tfrac{1}{2}\partial^2(\rho+i\sigma)) + G_C^+ \\
G^-(z) &= e^{-i\sigma} + G_C^- \\
J(z) &= \partial(\rho+i\sigma) + J_C \\
J^+(z) &= e^{\rho+i\sigma}J_C^+ = -e^{\rho+i\sigma+iH_C} \\
J^-(z) &= -e^{-\rho-i\sigma}J_C^- = -e^{-\rho-i\sigma-iH_C} \\
\tilde{G}^+(z) &= e^{iH_C+\rho} + e^{\rho+i\sigma}\tilde{G}_C^+ \\
\tilde{G}^-(z) &= e^{-iH_C}[-e^{-3\rho-2i\sigma}(p)^4 - \tfrac{i}{2}e^{-2\rho-i\sigma}p_a p_b\partial x^{ab} + e^{-\rho}(-\tfrac{1}{2}\partial x^m\partial x_m \\
&\quad - p_a\partial\theta^a - \tfrac{1}{2}\partial(\rho+i\sigma)\partial(\rho+i\sigma) + \tfrac{1}{2}\partial^2(\rho+i\sigma))] \\
&\quad - e^{-\rho-i\sigma}\tilde{G}_C^-\,.
\end{aligned}
\tag{1.6}
$$

These currents are given in terms of the left-moving bosons $\partial x^m, \rho, \sigma$, and the left-moving fermionic worldsheet fields p_a, θ^a, where $0 \leq m \leq 5; 1 \leq a \leq 4$. All fields are ordinary conformal fields, no spin fields occur in this formulation. The conformal weights of p_a, θ^a are 1 and 0, respectively. We define $p^4 \equiv \frac{1}{24}\epsilon^{abcd}p_a p_b p_c p_d = p_1 p_2 p_3 p_4$; and $\partial x^{ab} = \partial x^m \sigma_m^{ab}$ where $\sigma_m^{ab}\sigma_{nac} + \sigma_n^{ab}\sigma_{mac} = \eta_{mn}\delta_c^b$. Here lowered indices mean $\sigma_{mab} \equiv \frac{1}{2}\epsilon_{abcd}\sigma_m^{cd}$. Note that e^ρ and $e^{i\sigma}$ are worldsheet fermions. Also $e^{\rho+i\sigma} \equiv e^\rho e^{i\sigma} = -e^{i\sigma}e^\rho$. Here $J_C \equiv i\partial H_C, J_C^+ \equiv -e^{iH_C}, J_C^- \equiv e^{-iH_C}$. Both $\tilde{T}, G^\pm, J, J^\pm, \tilde{G}^\pm$ and the generators describing the compactification $\tilde{T}_C, G_C^\pm, J_C, J_C^\pm, \tilde{G}_C^\pm$ satisfy the twisted $N = 4$, $c = 6$, superconformal algebra (1.8), *i.e.* both $\tilde{T}$ and $\tilde{T}_C$ have $c = 0$. However, as seen in (1.8) and (1.1), c still appears in the twisted $N = 4$ and $N = 2$ algebras.

The other non-vanishing OPE's are $x^m(z,\bar{z})x^n(\zeta,\bar{\zeta}) = -\eta^{mn}\ln|z-\zeta|$; for the left-moving worldsheet fermion fields $p_a(z)\theta^b(\zeta) = (z-\zeta)^{-1}\delta_a^b$; and for the left-moving worldsheet bosons $\rho(z)\rho(\zeta) = -\ln(z-\zeta)$; $\sigma(z)\sigma(\zeta) = -\ln(z-\zeta)$. Right-movers are denoted by barred notation and have similar OPE's.

In this description, the physical states satisfy the cohomology

$$G_0^+\Phi^+ = 0; \qquad \tilde{G}_0^+\Phi^+ = 0; \qquad (J_0 - 1)\Phi^+ = 0. \tag{1.7}$$

For the IIB superstring on $AdS3 \times S3 \times K3$, the generators (1.6) are modified[8], and the cohomology (1.7) implies constraints on the AdS vertex operators[9]. For this case, we review in the last section the 16 supersymmetries. We show how they are all given by ordinary conformal fields, *i.e.* not spin fields, and how 8 of them act geometrically on functions of the group manifold, *i.e.* how 8 of them are manifest.

The twisted $c = 6$, $N = 4$ superconformal algebra is

$$\begin{aligned}
\tilde{T}(z)\tilde{T}(\zeta) &= (z-\zeta)^{-2}2\tilde{T}(\zeta) + (z-\zeta)^{-1}\partial\tilde{T}(\zeta)\,, \\
\tilde{T}(z)G^{+}(\zeta) &= (z-\zeta)^{-2}G^{+}(\zeta) + (z-\zeta)^{-1}\partial G^{+}(\zeta)\,, \\
\tilde{T}(z)G^{-}(\zeta) &= (z-\zeta)^{-2}2G^{-}(\zeta) + (z-\zeta)^{-1}\partial G^{-}(\zeta)\,, \\
G^{+}(z)G^{-}(\zeta) &= (z-\zeta)^{-3}\frac{(c=6)}{3} + (z-\zeta)^{-2}J(\zeta) \\
&\quad + (z-\zeta)^{-1}\tilde{T}(\zeta)\,, \\
G^{+}(z)G^{+}(\zeta) &= 0\,, \qquad G^{-}(z)G^{-}(\zeta) = 0\,, \\
\tilde{T}(z)J(\zeta) &= (z-\zeta)^{-3}(-\frac{(c=6)}{3}) + (z-\zeta)^{-2}J(\zeta) \\
&\quad + (z-\zeta)^{-1}\partial J(\zeta)\,, \\
J(z)J(\zeta) &= (z-\zeta)^{-2}\frac{(c=6)}{3}\,, \\
J(z)G^{\pm}(\zeta) &= \pm(z-\zeta)^{-1}G^{\pm}(\zeta)\,, \\
J(z)J^{\pm}(\zeta) &= (z-\zeta)^{-1}(\pm 2)J^{\pm}(\zeta)\,, \\
J^{+}(z)J^{-}(\zeta) &= -(z-\zeta)^{-2}\frac{(c=6)}{6} - (z-\zeta)^{-1}J(\zeta)\,, \\
\tilde{T}(z)J^{+}(\zeta) &= (z-\zeta)^{-1}\partial J^{+}(\zeta)\,, \\
\tilde{T}(z)J^{-}(\zeta) &= (z-\zeta)^{-2}2J^{-}(\zeta) + (z-\zeta)^{-1}\partial J^{-}(\zeta)\,, \\
J^{+}(z)G^{-}(\zeta) &= -(z-\zeta)^{-1}\tilde{G}^{+}(\zeta)\,, \qquad J^{-}(z)G^{-}(\zeta) = 0\,, \\
J^{-}(z)G^{+}(\zeta) &= -(z-\zeta)^{-1}\tilde{G}^{-}(\zeta)\,, \qquad J^{+}(z)G^{+}(\zeta) = 0\,, \\
J^{-}(z)\tilde{G}^{+}(\zeta) &= (z-\zeta)^{-1}G^{-}(\zeta)\,, \qquad J^{-}(z)\tilde{G}^{-}(\zeta) = 0\,, \\
J^{+}(z)\tilde{G}^{-}(\zeta) &= (z-\zeta)^{-1}G^{+}(\zeta)\,, \qquad J^{+}(z)\tilde{G}^{+}(\zeta) = 0\,, \\
G^{-}(z)\tilde{G}^{+}(\zeta) &= 0\,, \qquad G^{+}(z)\tilde{G}^{-}(\zeta) = 0\,, \\
\tilde{G}^{+}(z)\tilde{G}^{-}(\zeta) &= (z-\zeta)^{-3}\frac{(c=6)}{3} + (z-\zeta)^{-2}J(\zeta) \\
&\quad + (z-\zeta)^{-1}\tilde{T}(\zeta)\,, \\
G^{+}(z)\tilde{G}^{+}(\zeta) &= (z-\zeta)^{-2}2J^{+}(\zeta) + (z-\zeta)^{-1}\partial J^{+}(\zeta)\,,
\end{aligned}$$

$$G^-(z)\tilde{G}^-(\zeta) = (z-\zeta)^{-2}2J^-(\zeta) + (z-\zeta)^{-1}\partial J^-(\zeta),$$
$$\tilde{G}^+(z)\tilde{G}^+(\zeta) = 0, \quad \tilde{G}^-(z)\tilde{G}^-(\zeta) = 0$$
$$\tilde{T}(z)\tilde{G}^+(\zeta) = (z-\zeta)^{-2}\tilde{G}^+(\zeta) + (z-\zeta)^{-1}\partial\tilde{G}^+(\zeta),$$
$$\tilde{T}(z)\tilde{G}^-(\zeta) = (z-\zeta)^{-2}2\tilde{G}^-(\zeta) + (z-\zeta)^{-1}\partial\tilde{G}^-(\zeta),$$
$$J(z)\tilde{G}^\pm(\zeta) = \pm(z-\zeta)^{-1}\tilde{G}^\pm(\zeta). \tag{1.8}$$

3. GEOMETRIC ACTION OF SUPERSYMMETRIES

For the IIB superstring on $AdS3 \times S3 \times K3$, the worldsheet fields are denoted by F_a, E_a, K_{ab} and their bars. The supersymmetry generators are given by

$$q_a^- = \oint F_a$$
$$q_a^+ = \oint (e^{-\rho-i\sigma} F_a + iE_a)$$
$$\bar{q}_a^- = \oint \bar{F}_a$$
$$\bar{q}_a^+ = \oint (e^{-\rho-i\sigma} \bar{F}_a + i\bar{E}_a).$$

Together with the $SO(4) \times SO(4)$ generators $K_{ab}, \bar{K}_{ab}$, they close the $PSU(2|2) \times PSU(2|2)$ supergroup, which replaces the super-Poincare group since we are on an AdS space which has this isometry group. In what follows, we distinguish between the worldsheet fields and their zero moments F_a, E_a, K_{ab}, and give their action on a function $g(x, \theta, \bar{\theta})$ on the $PSU(2|2)$ supergroup manifold. The operators E_a, F_a, and K_{ab} represent the left action of e_a, f_a, and t_{ab} on g. In the above coordinates,

$$F_a = \frac{d}{d\theta^a}, \qquad K_{ab} = -\theta_a \frac{d}{d\theta^b} + \theta_b \frac{d}{d\theta^a} + t_{Lab}$$
$$E_a = \tfrac{1}{2}\epsilon_{abcd}\,\theta^b\,(t_L^{cd} - \theta^c \frac{d}{d\theta_d}) + h_{a\bar{b}} \frac{d}{d\bar{\theta}_{\bar{b}}}, \tag{1.9}$$

where we have introduced an operator t_L that generates the left action of $SU(2) \times SU(2)$ on h alone, without acting on the θ's. Here

$$g = g(x, \theta, \bar{\theta}) = e^{\theta^a f_a} e^{\frac{1}{2}\sigma^{pcd} x_p t_{cd}} e^{\bar{\theta}^{\bar{a}} e_{\bar{a}}} = e^{\theta^a f_a} h(x) e^{\bar{\theta}^{\bar{a}} e_{\bar{a}}}$$

$$t_{Lab}\, g = e^{\theta^a f_a} (-t_{ab}) h(x) e^{\bar{\theta}^{\bar{a}} e_{\bar{a}}},$$

and we found[9] (1.9) by requiring $F_a g = f_a g$, $E_a g = e_a g$, $K_{ab} g = -t_{ab} g$. Similar expressions hold for the right-acting generators $\bar{K}_{\bar{a}\bar{b}}$, $\bar{E}_{\bar{a}}$, and $\bar{F}_{\bar{a}}$. Therefore the 8 supersymmetries F_a and $\bar{F}_a$ act geometrically on the target space, that is as a simple derivative. And these supersymmetries are manifest. The other 8 given in terms of $E_a, \bar{E}_a$ are not.

REFERENCES

1. D. Diaconescu, G. Moore, and E. Witten, *A Derivation of K theory from M theory*, hep-th/0005091; *E(8) Gauge Theory, and a Derivation of K theory from M theory*, hep-th/0005090; G. Moore and E. Witten, *Self-Duality, Ramond-Ramond Fields, and K-theory*, JHEP 0005:032,2000, hep-th/0012279; D. Freed, and M. Hopkins, *On Ramond-Ramond Fields and K Theory*, JHEP 0005:044,2000; hep-th/0002027.

2. A.A. Tseytlin, Superstrings in AdS in Light-cone Gauge, hep-th/0009226; R. Kallosh and A.A. Tseytlin, Symplifying Superstring Action on AdS5 x S5, JHEP 9810:016,1998, hep-th/9808088; N. Drukker and D. Gross, *Green-Schwarz String in AdS5 x S5 : Semiclassical Partition Function*, JHEP 0004:021,2000, hep-th/0001204.

3. N. Berkovits, *A New Description of the Superstring*, hep-th/9604123.

4. N. Berkovits, *Super-Poincare Covariant Quantization of the Superstring*, JHEP 0004:018,2000, hep-th/0001035.

5. N. Berkovits, *Cohomology in the Pure Spinor Formalism for the Superstring*, JHEP 0009:046,2000, hep-th/0006003.

6. N. Berkovits, *Consistency of Super-Poincare Covariant Superstring Tree Amplitudes* JHEP 0007:015,2000, hep-th/0004171.

7. N. Berkovits, *Covariant Quantization of the Superstring*, hep-th/0008145.

8. N. Berkovits, C. Vafa, and E. Witten, *Conformal Field Theory of AdS Background With Ramond-Ramond Flux*, **JHEP 9903: 018, 1999**, hep-th/9902098.

9. L. Dolan and E. Witten, *Vertex Operators for AdS3 Background With Ramond-Ramond Flux*, **JHEP 11:003:1999**, hep-th/9910205.

10. N. Berkovits and O. Chandia, *Superstring Vertex Operators in an AdS5 x S5 Background*, Nucl.Phys.B596:185-196,2001, hep-th/0009168.

NEUTRINO PHYSICS AND COSMOLOGY

SUPERNOVA II NEUTRINO BURSTS AND NEUTRINO MASSIVE MIXING

David B. Cline*

1. INTRODUCTION OF THE NEUTRINO SPECTRUM FROM A SN II

The issue of whether or not neutrinos have masses is important for astrophysics and cosmology. Astrophysical considerations may represent the best hope for determining neutrino masses and mixings. In this paper, we examine how proposed neutral-current-based, supernova neutrino-burst detectors, in conjunction with the next generation water-Čerenkov detectors, could use a galactic supernova event to either measure or place constraints on the $\nu_{\mu,\tau}$ masses in excess of 5 eV[1,2]. Such measurements would have important implications for our understanding of particle physics, cosmology, and the solar neutrino problem and would be complementary to proposed laboratory vacuum-oscillation experiments.

A light neutrino mass between 1 eV and 100 eV would be highly significant for cosmology. In fact, if a neutrino contributes a fraction Ω_ν of the closure density of the Universe, it must have a mass $m_\nu \approx 92\, \Omega_\nu h^2$ eV, where h is the Hubble parameter in units of 100 km s^{-1} Mpc^{-1}. Reasonable ranges for Ω_ν and h then give 1 eV to 30 eV as a cosmologically significant range. A neutrino with a mass in the higher end of this range (*i.e.*, $10 \leq m_\nu \leq 30$ eV) could contribute significantly to the closure density of the Universe. The cosmic background explorer (COBE) observation of anistropy in the microwave background, combined with observations at smaller scales, and the distribution of galaxy streaming velocities, have been interpreted as implying that there are two components of dark matter: hot ($\Omega_{HDM} \sim 0.3$) and cold ($\Omega_{CDM} \sim 0.6$).

* Department of Physics and Astronomy, Box 951647 University of California Los Angeles Los Angeles, CA 90095-1647, USA

The Role of Neutrinos, Strings, Gravity, and Variable Cosmological Constant in Elementary Particle Physics, Edited by Kursunoglu *et al.*, Kluwer Academic/Plenum Publishers, New York, 2001

The hot dark matter (HDM) component could be provided by a neutrino with a mass of about 7 eV.**

Scheme	$\nu_\odot$	ν_{atmos}	LBL	SBL	SN ν's	BBN	SNN
I 3ν mixing No LSND	Yes $\nu_e \rightarrow \nu_{\mu,\tau}$	Yes $\nu_\mu \rightarrow \nu_\tau$	Yes $\nu_\mu \rightarrow \nu_\tau$	No $\nu_\tau \rightarrow \nu_e$ $\nu_\mu \rightarrow \nu_e$ τ appearance?	√	OK	OK
II 4ν mixing $\nu_\mu \rightarrow \nu_\tau$ (Doublet) $\nu_\odot \rightarrow \nu_s$ LSND	Yes $\nu_e \rightarrow \nu_s$ (No extra N.C. signal)	Yes $\nu_\mu \rightarrow \nu_\tau$	Yes $\nu_\mu \rightarrow \nu_\tau$	No? $\nu_e \rightarrow \nu_\tau$ τ appearance? $\nu_e \rightarrow \nu_\tau$? $\nu_e \rightarrow \nu_\mu$?	√ Hot ? Maybe Maybe	D/N ν_e- spectrum	?? Good or bad ??
III 4ν mixing $\nu_\mu \rightarrow \nu_s$ Doublet No LSND	Yes $\nu_e \rightarrow \nu_\mu$	Yes $\nu_\mu \rightarrow \nu_s$	Yes $\nu_\mu \rightarrow \nu_s$	No $m_{\nu\tau}$ $\nu_\mu \rightarrow \nu_\tau$?	√ ? 	? T of F	r-process constraint

2. EFFECTS OF NEUTRINO MIXING FOR SN II

SN II emit all types of neutrinos and anti-neutrinos with about the same luminosity per neutrino on anti-neutrino species. The expected energy spectrum is very different as can be seen in Fig. 2. Table 1 gives a set of optionsfor mixing and tests.

The possible different models of neutrino mass is shown in Fig. 2. These different mass structures have important consequences for the MSW effect in the SN II and the search for the correct neutrino mass spectrum.

3. A NEW ANALYSIS OF SN1987A AND THE LMA SOLAR NEUTRINO SOLUTION

Because the ν_x spectrum is expected to be harder than the ν_e or $\bar{\nu}_e$ (Fig. 3), one signature for neutrino oscillation is to observe a hard component in the ν_e or $\bar{\nu}_e$ spectrum

** While current cosmological data do not favor this, it is not completely excluded.

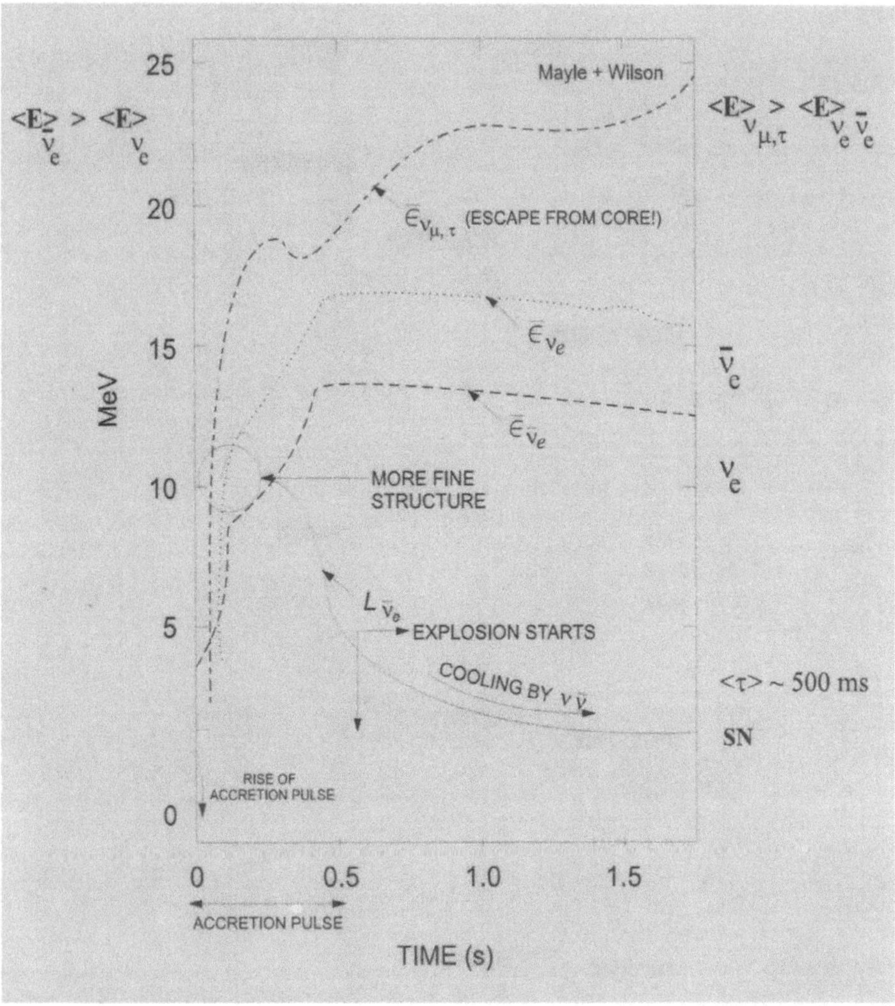

Fig. 1. The expected neutrino luminosity and average energy of the different species following the work of Mayle and Wilson..

from the process $\nu_x \to \nu_e$ or $\bar{\nu}_x \to \bar{\nu}_e$ [2]. We can define the neutrino flux in the following way:

$$\langle \nu_e \rangle = (1-p)\langle \nu_e \rangle_0 + p\langle \nu_x \rangle_0 \ , \tag{1}$$

$$\langle \bar{\nu}_e \rangle = (1-\bar{p})\langle \bar{\nu}_e \rangle_0 + \bar{p}\langle \bar{\nu}_x \rangle_0 \ , \tag{2}$$

where $\langle \nu_e \rangle_0$, $\langle \bar{\nu}_e \rangle_0$, $\langle \nu_x \rangle_0$, $\langle \bar{\nu}_x \rangle_0$ denote the unmixed neutrino spectra and $p, \bar{p}$ the mixing fraction.

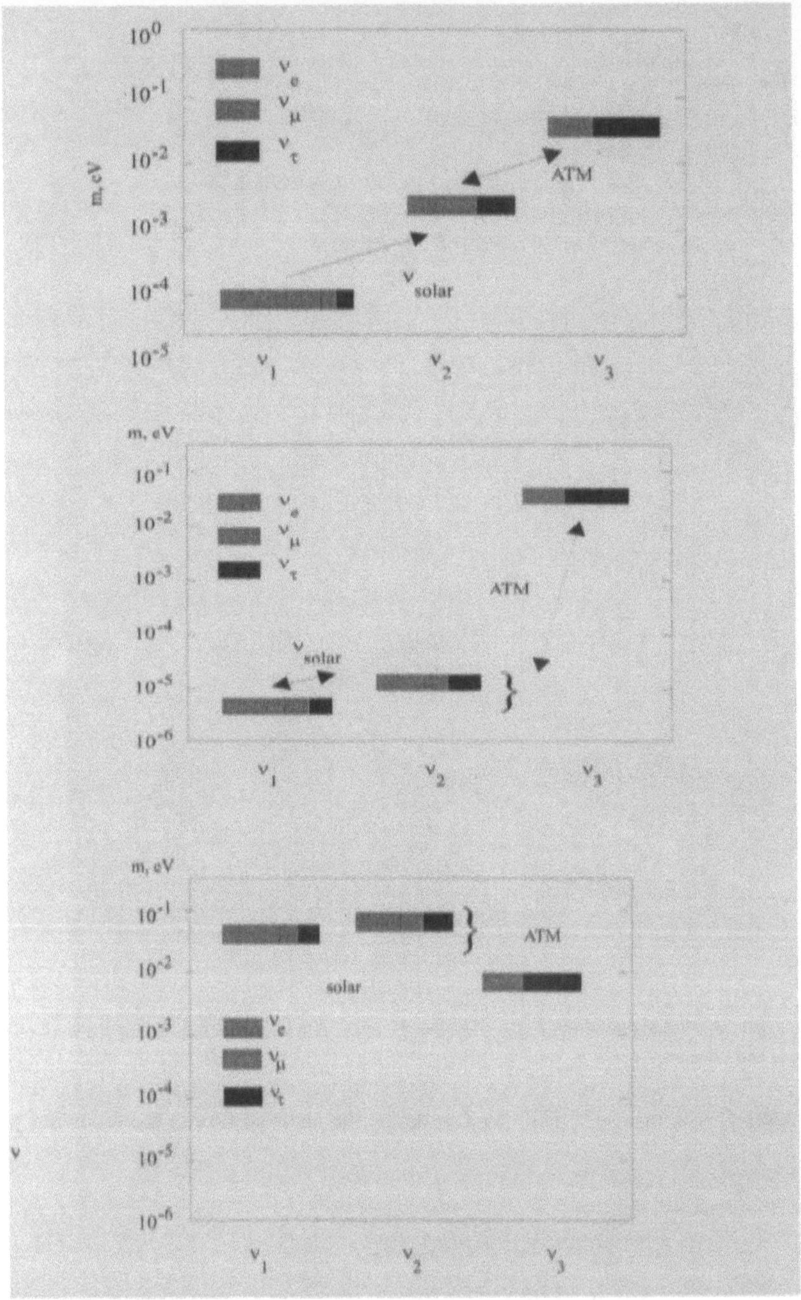

Fig. 2a, 2b, 2c. **Different schemes for the Neutrino Mass spectrum adopted from Ref. 3.**

First we comment on the Kamiokande and IMB event populations:

1. The IMB detector had a strong bias against low energy events.
2. The mass of the IMB detector was about three times larger than the Kamiokande detector and thus was more sensitive to higher energy neutrino events that are less probable.
3. The Kamiokande detector had excellent low energy properties, as was later demonstrated by the observation of solar neutrinos.
4. Some of the pmt's for the IMB detector were not operational during the recording of SN1987A events.

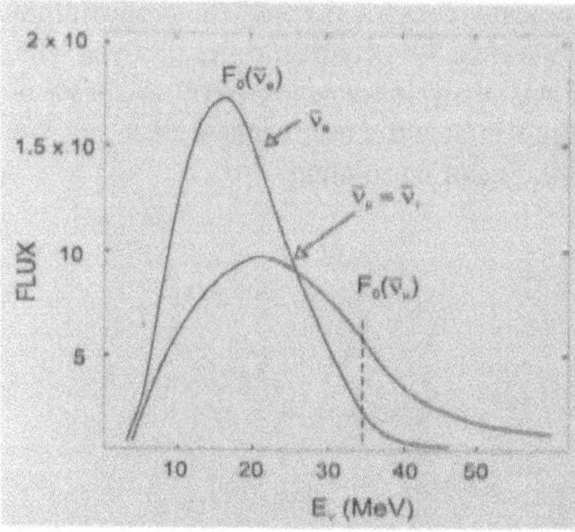

Figure 3. Expected $\bar{\nu}_e$ and $\bar{\nu}_x$ spectra from an SNII explosion.

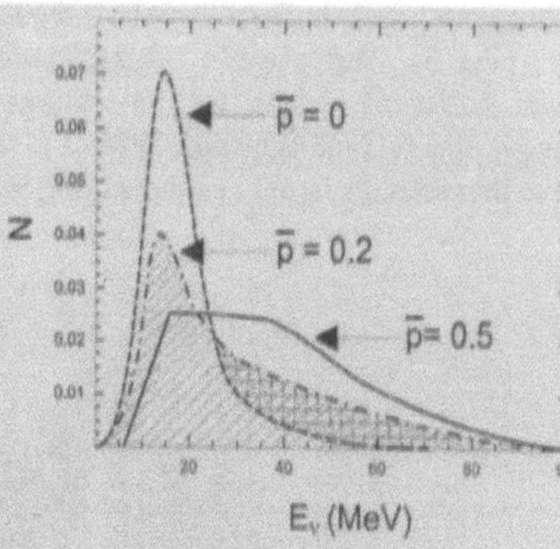

Figure 4. Events distribution expected for different values of the mixing $\bar{p}$.

In the Krauss analysis, an attempt was made to incorporate both populations of events by correcting the model as shown in Fig. 5. While this showed an acceptable fit, the mean temperature of the distribution was 4.5 MeV, which was higher than would be expected normally.

Smirnov, Spergel, and Bahcall tried another analysis [7]. They fit the combined Kamiokande and IMB data with a model that allowed for $\nu_x - \nu_e$ mixing as given by formula (2). They did not correct for the obvious differences in the Kamiokande and IMB event population (threshold different detector ,masses, possible differences in detection efficiency, etc.) but just added up the integral of the total event energies. They gave an exhaustive discussion of the different types of temperature distributions that may occur for the different neutrino flavors from the supernova emission. They expressed their results as a function of the mean energy of the ν_x neutrinos. Since most models gave this value to be 22 MeV or greater, we use that value in the results shown here. In Fig. 5, we have replotted the results of this analysis for the 95% confidence level reported in the paper. Note that most of the LMA and all of the vacuum oscillation (or "just so") is excluded.

One can be critical of this analysis due to the fact that no attempt was made to correct for the different experimental conditions in the Kamiokande and IMB experiments.

However, these results may well be a conservative lower limit, since corrections for the experimental differences will decrease the impact of the high energy events in IMB, as shown by the Krauss analysis.

3a. A NEW ANALYSIS OF THE SN1987A DATA FOR NEUTRINO MIXING

Because of the problems of comparing the two populations of events illustrated in points 1 through 4 above, we propose that a sensible analysis should use the data set with the least bias. Based on the Kamiokande data alone, while this set has no event with an energy above 40 MeV, there is no reason why the detector would not have recorded such events had they been produced. In Fig. 6, we show the Kamiokande data and the predictions of neutrino mixing (Fig. 3). The case $\bar{p} = 1/2$ is excluded to at least 99% confidence level. Even with a lower statistical sample, the conclusions of this analysis are as powerful as those of Smirnov, Spergel, and Bahcall. Table 2 gives the Kolmogurov test for these data [8]. We believe these results largely exclude the LMA solar neutrino solution.

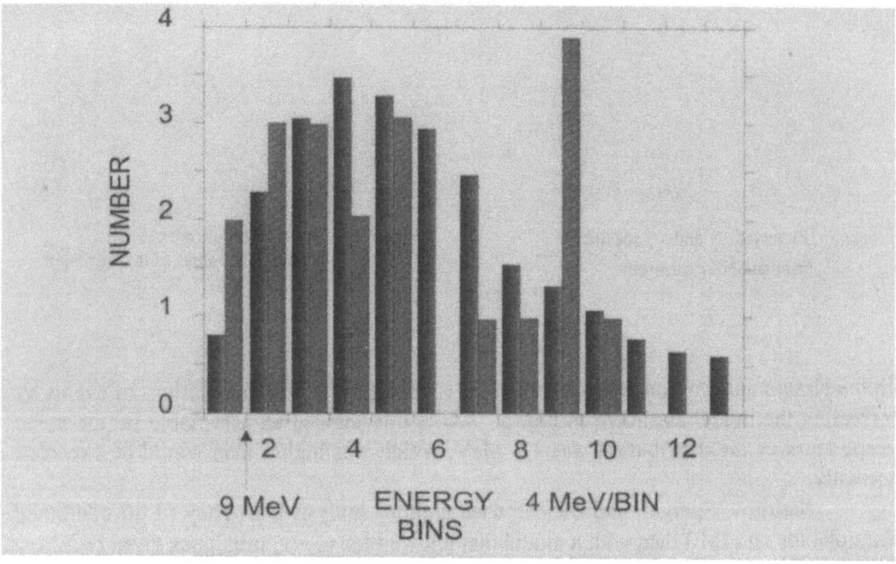

Figure 5. Krauss analysis of SN1987A data. The hatched events are the combined data and the solid blocks are the results of a model that incororates the effects of the detectors, etc. [6]. Neutrino mixing was not assumed.

4. POSSIBLE DETECTION OF THE DIFFERENT RELIC NEUTRINO FLUX FROM PAST SN II

Another kind of relic neutrinos are the neutrinos that arise from the integrated flux from all past type-II supernovae. Figure 8 shows a schematic of these (and other) fluxes. These fluxes could be modified by transmission through the SNII environment, as discussed recently. [1]

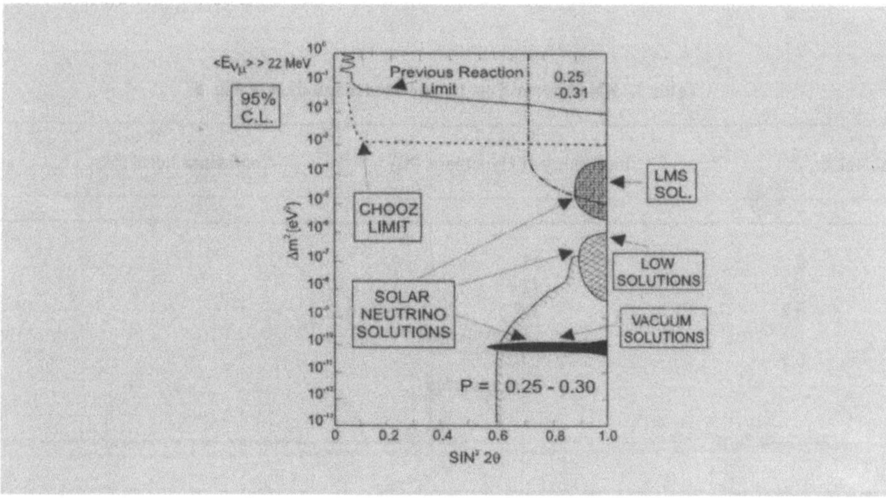

Figure 6. Limits on $\bar{\nu}_e \rightarrow \bar{\nu}_\mu$ from SN1987A (from Ref. [7]).

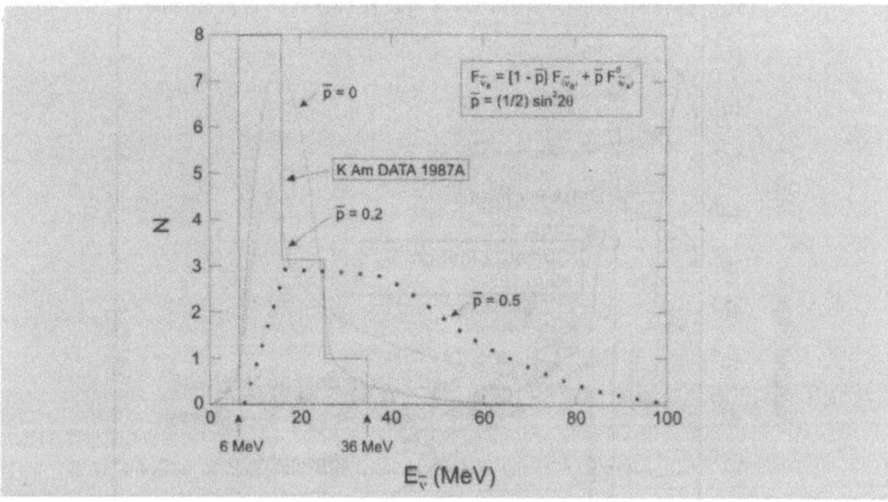

Figure 7. Comparison of the Kamiokande data with the neutrino oscillation models.

The detection of $\bar{\nu}_e$ from the relic supernovae may someday be accomplished by the SK detector. It would be as interesting to detect ν_e with an ICARUS detector, as illustrated in Table 3. High-energy ν_e would come from $\nu_{\mu,\tau} \rightarrow \nu_e$ in the supernova. A window of detection occurs between the upper solar neutrino energy and the atmospheric neutrinos,as first proposed by D. Cline and reported in the first ICARUS proposal (1983-1985). The ideal detector to observe this is a large ICARUS liquid-argon detector [1].

Table 2. Kolmogurov Test for the Model and data in Fig. 8.

Parameter, $\bar{p}$	Probability of Hypothesis (%)	Confidence Level (%)
0	58	
	42	
0.2	3.6	
	96.4 excluded	
0.5	0.02	
	> 99 excluded	

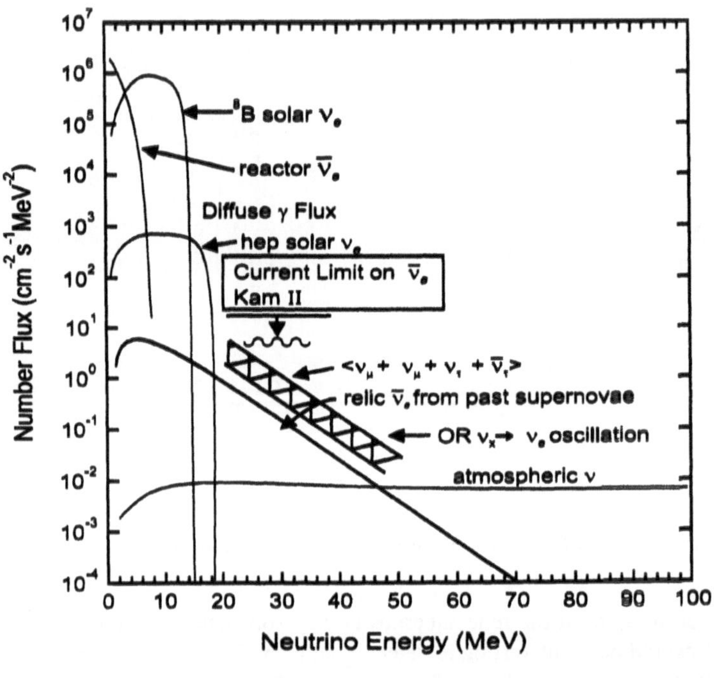

FIGURE 8. Relic neutrinos from past supernova. Note: $\nu_x \to \nu_e$ in the supernova can boost the energy of the ν_e if we find $<E\nu_e> \gg <E\bar{\nu}_e>$. This will be a signal for neutrino oscillation in supernovae! and measure $\sin^2 \theta x_e$. [1]

Table 3. Detection of $\nu/\bar{\nu}_e$ relic neutrino flux from time integrated SNII.

1.	Relic $\nu/\bar{\nu}_e$ from all SNII back to $Z \sim 5$: $<E_\nu> \sim 1/(1+Z)<E_\nu>$
2.	Detection would give integrated SNII rate from Universe - Window of detetion [D. Cline, ICARUS proposal, 1984]
3.	Neutrino oscillations in SNII would give $\nu_x \to \nu_e$ with higher energy than $\bar{\nu}_e$
4.	Detect $\bar{\nu}_e$ with SK or ICARUS. Attempt to detect ν_x/ν_e detection.

5. THE OMNIS DETECTOR CONCEPT AND OTHER SUPERNOVA NEUTRINO DETECTORS

Recently there has been real progress in supernova simulations giving an explosion. These calculations give interesting predictions for the neutrino spectra. Detectors like the SK and SNBO/OMNIS may be able to detect such effects, however the SNBO/OMNIS detector may be of crucial importance for this study [9]. Using these various detectors, it should be possible to detect a finite neutrino mass. The characteristics of this detector are listed in Table 4.

In this analysis, we have assumed the existence of a very massive neutral-current detector (the SNBO/OMNIS), which we discuss next [9]. By using these different detectors it will be possible to measure the μ or τ neutrino masses, as shown in Table 3, which could determine a mass to ~ 10 eV [9]. To go to lower mass, we need to use the possible fine structure in the burst; we have shown that it may be possible to reach ~ 3 eV with very large detectors in this case. The detection of two-neutron final stateswould be useful for Pb detection. [10].

Table 4. Properties of the Pproposed) OMNIS/SNBO Detector

Targets:	NaCℓ (WIPP site) Fe and Pb (Soudan and Boulby sites)
Mass of Detectors:	WIPP site ≥ 200 ton Soudan/Boulby sites ≥ 200 ton
Types of Detectors:	Gd in liquid scintillator
	^{6}Li loaded in the plastic scintillators that are read out by scintillating-fiber-PMT system

The major problem of supernova detection is the uncertain period of time between such processes in this Galaxy. In addition, complimentary detectors should be active when the supernova goes off in order to gain the maximum amount of information possible about the explosion process and neutrino properties.

Lacking an ideal observatory, a group of us have been studying a very large detector, SNBO/OMNIS. Table 4 gives some of the guidelines for this detector. We have located a possible site for the observatory near Carlsbad, NM, which is the WIPP site (shown in Fig. 9). We believe that expectation that the background for a galactic supernova is much smaller than the signal at this site. A schematic of the NBO/OMNIS detector is shown in Fig. 9.

The basic idea of OMNIS/SNBO is to develop a long lived inexpensive SN II detector that detects υ_τ and υ_μ neutrinos by the neutral current process. The Carlsbad site gives a unique long-lasting underground laboratory.

In Table 5 we give the rates of different detectors for a galactic supernova and Figure 10 shows the rates by neutrino species [1].

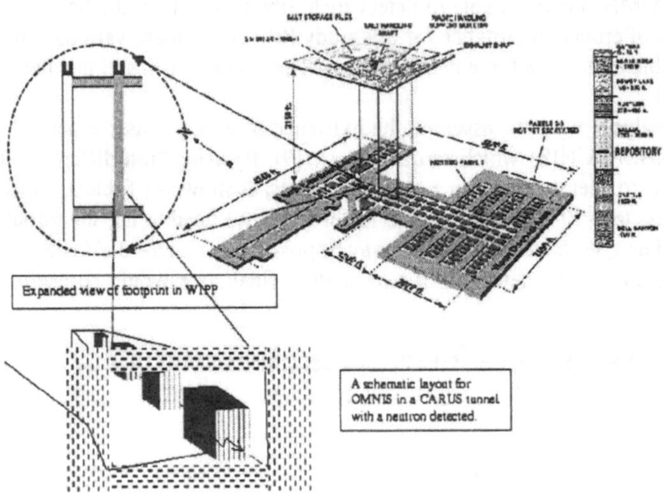

Figure 9. Schematic layout for OMNIS in an ICARUS tunnel with a neutron detected.

TABLE 5: YIELDS OF SUPERNOVA NEUTRINO DETECTORS

Detector	Target Material	Fiducial Mass (Ton)	Target Element	Yield (υ_e)	Yield ($\overline{\upsilon}_e$)	Yield ($\upsilon_\mu+\upsilon_\tau+\overline{\upsilon}_\mu+\overline{\upsilon}_\tau$)
Super K	H_2O	32000	p, e, O	180	8300	50
LVD	CH_2	1200	p, e, C	14	540	30
SNO	H_2O	1600	p, e, O	16	520	6
SNO	D_2O	1000	d, e, O	190	180	300
OMNIS	*Fe*	8000	*Fe*	20‡	20‡	1200‡
OMNIS	*Pb*	2000	*Pb*			
no osc.				110‡‡	40‡‡	860‡‡
$\upsilon_\mu \rightarrow \upsilon_e$ osc.				≤4420 ‡‡	40‡‡	≤ 640‡‡

‡ Assumes same efficiency as in Smith 1997
‡‡ Assumes a single neutron detection efficiency of 0.6

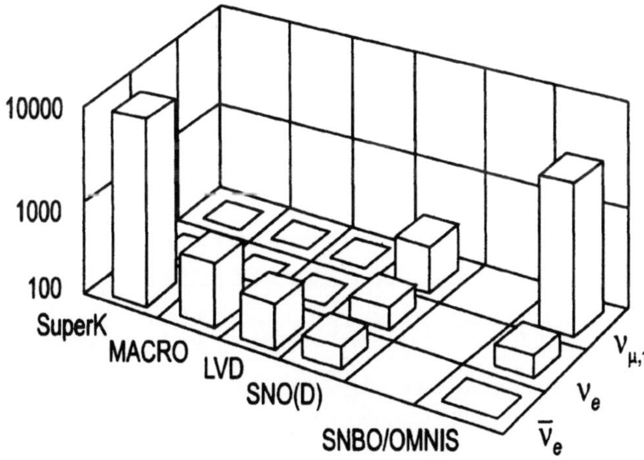

FIGURE 10. Comparison of world detectors (event numbers for supernovae at 8 kpc).

REFERENCES

1. Cline, D. B. "Search for Relic Neutrinos and Supernova Bursts", in *Proceedings, Eighth Intl. Wksp. on Neutrino Telescopes* (Venice, Feb. 23-26, 1999), ed. M. Baldo Ceolin, 1999, Vol. II, pp. 309-320.

2. R. Mayle, J. R. Wilson, and D. B. Schramm, Astrophys. J. 318 (1987) 288.

3. Figure 3 is adapted from the paper A. Dighe and A. Yu Smirnov, Phys. Rev. D. 62, 033007.

4. Kamiokande Collaboration, K. S. Hirata et al., Phys. Rev. Lett. 58 (1987) 1490.

5. IMB Collaboration, R. M. Bionata et al., ibid. 1494.

6. M. Krauss, Nature (London) 329 (1987) 689.

7. A. Yu. Smirnov, D. N. Spergel, and J. N. Bahcall, Phys. Rev. D

8. S. Otwinowski, UCLA, private communication (2000).

9. Cline, D. B., Fuller, G., Hong, W. P., Meyer, B., and J. Wilson, Phys. Rev. D50, 720 (1994).

10. G. M. Fuller, W. C. Haxton, and G. C. McLaughlin, Phys. Rev. D 59 (1999) 085005 (and Refs. therein)

A NEUTRINO COMPONENT OF ULTRA HIGH ENERGY COSMIC RAYS?

G. Domokos and S. Kovesi-Domokos*

1. INTRODUCTION

Ultrahigh energy cosmic rays present us with an interesting puzzle: they should not reach us. The reason is very simple. The Universe is filled with radiation. The largest photon density is in the cosmic microwave background radiation (CMBR), about 400 photons/cm^3. When a high energy proton interacts with a photon in the CMBR, pion production takes place. This process begins to be significant around proton energies of the order of 10^{19}eV. In the center of mass this corresponds to an energy where photoproduction proceeds predominantly through the excitation of the Δ resonance. The inelasticity of the process is about 25 %; thus, after a few interactions the proton loses a substantial part of its energy. This is the celebrated *Greisen - Zatsepin - Kuzmin (GZK) cutoff* of the cosmic ray spectrum [1]. As a consequence of this, ultra high energy (UHE) protons have to originate not further than about 50Mpc from us in order to survive the journey without substantial energy losses, see [2].

Here is the puzzle.

- By now, about 100 events have been observed of energies $E \geq 5 \times 10^{19}$eV.
- Various analyses appear to conclude that there are no candidate sources within the "GZK sphere": a sphere of radius 50Mpc around the solar system.

The situation is reviewed in a recent article by Bhattacharjee and Sigl, [3] as well as in the Erice lectures of Stecker [4]. It has to be emphasized that particle physics effects are unlikely to alter the photoproduction process at the relevant energies: the CMS energies involved are of the order of a few GeV. This is an

* Department of Physics and Astronomy, The Johns Hopkins University, Baltimore, MD 21218, skd@jhu.edu

The Role of Neutrinos, Strings, Gravity, and Variable Cosmological Constant in Elementary Particle Physics, Edited by Kursunoglu *et al.*, Kluwer Academic/Plenum Publishers, New York, 2001

energy region which has been studied for several decades now and it is very well understood. Likewise, the requirements for an astrophysical source to accelerate protons up to a given energy – in this case, at least up to a few times 10^{20}eV – are well understood, see [5].

Several solutions have been proposed to this conundrum: most of them are outlined in the Proceedings of the *Workshop on Observing Giant Cosmic Ray Air Showers* [6] and Stecker's Erice lecture notes just cited.

In what follows, we describe our own proposal [7] which was first outlined in a rather rudimentary form several years ago [8].

2. NEW PHYSICS?

The solution to the puzzle of trans-GZK cosmic rays is based on some recent developments in string theory [9]. (For a recent review, see [10].) The basic idea is the following. *Any* presently known string theory has to live in a space-time of dimensionality $D > 4$. Presumably, the extra dimensions are either compactified or our universe is a stable submanifold of the D dimensional space (a brane). In either case, the D-dimensional Planck mass, M_D, and the Planck mass in our universe are not necessarily equal: the D-dimensional Planck mass can be considerably lower than the 4-dimensional one, perhaps something like a few tens of TeV. In such a scenario, unification of the interactions takes place at energies around M_D and string excitations become noticable in the same energy range. If this scenario is a viable one, observable effects begin to surface in the energy range of the order of M_D.

It has to be emphasized that no phenomenologically viable string model has been constructed so far. Even if one of the models constructed so far were viable, experimental tests would require an understanding of the symmetry breaking, the construction of unitary amplitudes, *etc.* In such a situation, one has to try and extract robust features of string models. One can then rely upon reasonable guesses of how those features affect the behavior of the various measurable quantities once $\sqrt{s}$ becomes close to M_D. We believe that some of such robust features are the following.

- The density of states is a rapidly rising function of the mass (equivalently, of $\sqrt{s}$). One has exact asymptotic estimates for the level densities, $D(N)$, stemming from the fact that string models are conformal field theories. One has:

  $$\ln D(N) \sim \sqrt{N} + \text{const},$$

  where N is the Nth level of excitation. In the s channel, therefore, $N \sim \alpha' s$. There is evidence that at lower excitations the level density is rising even more rapidly. For instance, in the Ramond sector of the superstring the level density of the first few excited states can be interpolated by a function $\propto \exp(CN)$, where C is a constant of order one.

If all the resonances were purely elastic, this would mean an exponential rise of the cross section, since at resonance the phase shift equals to $\pi/2$. Clearly, this can hold only in a transition region: unitarity must asymptotically limit the cross section to a more modest growth. However, it is reasonable to expect that in the region where string effects take over from local field theory, the transition is rapid.

- The Regge slope, (α') , the unification mass and M_D are equal to each other within factors of order unity. Hence, the onset of the string regime also means a unification of interactions: all interactions described by the model become of equal strength. (For a discussion see [10].)

Given such a scenario, one can understand, at least qualitatively, the existence of trans-GZK cosmic rays. The CMS energy, $\sqrt{s}$, in an interaction between an incoming particle and a photon of the CMBR is of the order of a few GeV, depending on the rest mass of the particle. Hence, the Standard Model rules: protons lose energy by photoproduction, neutrinos penetrate without any interaction, *etc.* The situation is, however, drastically different once a particle reaches the atmosphere. If one views an "air nucleus" merely a collection of almost free nucleons (a picture of some validity), one finds that a trans-GZK cosmic ray of a laboratory energy somewhere above 10^{18}eV colliding with a nucleon has a CMS energy of several hundred TeV. Consequently, if some type of low scale string model is Nature's preference, one is already in the string/unification regime. Protons do not get to us, but in the string regime there is no essential difference between a quark-quark and a neutrino-quark interaction. The incoming neutrino gives rise to an air shower (roughly) in the same way as a quark-quark interaction would. This picture remains valid as long as the CMS energy in the interactions remains above M_D. Once the CMS energy in a collision drops below the unification mass, the shower continues to evolve according to the Standard Model.

Is this scenario testable by observing trans-GZK air showers?

In order to understand the answer in qualitative terms, let us recall that in a "normal", proton induced air shower, the lepl'tonic (electron-photon) component is, in essence, a secondary one. Photons arise mostly from the decay $\pi^0 \rightarrow \gamma\gamma$; those, in turn, initiate an electromagnetic shower. There is hardly any feedback to the hadronic component, since photoproduction cross sections are small. Hadronic interactions are typically of high multiplicity, hence fluctuations in the shower are small.

By contrast, in the trans-GZK showers when the transition to the Standard Model occurs, one has comparable numbers of quarks and leptons: there is a substantial primary lepton component, carrying something like 1/2 of the primary energy. As a consequence, in the Standard Model phase, there are large fluctuations, since in the interactions of electrons and photons the multiplicity is typically 2. One cannot possibly expect large deviations in the average development of the showers: at unification, *all* elementary cross sections should be of a comparable magnitude and close to a quark-quark cross section.

This suggests that one should study the fluctuations in the longitudinal development of the trans-GZK showers in order to distinguish between our scenario and some others in which, for instance, protons are produced locally or exotic neutral hadrons initiate the trans-GZK showers.

At this stage, there are large uncertainties in implementing this picture: in particular, little is known about the value of the unification mass. A one dimensional MC program was written by P. Mikulski [11] in approximation A in order to study this question. The unification mass was left as a free parameter which then was varied within reasonable limits.

In the next section, we discuss the results of such a simulation. The average development profiles are presented in real space as observed by orbiting detectors or by Fly's Eye type ones on the ground.

3. RESULTS

For the purpose of exploring the conjectured "new physics" described in ref. [7], it was assumed that as long as $\sqrt{s} \geq M_D$, an interaction produces an equal number of leptons and quarks. Once the primary energy falls below the string scale, M_D, the shower evolves according to Standard Model physics. It was found that a step function-like onset of the "new physics" gives a description indistinguishable from an exponentially rising level density which for $\sqrt{s} \gg M_D$ levels off due to unitarity corrections. Hadronization of the quarks both in the string and Standard Model regimes was handled by means of a conventional splitting algorithm. Due to uncertainties in the models, at present no predictions can be made about how fast the various coupling constants are running with energy. It is expected that they run faster than in the Standard model, but the exact functional form of the β functions depends on the spectrum of Kaluza-Klein excitations and on the specific string theory. For this reason, we extrapolated the strong cross sections into the energy range of $\sqrt{s} \approx 600\text{TeV}$ using the fit of Block *et al.* [12] and worked with half of that value as the cross section in the string regime. Again, the results do not depend critically on the precise magnitude of the cross section in the string regime.

3.1. Average behavior

For the purpose of presenting the results, an exponential atmosphere was used with a scale height, $h_0 = 6.4$km. Due to the fact that all relevant showers take place in the upper atmosphere, this gives an adequate approximation to Shibata's parametrization [13]. The geometry of a shower has to be characterized in a manner independent of its relative position with respect to the detector. We chose the impact parameter, b to characterize the geometry of the shower. (One recalls that b is the distance of closest approach of the shower axis measured from the center of the Earth.) In this work we present results only for $b > R_\oplus$ and write $b = R_\oplus + h$, where, obviously, h is the distance of closest approach measured from the surface of the Earth. Horizontal distances, by definition, mean distances measured along the shower axis, taking the point of closest approach as zero distance.

In Figures 1 and 2, we display the average "anomalous", *i.e.* neutrino induced, shower profiles at various heights above the surface of the Earth choosing, as an example, $\sqrt{S}$ = 30TeV and $\sigma_\nu = \sigma_p/2$. All simulations were carried out for a primary energy $E_0 = 2 \times 10^{20}$eV, corresponding to $\sqrt{s} \approx$ 600TeV. The number of electrons at the shower maximum, N_{max}, is only weakly dependent on the value of M_D ($N_{max} \approx 3 \times 10^{10}$) and thus, the average profiles were normalized to this quantity. At higher altidudes the

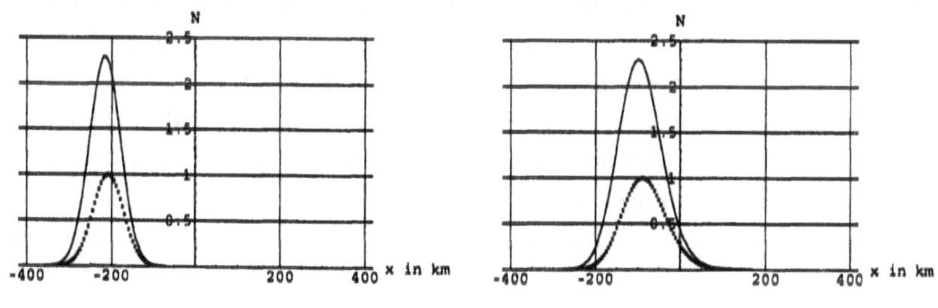

(a) 15 km

(b) 20 km

Figure 1. Average shower profiles of ν induced showers at h=15km and 20km above the Earth (dotted line). For comparison, the average profile of a proton induced shower of the same primary energy is shown (solid line).

shower development is considerably broader due to the low density of air. This is evident from the profile at h=25km.

3.2. Fluctuations

The average shower profiles displayed in Fig. 1 and Fig. 2 reveal an interesting feature. Even though the average shower profile of a ν induced shower is

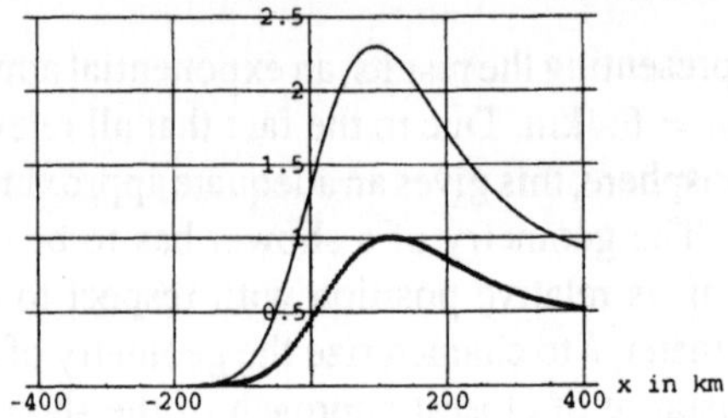

Figure 2. Average shower profile at h=25km. Thick line: Neutrino induced shower; solid line: proton induced shower.

different from a "normal", proton induced one, it is hard to distinguish between the two types on an event-by-event basis. While it is true that *on average* $N_{\max}$ in a proton induced shower is roughly twice as large as it is in a ν induced one and the development of the ν induced showers is somewhat slower than that of the proton induced ones, fluctuations are likely to smear out such differences in any given shower.

A statistical analysis of a sample of showers, however, should reveal a significant difference between ν induced and proton induced showers. One recalls that, from a qualitative point of view, smaller cross sections and smaller average multiplicities in individual interactions lead to larger fluctuations in the shower development. In the present model of precocious string unification, any particle produced is a lepton or quark with a probability 1/2 as long as the energy is larger than M_D. A lepton interacts producing a multiplicity of 2 (counting leptons and photons on an equal footing), whereas a quark, after hadronization produces high multiplicities in each interaction. (At the energies considered, a muon emits a Bremsstrahlung photon almost at the same rate as an electron.) Consequently, there is a low multiplicity component in the shower at least for the first few generations. This is absent in proton induced showers: there, the source of the leptonic component is the decay, $\pi^0 \to \gamma\gamma$. (At the present level of accuracy, photoproduction of pions and other mesons can be neglected.) This qualitative expectation is borne out by the simulation. In the following Figures, the distribution of X_{max} has been plotted for a given characteristic energy of the onset of the "new physics", $\sqrt{S}$. Following the previous discussion, the energy of onset was just put equal to M_D and the plots were constructed accordingly. Figure 3 displays the distribution of $X_{\max}$, the position of the shower maximum, for various values of S, the characteristic scale. For comparison, the same distribution is shown for "normal", proton induced showers in the absence of "new physics". Clearly, the distribution in X_{max} for all the showers containing the "new physics" is considerably broader than in a proton induced shower. One also observes that given sufficient

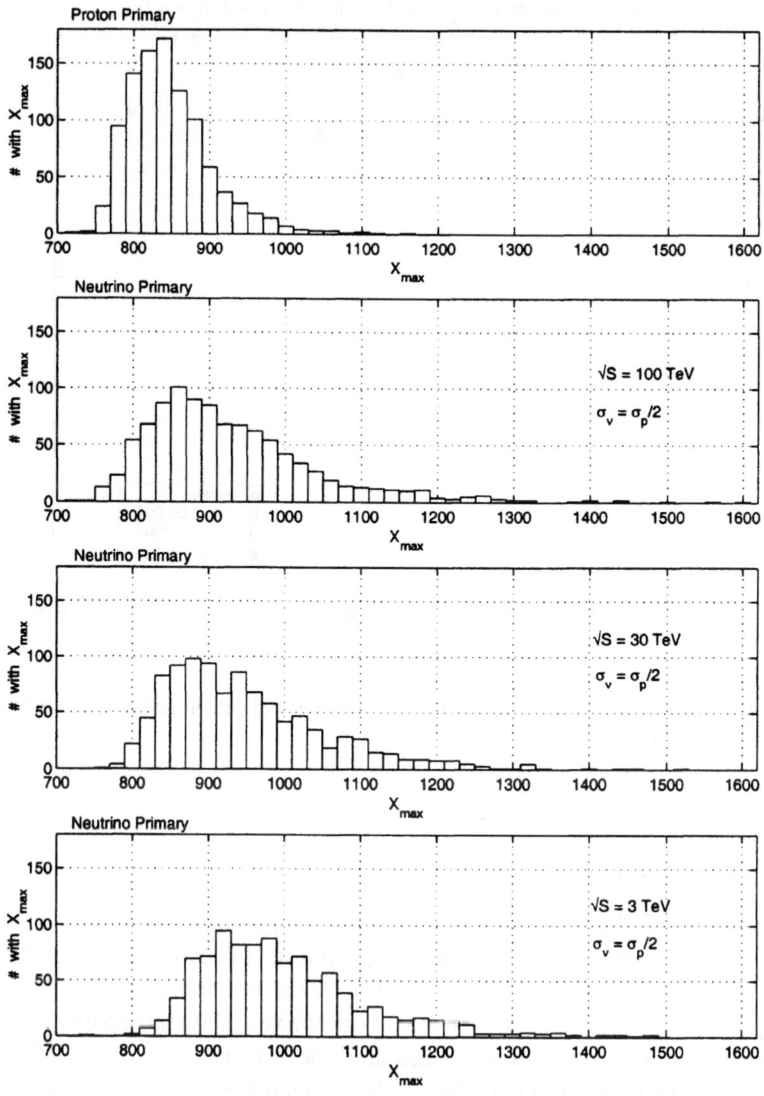

Figure 3. Distribution of shower maxima as a function of column density.

statistics, the distribution in $X_{\rm max}$ gives a hint about the magnitude of M_D: lower characteristic energies give rise to longer tails in the distribution. This is clearly borne out if one plots the second central moments of the distributions: there is a clear difference between proton and neutrino induced showers, see Fig. 4.

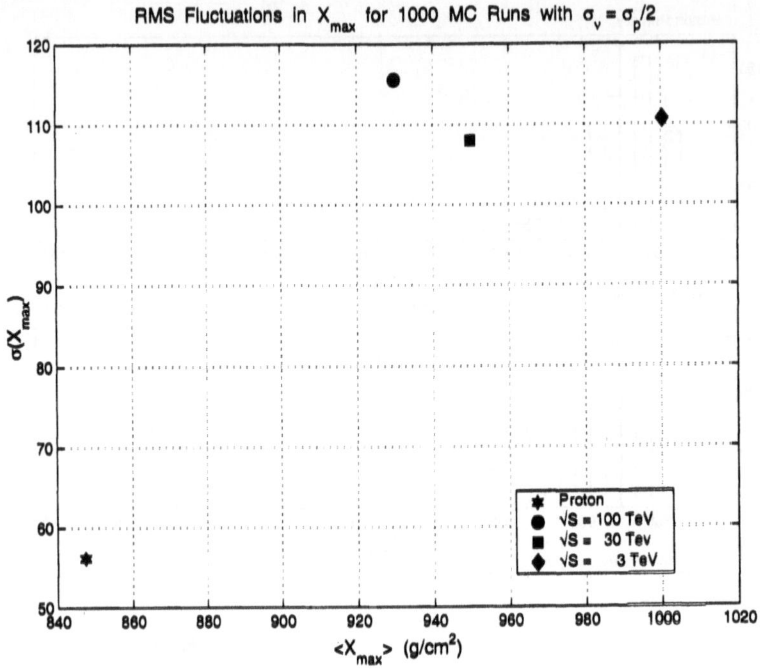

Figure 4. Standard deviations of the X_{max} distributions of proton and neutrino induced showers at various values of M_D.

4. DISCUSSION

The results of the simulation are rather pleasing from at least two points of view. First, the average longitudinal profile of a shower is quite similar to the profile of a hadron induced shower. This is important: the highest energy Fly's Eye and HIRES events are hadronlike. Thus our results, based on "new physics" at least do not contradict the available data. Second, the scenario outlined here is "falsifiable". The "smoking gun" – if there is one – will be the presence of anomalously large statistical fluctuations in the shower development. There are other features as well which single out showers induced by physics beyond the Standard Model. For instance, one expects high energy muons in the shower core: those are the prompt muons produced while the CMS energy was substantially above M_D. By contrast, in conventional, proton induced showers, muons arise from various meson decays; hence, they are spread rather widely away from the core. At present, such muons are hard to detect by existing detectors. However, future detectors may be able to ascertain their presence.

In closing, we wish to thank the organizers of this successful and stimulating meeting. We were happy to meet old friends and new ones. The informal atmosphere of the *Orbis Scientiæ* meetings stimulate discussions and – we hope – they lead to the generation of new ideas.

References

[1] K. Greisen, Phys. Rev. Lett. **16**, 748 (1966). G.T. Zatsepin and V.A. Kuzmin, Pis'ma Zh. Eksp. Teor. Fiz. **4**, 114 (1966).

[2] F.W Stecker, Phys.Rev. Letters, **21**, 1016 (1968).

[3] P. Bhattacharjee and G. Sigl, Phys. Reports **327**, 109 (2000)

[4] F.W. Stecker, astro-ph/0101072.

[5] A.M. Hillas, Ann. Rev. Astron. Astrophysics, **22**, 425 (1984).

[6] J.F. Krizmanic, J.F. Ormes, and R.E.Streitmatter, (editors), American Institute of Physics, New York (1998).

[7] G. Domokos and S. Kovesi-Domokos, Phys. Rev. Letters, **82**, 1366 (1999.

[8] G. Domokos and S. Nussinov, Phys. Letters **B187**, 372 (1987).

[9] P. Horava and E. Witten, Nuc. Phys. **B460**, 506 (1996). J. Lykken, Phys. Rev. **D54**, 3693 (1996). I. Antoniadis and M. Quiros, Phys.Lett. **B392**, 61 (1997).

[10] P. Nath, hep-ph/0011177

[11] P.T. Mikulski, dissertation, The Johns Hopkins University, (2000) and in preparation.

[12] M.M. Block, F. Halzen and T. Stanev, Phys. Rev. **D 62**, 077501 (2000).

[13] T.K. Gaisser, "Cosmic Rays and Particle Physics", Cambridge University Press, New York (1990). Sec. 3.5.

PROBING THE NATURE OF NEUTRINOS AT PION FACTORIES

Ali R. Fazely*

1. INTRODUCTION

Searches for neutrino oscillations and other neutrino induced reactions, as a comprehensive test of the SM of particle physics are to be performed at the Spallation Neutron Source (SNS) under construction at the Oak Ridge National Laboratory (ORNL). The principal motivation of these measurements are to perform a class of experiments to rigorously test the validity of the SM of particle physics using neutrinos as probes. The experiment will search for the appearance of $\bar{\nu}_e$ from $\bar{\nu}_\mu$ produced in the beam-stop from muons decaying at rest. It will also search for $\nu_\mu \rightarrow \nu_e$ oscillations with super high sensitivity. The detector will also be sensitive to the disappearance of ν_e's when the second proton storage ring is complete. Recent reported neutrino oscillation signal observed by the Liquid Scintillator Neutrino Detector[1,2] (LSND) can also be scrutinized with an improved sensitivity of about two orders of magnitude. Furthermore, measurements will yield high statistics neutrino-electron scattering data and due to the fast-spill nature of the beam, a precise measurement of the $\sin^2\theta_w$ will be possible. These measurements will be performed using the newly approved high-intensity short beam-spill spallation source. The experiment will yield data with high statistical precision and small systematic errors due to the precise measurement of the known ν_e+^{12}C$\rightarrow$^{12}N$_{g.s.}$+e$^-$ cross section. Measurements will also be made of other neutrino nucleus cross sections with super high precision which are of interest to fundamental physics as well as nuclear physics and astrophysics.

Neutrino oscillations phenomenon is a class of lepton-number violating processes which can be utilized to test the validity of the SM of the electroweak sector in particle physics. Neutrino oscillations can only occur if the physical neutrinos are not pure quantum mechanical systems, but rather a superposition of different mass

* Ali R. Fazely, Southern University, Baton Rouge, Louisiana 70813

The Role of Neutrinos, Strings, Gravity, and Variable Cosmological Constant in Elementary Particle Physics, Edited by Kursunoglu *et al.*, Kluwer Academic/Plenum Publishers, New York, 2001

eigenstates. This immediately implies that at least of the neutrino flavor eigenstates has to be massive. This is obviously in contradictions with the SM where the neutrinos are assumed to be massless.

The major advances in particle physics in the last twenty five years have been the success of the electroweak model of Weinberg, Salam and Glashow unifying the electromagnetic and the weak interactions. The SM describes experimental data to within a high degree of precision, apart from masses and mixing angles, in terms of a single phenomenological parameter $\sin^2\theta_w$. With a value of $\sin^2\theta_w = 0.230 \pm 0.005$. Measurements of the left-right cross section asymmetry in Z boson production by the polarized e^+e^- collision using the SLD detector[3] has determined the value of the effective weak mixing angle to be $\sin^2\theta_w = 0.2292 \pm 0.0009(stat) \pm 0.0004(syst)$ at high Q^2's. At low Q^2 the situation is not clear and a stopped pion facility such as SNS with its fast spill could shed important light on this subject.

2. NEUTRINO OSCILLATIONS

In the past two decades, questions regarding the masses of neutrinos and the extent of mixing of different flavor eigenstates have received a great deal of attention. Evidence for neutrino oscillations from the LSND[1,2] and flux deficit in solar neutrino experiments[4,5] have intensified the debate. The atmospheric neutrino experiments with recent results from Super K[6] has also reported deficit in the ν_μ flux consistent with neutrino oscillations. Although LSND, solar neutrino, and atmospheric neutrino results do not point to the same conclusion regarding neutrino masses and mixing within the traditional two-neutrino mixing analysis approach, one could reconcile these two measurements with more sophisticated analysis of three-neutrino mixing with an extra "sterile" neutrino. It is, therefore, essential to check the validity of these measurements experimentally. Amongst these measurements, LSND by far is the most intriguing. The LSND collaboration has reported 22 events consistent with $\bar{\nu}_e$ appearance resulting from $\bar{\nu}_\mu \rightarrow \bar{\nu}_e$ oscillations. The most attractive aspect of the LSND experiment is the full understanding of the neutrino flux and systematics of the experiment. Fig. 1 shows the energy spectrum for oscillations and Fig. 2 shows the areas of probable neutrino oscillations in the customary δm^2 vs. $\sin^2 2\theta$ notation[7].

2.2. Decay-at-Rest Neutrino Oscillations

At SNS pion are copiously produced in the Hg target at a rate of approximately 0.1 π^+ per incident proton. The majority of these pions (99%) come to rest inside the target and decay. The decay chain produces the well-known neutrino spectrum shown in Fig. 3. We, thus, have:

$$\pi^+(\tau = 26\text{ns}) \rightarrow \mu^+ + \nu_\mu\ , \qquad \mu^+(\tau = 2.2\mu\text{s}) \rightarrow \text{e}^+ + \bar{\nu}_\mu + \nu_e \tag{1}$$

The search is done for the neutrino flavor $\bar{\nu}_e$. Note that the this flavor is missing in the above decay scheme. Only the negative pions that are not captured and decay can produce this flavor of neutrinos. We estimate this number to be suppressed by 10^4. Therefore, if $\bar{\nu}_e$ is observed in a detector, then the most likely conclusion is that $\bar{\nu}_\mu$'s have oscillated into $\bar{\nu}_e$'s.

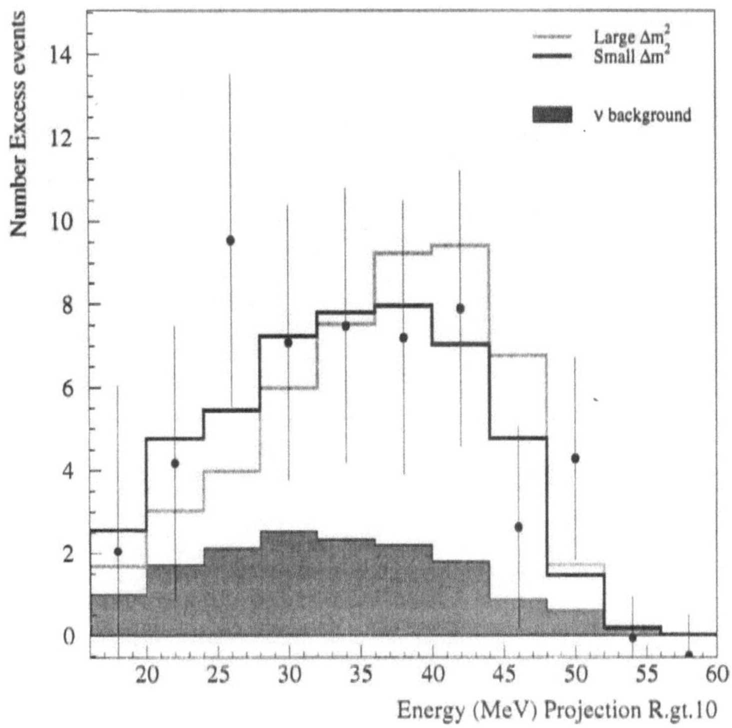

Figure 1: Electron energy spectrum for the LSND signal. There are two calculations indicating low and high mass solutions.

The detection method for this mode of neutrino oscillation is the detection of a positron followed by a neutron. The reaction inside a scintillator detector with H as the target for $\bar{\nu}_e$ is the usual inverse β-decay process;

$$\bar{\nu}_e + \mathrm{p} \rightarrow \mathrm{e}^+ + \mathrm{n} \tag{2}$$

The positron has a Michel energy spectrum and provides the primary trigger. The detection of neutron is usually done in a so-called slow coincidence technique such as LSND. In this method a gate of a few hundreds of μ-seconds is opened to allow for the neutron to thermalize and a capture process on H or an additive such as Gd would produce low energy γ's constituting the signature for the slow coincidence. In a fast coincidence method one looks for recoil protons which can only be done in a fine modular detector such as the one we are describing here. As the neutron slows down, it collides with protons in the scintillator and produces light. This method is background free. Note that these two methods are not mutually exclusive and a gold plated event would, ideally, have both signatures in coincidence with the primary positron.

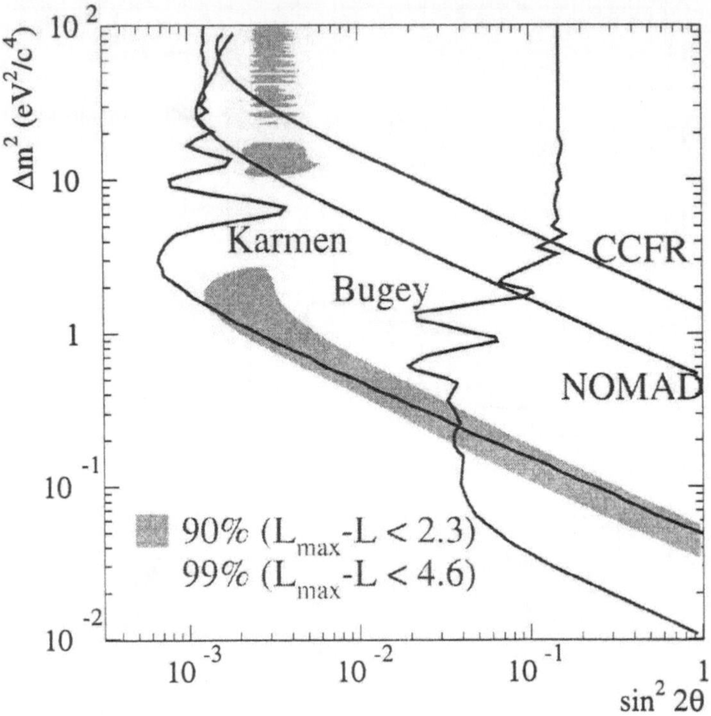

Figure 2: Probable region of $\bar{\nu}_\mu \to \bar{\nu}_e$ oscillation for 2.3 σ and 4.6 σ. Other neutrino oscillations experimental limits are also displayed.

2.2. Decay-in-flight Neutrino Oscillations

Given a pion decay-in-flight (DIF) source at SNS one can perform a DIF neutrino oscillation study.

2.2.1. $\nu_\mu \to \nu_e$ Oscillations Analysis Using ^{12}N, inclusive states

In this oscillation search, electron-like events, above 55 MeV of total energy are considered. In this mode of data analysis identified particles are free from the Michel electrons. The LSND reports 10.5 and 10.1 in two different analyses for the inclusive states[8] in ^{12}N.

2.2.2. $\nu_\mu \to \nu_e$ Oscillations Analysis Using the Ground State of ^{12}N

The experiment in the pion decay-in-flight (DIF) region, is mainly sensitive to reactions of ν_e form $\nu_\mu \to \nu_e$ oscillations in the beam. The ν_e's are detected by the inverse beta decay reaction. The detector is also sensitive to detecting positrons form β^+ decay of the $^{12}N_{g.s.}$ with an end point energy of 16.826 MeV. This con-

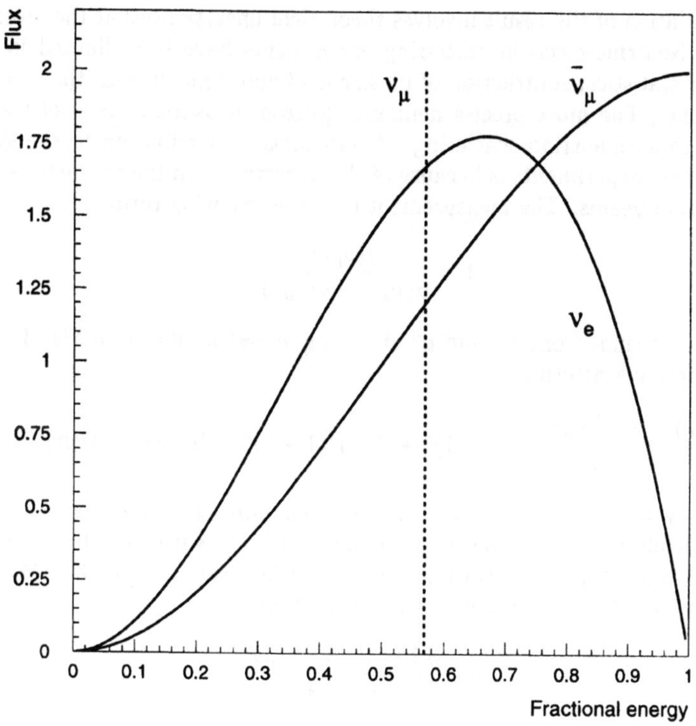

Figure 3: Neutrino energy spectra for pion and muon decay-at-rest

stitutes a unique signature for the DIF oscillations search. The LSND experiment observed a few such events above 55 MeV with virtually zero background[9,10].

3. PRECISION $\sin^2\theta_w$ MEASUREMENT

The Large Cerenkov Detector[11] (LCD) at LAMPF's Proton Storage Ring (PSR) was proposing to build a 10-kilo ton water Cerenkov detector and measure the $\sin^2\theta_w$ within a 1% error. Although this experiment was never built, the opportunity is being provided for a similar measurement at SNS. With much higher neutrino flux at SNS than that of the PSR at LAMPF (10 to 40 times), we have a unique facility to build a laboratory to fulfill many of the goals set out by the LCD.

3.1. Neutrino-electron elastic scattering theory

Measurements of the masses of the intermediate vector bosons made in the past few years provide the most accurate determination of the parameter $\sin^2\theta_w$. The prediction of the SM that $\sin^2\theta_w$ as the sole parameter must, however, be tested with neutrinos at low Q^2. Deep inelastic scattering on nucleons has proved to be the most accurate method so far but has the problem common to hadronic processes that calculation of the result involves theoretical uncertainties at the level of a few percent. Neutrino-electron scattering experiments have been limited in the past by lack of statistics, subtraction of background and difficulties in the neutrino flux calculations. The most precise neutrino-electron measurements is of the ratio of neutrino to antineutrino scattering. A substantial contribution to the systematic error in these experiments is because of the uncertainty in the characteristics of the two neutrino beams. The measurement is of the following ratio,

$$R = \frac{\sigma(\nu_\mu e)}{\sigma(\nu_e e) + \sigma(\bar{\nu}_\mu e)} \tag{3}$$

With neutrinos from π^+ and μ^+ decaying at rest as shown in Fig. 3. The cross section for $\bar{\nu}_\mu e$ scattering is

$$\frac{d\sigma(\nu_\mu e)}{dy} = \frac{G_\mu{}^2 m_e E_\nu}{2\pi}(2s^2-1)^2 + (2s^2)^2(1-y)^2 - 2s^2(2s^2-1)(m_e/E_\nu)y \tag{4}$$

where $y = (E_e - m_e)/E_\nu, s^2 = \sin^2\theta_w$. For $\nu_e e$ scattering, replace $2s^2 - 1$ by $2s^2 + 1$, while for $\nu_\mu e$ scattering interchange $2s^2 - 1$ with $2s^2$. Because m_e/E_ν is small, we can ignore the term linear in y, and integrate over y, the SM expression for the ratio of the cross sections may be written:

$$R = \frac{3}{4}\frac{1 - 4s^2 + 16/3s^4}{1 + 2s^2 + 8s^4} \tag{5}$$

where $R = 0.13$ for $\sin^2\theta_w = 0.23$. When $\sin^2\theta_w$ increases by a given relative amount, the ratio R increases by a larger fraction (2.08 with a 10 MeV electron energy cut, 1.66 with no energy cut).

We, therefore, can make a precision measurement of the ratio of neutrino electron scattering with neutrinos from pion decay and muon decay simultaneously. Measurement of the ratio R above, with an accuracy of 2% will yield $\sin^2\theta_w$ with a total of approximately 1%.

4. OTHER TESTS OF THE SM

4.1 Limits on $\pi^o \to \nu\bar{\nu}$, and $\eta \to \nu\bar{\nu}$, Decays

The observation of the decays $\pi^o \to \nu\bar{\nu}$ and $\eta \to \nu\bar{\nu}$ would imply physics beyond the SM. Momentum and angular-momentum conservation require that the decay ν and $\bar{\nu}$ possess the same helicity. Therefore, in the standard electroweak interaction model of Weinberg-Salam, where the neutrinos are assumed to be massless and purely left handed (antineutrinos purely right handed) the above-mentioned processes are strictly forbidden. One also can argue that since π^o is a pseudo-scalar meson (0^-), and vacuum is 0^+, this transition is strictly forbidden, within the V-A interaction. These decays, however, can proceed if pseudo-scalar weak interaction exists, or both right and left-handed neutrinos exist and/or the total lepton number is not conserved. The information derived from $\pi^o \to \nu\bar{\nu}$ and $\eta \to \nu\bar{\nu}$ are complementary because the former is sensitive only to the isovector neutral-current interactions while the latter is sensitive to the isoscalar neutral-current interactions[12].

Furthermore, π^o decays involve only u- and d-quarks while η decays additionally involve the s-quark and perhaps other heavier quarks[12].

Equally interesting are the limits on decays $\pi^o \to \nu_e \nu_\mu$ and $\eta \to \nu_e \nu_\mu$, which violate individual lepton number conservation in the neutral current sector.

An experimental upper limit, $\Gamma(\pi^o \to \nu_e \bar{\nu}_e)/\Gamma(\pi^o \to \text{all}) \leq 2.4 \times 10^{-5}$ (90% CL), was set by Herczeg and Hoffman[13] who used the $K^+ \to \pi^+ \nu \bar{\nu}$ data and focused on the $K^+ \to \pi^+(108\text{MeV})\nu\bar{\nu}$ region of the decay spectrum. Hoffman[13] has further set limits for the branching ratio for $\pi^o \to \nu\bar{\nu}$ by using the data from several beam-dump experiments. Similar limits were obtained by Dorenbosch et al.[14]. Other limits from $K^+ \to \pi^+\nu\bar{\nu}$ data sets an inclusive tighter limit[15] of $\Gamma(\pi^o \to \nu_e \bar{\nu}_e)/\Gamma(\pi^o \to \text{all}) \leq 8.3 \times 10^{-7}$ (90% CL). These limits can be improved by two orders of magnitude at the SNS.

4.2. Strange quark content of the nucleon, νp $\to$ νp

The semileptonic νp elastic scattering is a simple reaction and provides a powerful tool to measure the strange quark content, Δs of the proton and its relative contribution to the proton spin. Cross sections for these reactions are well known, within small theoretical uncertainty, and small radiative corrections, characteristic of neutral-current neutrino scattering. Dependence on the the nucleon form factors are all known from electron scattering measurements, except for the strange form factor, by using results from electron scattering and CVC to describe the vector part of the neutral current. The axial current is parameterized in a similar way.

The only two observables in νp $\to$ νp are the recoil proton kinetic energy and the angle of the recoil proton with respect to the incident neutrino. We assume that the recoil neutrino and the recoil proton polarization are not measurable in a practical way. The proton kinetic energy is given by $T_p = Q^2/2m_p$, where Q^2 is the momentum transfer squared of the reaction and m_p is the proton mass. The angle of the recoil proton in the lab frame can be expressed as $\cos(\theta) = (1 + m_p/E_\nu)/(1 + 2m_p/T_p)^{1/2}$, where E_ν is the neutrino energy. At SNS with a fine segmented detector, we will be able to measure the proton angle at high momentum transfers and this together with the proton kinetic energy will provide the signature for νp elastic scattering.

All the form factors relevant to νp $\to$ νp scattering are well understood, and the third, G_1 is the axial vector. All these form factors are functions of Q^2, therefore,

$$F_1(Q^2 = 0) = 0.034 - F_1{}^s/2\,, \qquad F_2(Q^2 = 0) = 1.017 - F_2{}^s/2 \tag{6}$$

Note that $F_1{}^s = 0$ at $Q^2 = 0$ and $F_2{}^s$, the weak strange magnetism, has been measured by HAPPEX[16] at JLAB in the forward direction and the SAMPLE experiment[17] at BATES in the backward direction. The measured value is $G_M^Z = 0.34 \pm 0.09 \pm 0.04 \pm 0.05$ n.m.(nuclear magnetons) at $Q^2 = 0.1(\text{Gev/c})^2$. $G_1(0) = -g_A/2 + G_1{}^s/2$, where $g_A = 1.26$ from neutron beta decay and $G_1{}^s = \Delta s$ is the strange quark contribution to the spin of the proton. We assume only first class current and therefore, F_3 and G_2 are zero and $G_3 = 0$, because the mass of the neutrino is very light compared to the electron mass.

The cross section at low Q^2's is:

$$\frac{d\sigma}{dQ^2} = \frac{G_F^2}{2\pi}[(-0.63 + \frac{G_1^s}{2})^2(1 + \frac{Q^2}{4E_\nu^2}) + 0.001156(1 - \frac{Q^2}{4E_\nu^2})] \tag{7}$$

In the above expression, the only unknown is G_1^s. A measurement of the above cross section will determine this strange quark contribution to the proton spin.

Fig. 4 shows a calculation for νp elastics for bound and free protons. It also shows the νn results. Note that in νn, the trigger requirement and the associated energy of neutron collision throughout the detector provide important tools to identify neutrons and protons in a fine segmented detector.

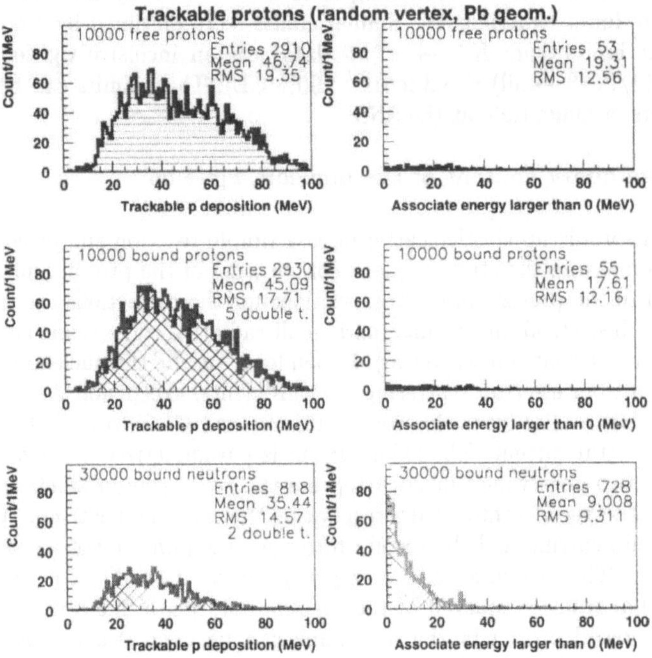

Figure 4: Calculated proton and neutron energy spectra for bound and free protons and neutrons.

4.3. Nuclear Physics Program

A segmented detector provides the ideal laboratory to measure neutrino-nucleus cross sections for nuclei of interest in astrophysics or nuclear physics. Measurement of such cross sections are important in order to get a better understanding of, for example, the dynamics of super novae. Some nuclei may have larger or smaller cross sections than what calculations predict and may provide tools by which we can design more efficient neutrino detectors.

Fig. 5 shows the electron energy spectra for Al, Si, iron and lead. Our calculations show that given sufficient mass of a given nuclear target, a better than a 10%

cross section measurement is possible. As mentioned above these measurements will have important astrophysical consequences.

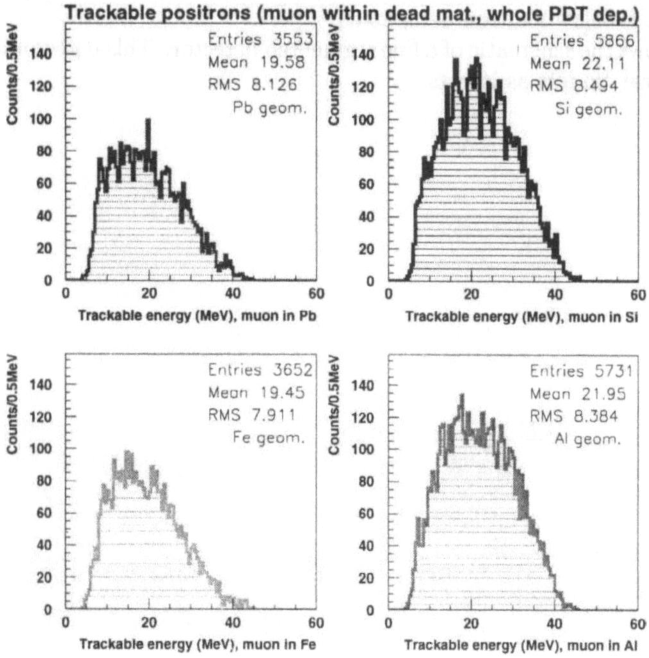

Figure 5: Calculated electron spectra for neutrino charged-current reactions on Al, Si, Fe and Pb.

5. THE DETECTOR

We believe the best and the most suitable place to extend the sensitivity of LSND and KARMEN measurements, is the SNS being built at ORNL. The 1 GeV beam energy at SNS would be perfect for copious pion production in an Hg target. This energy being higher than LANSCE by about 20% provides higher pion flux and is also well below kaon production threshold, thus keeping the oscillation signal free from neutrinos resulting from K-decay. The pion decay at rest neutrino spectrum is well known (Fig. 3) and pion production at these energies have been measured and/or can be calculated reliably. Furthermore, The high-Z Hg target insures higher π^- and μ^- capture rate than Cu beam-stop at LANSCE, thus reducing the beam-related $\bar{\nu}_e$'s.

A segmented detector can be designed containing long-thin plastic modules 5 meters long by 1 cm by 20 cm cross section. These modules will be arranged in x layers to provide calorimetry followed by x-y planes of proportional drift tubes (PDT) for tracking. Each optically-isolated module will be read out with green wave-shifter fibers at both ends. The outside of each module will be painted with Gd_2O_3 paint for slow neutron detection.

Fig. 6 shows the schematic of a fine segmented detector. This detector is designed to have nuclear targets as inserts.

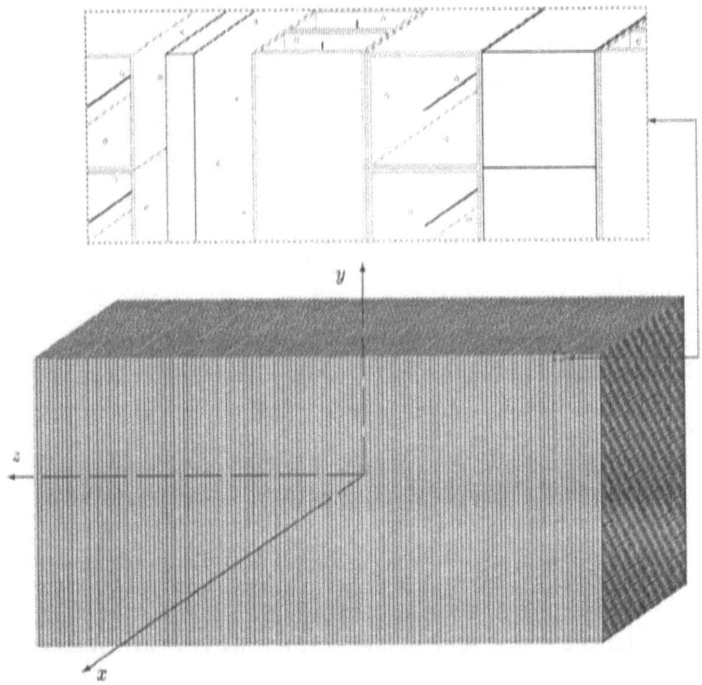

Figure 6: Schematic view of a fine segmented neutrino detector at the SNS

The modular scintillator design has also the capability of providing particle identification. This is due to observation of ionization of particle via dE/dx technique. At energies below 100 MeV the electrons and protons are well separated and an electron-proton rejection of 10^{-4} is achievable as it was done in E645[18] at LAMPF. A segmented detector also provides the capability of fast neutron detection for the DAR neutrino oscillations measurement.

6. REFERENCES

1. C.A. Athanassopoulos, *et. al.*, Phys. Rev. Lett. **75**, 2 650 (1995).
2. C.A. Athanassopoulos, *et. al.*, Phys. Rev. Lett. **77**, 3082 (1996).
3. K. Abe, *et. al.*, Phys. Rev. Lett. bf73, 25 (1994).
4. P. Anselmann, *et. al.*, Phys. Lett. **327B**, 377 (1994).
5. GALLEX Collaboration, Phys. Lett. **285B**, 376 (1992).

6. Y. Fukuda, *et. al.*, Phys. Rev. Lett. **81**, 1562 (1998)
7. Bill Louis, private communication.
8. C.A. Athanassopoulos, *et. al.*, Phys. Rev. Lett. **81** 1774 (1998).
9. C.A. Athanassopoulos, *et. al.*, Phys. Rev. C**58** 2489 (1998).
10. A. R. Fazely, LSND Technote 108 (1998).
11. Large Cerenkov Detector LAMPF Proposal, LA-11300-P, (unpublished).
12. P. Herczeg, Proceedings of the Workshop on Production and Decay of Light Mesons, ed. P. Fleury, World Scientific Publishers, Singapore(1988).
13. C.M. Hoffman, Phys. Lett. B**208**,149 (1988).
14. J. Dorenbosch, *et al.*, Z. Phys. C**40**, 497 (1988).
15. M. S. Atiya, *et al.* Phys. Rev. Lett. 66 2189 (1991).
16. K. A. Aniol, *et al.* Phys. Rev. Lett. 82 1096 (1999).
17. D. T. Spayde, *et al.* Phys. Rev. Lett. 84 1106 (2000).
18. L. S. Durkin, *et al.* Nucl. Instr. and Meth. A277 386 (1989).

INCLUSIVE NEUTRINO AND ANTINEUTRINO REACTIONS IN IRON AND AN INTERESTING RELATION

Stephan L. Mintz*

1.INTRODUCTION

Neutrino reactions in nuclei are very interesting for a number of reasons. They present us with the very few opportunities which we have for studying neutrino reactions in a laboratory setting under controlled conditions. In particular iron is an unusually suitable reaction to study. Virtually every neutrino experiment has some background from iron used in shielding or for tanks and other equipment. This is particularly true for the KARMEN experiments[1] which make use of a 6000 ton steel shield for their calorimeter. The KARMEN collaboration[2] considers neutrino reactions in iron to be one of its most serious backgrounds. They have extracted some results for electron neutrino cross sections on ^{56}Fe averaged over the Michel spectrum which have been already published[3,4].

The interest in neutrino reactions in iron however is not merely as a background. There has been an experiment involving high energy neutrino and antineutrino reactions in iron[5]. Furthermore there is a new experiment[6] for neutrino reactions in iron which will start in the near future and will run from threshold to about 3 GeV. These experiments both planned and already performed will provide useful information on how present calculations, which seem to work well for neutrino reactions in ^{12}C up to a few hundred MeV might be extended. In particular it may be possible to make use of already determined structure functions to extend our present calculations.

Besides obtaining the cross sections for the reactions $\nu_e +^{56} Fe \longrightarrow e^- + X$, $\nu_\mu +^{56} Fe \longrightarrow \mu^- + X$, $\bar{\nu}_e +^{56} Fe \longrightarrow e^+ + X$, and $\bar{\nu}_\mu +^{56} Fe \longrightarrow \mu^+ + X$, we also obtain averaged cross sections for both the Michel spectrum electron neutrinos for the electron neutrino reaction, and also the LAMPF decay-

* Physics Department,Florida International University,Miami, FL. 33157

The Role of Neutrinos, Strings, Gravity, and Variable Cosmological Constant in Elementary Particle Physics, Edited by Kursunoglu *et al.*, Kluwer Academic/Plenum Publishers, New York, 2001

in-flight muon neutrino spectrum for the muon neutrino reaction. We do these averages only for the neutrino reactions. This is because the electron antineutrino spectrum is very suppressed for both LAMPF and KARMEN. There is still a substantial muon antineutrino contribution to experiments at both of these facilities but at Michel spectrum energies, this leads only to neutral current reactions. For the decay in flight LAMPF muon neutrino experiment, the muon antineutrino spectrum is also suppressed.

We expect a priori that the neutrino cross sections in iron will be relatively large. The cross section increases generally with the number of neutrons available for interaction and a similar effect is noted for total muon capture rates in nuclei, which increase with the number of protons. Thus the background contributions could be significant for variety of experiments as already noted.

In section II of this paper we shall obtain the matrix elements needed to calculate these four inclusive neutrino reactions in ^{56}Fe. In section III we shall present our results. Finally in section IV we shall discuss these results and and consider how they might be usefully connected to higher energy data for neutrino reactions in iron already in existence[5].

2. MATRIX ELEMENTS

We make use of a phenomenologically based method of calculation which employs total muon capture rate data to determine the parameters necessary to calculate the inclusive neutrino cross sections. This method has been used in several calculations which have previously appeared in the literature[7,8,9,10,11,12]. It gives results in agreement with experiment for the ^{12}C case[1,13,14] and for the limited results available[15] for the ^{127}I case. We present a summary of this method in what follows.

We have four reactions to consider here. However it is sufficient to explicitly show only the $\nu_e +^{56} Fe \longrightarrow e^- + X$ case. The muon neutrino reaction can be obtained from the electron neutrino case by simply replacing the electron mass and four-momentum by the corresponding muon quantities. The antineutrino reactions can be handled in much the the same way. The currents are replaced by their hermitian conjugates and the threshold and final state interaction is adjusted. There are also some changes in the matrix elements which will be discussed shortly but the basic structure is the same.

We write the transition matrix element for the electron neutrino case, $\nu_e +^{56} Fe \longrightarrow e^- + X$, as:

$$M_{ki} = \frac{G}{\sqrt{2}} cos\theta_C \overline{u}_e \gamma^\lambda (1 - \gamma_5) u_\nu < k|J_\lambda(0)|^{56}Fe > \tag{1}$$

where k is a particular final state and:

$$J_\mu(0) = V_\mu(0) - A_\mu(0). \tag{2}$$

The formulation of Eq.(1) is sufficient for our purpose because the incident neutrino energies are relatively low for the reactions being consider here in comparison with the mass of the $W^\pm$ boson.

We may therefore write the cross section in a completely standard way as:

$$\sigma_c = \sum_k \frac{m_\nu}{2ME_\nu} \int d^3P_e \, |M_{ki}|^2 \, \frac{m_e}{E_e(2\pi)^3} \frac{d^3P_k}{2E_k(2\pi)^3} (2\pi)^4 \delta^4(P_k + p_e - p_\nu - P_i). \tag{3}$$

Here k stands for the possible final hadronic states,P_i^μ is the initial four momentum of the ^{56}Fe nucleus ,P_k^μ is the four momentum of the k state, and p_e^μ and p_ν^μ are the electron and neutrino four momenta respectively. The quantity $|M_{ki}|^2$ which appears in Eq.(3), is given by:

$$|M_{ki}|^2 = \frac{G^2 \cos^2\theta_C}{2m_\nu m_\nu} L^{\sigma\lambda} < k|J_\sigma(0)|^{56}Fe >< k|J_\lambda(0)|^{56}Fe >^* . \tag{4}$$

The quantity, $L^{\sigma\lambda}$, is the lepton tensor appropriate to this process and is given by:

$$L^{\sigma\lambda} = p_e^\sigma p_\nu^\lambda - p_e \cdot p_\nu g^{\sigma\lambda} + p_\nu^\sigma p_e^\lambda - \epsilon^{\alpha\sigma\beta\lambda} p_{e\alpha} p_{\nu\beta}. \tag{5}$$

Since we shall be working with average quantities, we assume an average nuclear excitation of δ given by:

$$M_x - M_i = \delta \tag{6}$$

where δ is clearly a function of the incoming neutrino energy. We also assume on the basis of our knowledge of individual states that the interaction is largely in the forward direction and that:

$$< E_e > \simeq E_\nu - \delta. \tag{7}$$

and:

$$< \vec{p}_e > \simeq \sqrt{(E_\nu - \delta)^2 - m_e^2}. \tag{8}$$

In general the value of δ will certainly increase with increasing incident neutrino energy. However we expect that above the giant dipole that this increase should be slow. For ^{56}Fe, the giant dipole resonance occurs at approximately[16] 18 MeV. For our case of interest there is differential cross section data for electron scattering from ^{56}Fe available[16,17] for incident electrons from 150 MeV to 250 MeV at a scattering angle of 35 degrees. An analysis of this data indicates that $\delta(E_\nu)$ in the region immediately above the giant dipole region is given by:

$$\delta(E_\nu) = .0215(E_\nu - m_l) + 17.58 \tag{9a}$$

where of course E_ν is the incident neutrino (or antineutrino) energy and m_l is the mass of the outgoing lepton. Below the giant dipole resonance we have used a linear delta which joins smoothly with Eq.(9a) at 35 MeV. Because the thresholds are slightly different for the neutrino and antineutrino cases this leads to:

$$\delta(E_\nu) = .6015(E_\nu - m_l) - 2.747 \qquad (9b)$$

for the neutrino reaction and:

$$\delta(E_{\bar{\nu}}) = .5845(E_{\bar{\nu}} - m_l) - 2.16 \qquad (9c)$$

for the antineutrino case.

Equations (9a) ,(9b), and (9c) enable us to obtain $< E_e >$ and $< \vec{p}_e >$ in a straightforward way. Using these quantities we obtain the cross section as:

$$\sigma_c = \frac{G^2 cos^2\theta_C}{2ME_\nu} \int d\Omega_e \sum_k < k|J_\sigma(0)|^{56}Fe >< k|J_\lambda(0)|^{56}Fe >^* L^{\sigma\lambda}$$
$$\times \frac{< |\vec{p_e}| >}{2M - 2E_e + 2E_e \cos\theta_e \frac{<E_e>}{<|\vec{p_e}|>}}. \qquad (10)$$

The hadronic part of eq.(10) may be replaced as:

$$< k|J_\sigma(0)|^{56}Fe >< k|J_\lambda(0)|^{56}Fe >^* \equiv Q_{\lambda\sigma}(P_i, < q >) \qquad (11)$$

which is a tensor. We have previously shown[7,8,9,10,11,12] that this tensor may be reduced to the simplier form:

$$Q^{\mu\nu} = \alpha g^{\mu\nu} + \frac{\beta}{M^2} P_i^\mu P_i^\nu. \qquad (12)$$

The cross section then becomes:

$$\sigma_c = \frac{G^2 \cos^2(\theta_C)}{4\pi} \frac{< |\vec{p}_e| >< E_e > D}{M(M + E_\nu)} \qquad (13)$$

where:

$$D = \beta - 2\alpha \qquad (14)$$

and an impulse approximation based calculation[18] gives $D(q^2)$ as:

$$D = a_o - b_o q^2. \qquad (15)$$

We assume this simple q^2 dependence for D, which we may also write as $D = a_o + b_o|q^2|$. We note that even without an impulse approximation based calculation this would merely be the first two terms in a general expansion of D.

Thus our result depends upon two parameters,a_o and b_o. These we need to determine phenomenologically. This can often be done by making use of total muon capture rates and sometimes by the use of electron scattering if there is enough available. The total muon capture rate is available for the process,$\mu^- + ^{56}Fe \longrightarrow \nu_\mu + X$. However it is clear that this is not the desired muon capture needed here. The muons react on the protons whereas the neutrinos react on the neutrons. Thus for the neutrino reactions described here muon capture results on a nucleus consisting of 30 protons and 26 neutrons is needed. This would be ^{56}Zn which is, of course, not stable. Because the antineutrinos react on the protons, total muon capture data for the reaction $\mu^- + ^{56} Fe \longrightarrow \nu_\mu + X$ is directly appropriate for obtaining the parameters for the antineutrino case. We can also see that the total muon capture in ^{56}Zn is what we want for the neutrino case immediately from charge symmetry[12]. This follows from the relation:

$$e^{i\pi I_2} J_\mu(0) e^{-i\pi I_2} = -J_\mu(0)^\dagger \tag{16}$$

and that a rotation of π about the I_2 axis essentially changes the sign of the value of I_3.

We therefore make use of a method we derived for the ^{127}I case[9] to create a muon capture rate for ^{56}Zn. We have used this method on a few occassions and it has been seen to provide reasonable results[9,12]. To do this we make use of a calculation by Goulard and Primakoff[18,19] which predicts total muon capture rates with remarkable accuracy, generally to within a few percent. The general form of the Goulard-Primakoff relation is given by:

$$\Gamma_{TOT} = Z_{eff}^4 G_1 \left[1 + G_2 \frac{A}{2Z} - G_3 \frac{A-2Z}{2Z} - G_4 \left(\frac{A-Z}{2A} + \frac{A-2Z}{8AZ} \right) \right] \tag{17}$$

where $G_1 = 261, G_2 = -0.040, G_3 = 0.26$ and $G_4 = 3.24$ and where these parameters are the most recent ones[21]. On the other hand following a method of calculation[7,8,9,10,11,12] similar to that leading to Eq.(12), we obtain:

$$\Gamma_{TOT} = \frac{C|\Phi(0)|^2 G^2 \cos^2\theta_C < E_\nu >^2 D}{8\pi M_i (M_i + m_\mu)}. \tag{18}$$

Equations (17) and (18) describe the same process and must be equal for the same nucleus. We therefore take the ratio of Eqs. (17) and (18) for the ^{56}Fe nucleus and the ^{56}Zn nucleus respectively and find:

$$\frac{(\Gamma_{TOT}^{GP})^{Zn}}{(\Gamma_{TOT}^{GP})^{Fe}} = \frac{(\Gamma_{TOT}^{MP})^{Zn}}{(\Gamma_{TOT}^{MP})^{Fe}}. \tag{19}$$

In Eq.(19), GP stands for the Goulard-Primakoff result of Eq.(17) and MP stands for the result given by Eq.(18). Because $C = (\frac{Z_{eff}}{Z})^4$ and $|\Phi(0)|^2 = Z^3\alpha^3(1 + m_\mu/M)^{-3}$ in Eq.(19),there is considerable cancellation in Eq.(19). The remaining expression is easily evaluated and the result is:

$$\frac{D^{Zn}}{D^{Fe}} = \frac{Z^{Zn}}{Z^{Fe}} \times 1.802 = 2.08. \qquad (20)$$

The total muon capture rate for ^{56}Fe is known [20]. The most recent measurement is $4.53 \pm 0.10 \times 10^6$ $sec.^{-1}$. Making use of the known value for $C = .322$, and the results of Eq.(20) we obtain:

$$D^{Zn} = 5.18 \times 10^{11} \; MeV^2 \qquad (21)$$

at q^2 appropriate to muon capture. We note that this is for the case of neutrino reactions only. For the antineutrino reactions being considered here, D^{Fe} is obtained directly from the total muon capture rate for iron. For q^2 appropriate to muon capture this number is:

$$D^{Fe} = 2.49 \times 10^{11} \; MeV^2. \qquad (22)$$

We thus need additional information to fully determine D as can be seen from Eq.(15). If this were the ^{12}C case we would rely on electron scattering data. However this is not available over a sufficient range of q^2 and so we make use of an impulse approximation result[18,19] which yields a value for $\tilde{D} = \frac{D(q^2)}{D(0)}$ given by:

$$\tilde{D} = \frac{[1 - (\frac{(A-Z)}{2A})\delta(\vec{q}^2)]}{[1 - \frac{(A-Z)}{2A}\delta(0)]} \qquad (23a)$$

where:

$$\delta(\vec{q}^{\,2}) = (\frac{d}{r_o})^3(1 - \frac{\vec{q}^{\,2}d^2}{10}). \qquad (23b)$$

with $r_o = \frac{R}{A^{\frac{1}{3}}}$, $\frac{d}{r_o} \simeq 1.5$, and where from Eq.(15) we may write:

$$\tilde{D} = (1 - \frac{b_o}{a_o}q^2) \equiv (1 - b'_o q^2). \qquad (23c)$$

Evaluating Eq.(23c) we obtain:

$$\tilde{D} = 1 - .328\frac{q^2}{m_\mu^2} \qquad (24a)$$

for the cases of neutrino reactions and

$$\tilde{D} = 1 - .5185\frac{q^2}{m_\mu^2} \qquad (24b)$$

for the cases of the antineutrino reactions. and finally:

$$D(q^2) = 4.202 \times 10^{11}(1 - .328\frac{q^2}{m_\mu^2}) \qquad (25a)$$

for the neutrino reaction reactions and

$$D(q^2) = 1.819 \times 10^{11}(1 - .5185\frac{q^2}{m_\mu^2}) \qquad (25b)$$

for the antineutrino reactions. Eqs.(25a) and (25b) as noted are largely determined by total muon capture rates. We expect them to fail as neutrino energies increase. Some rough calculations indicate that this will occur for energies from 150 to 200 MeV above threshold. We note that $\delta(\vec{q}^{\,2})$ is unrelated to the δ of Eq.(6). Thus we are now able to evaluate Eq.(13) and to obtain the desired cross sections.

3. RESULTS

We are now able to evaluate the cross section,Eq.(13), for the four reactions of interest here. In figure 1 we plot the cross section for the reaction,$\nu_e +^{56} Fe \longrightarrow e^- + X$, as a function of incoming electron neutrino energies from threshold to 240 MeV. In figure 2 we plot the cross section for the reaction $\nu_\mu +^{56} Fe \longrightarrow \mu^- + X$ as a function of incoming muon neutrino energy from threshold to 300 MeV. In figure 3 we plot the cross section for the reaction $\bar{\nu}_e +^{56} Fe \longrightarrow e^+ + X$ as a function of the electron antineutrino energy from threshold to 240 MeV. Finally in figure 4 we plot the cross section for the reaction $\bar{\nu}_\mu +^{56} Fe \longrightarrow \mu^+ + X$, as a function of the muon antineutrino energy from threshold to 300 MeV.

In all of the figures there is a point at which curves representing $\delta(E_\nu)$ below the giant dipole resonance and given by Eqs.(9b) for the neutrino case and (9c) for the antineutrino case join with the curve for $\delta(E_\nu)$ above the giant dipole resonance given by Eq.(9a). A more precise fitting of the available data has made this joining smoother for all of the reactions considered here than in previous calculations.

For the cases of the two neutrino reactions considered here, namely, $\nu_e +^{56} Fe \longrightarrow e^- + X$, and $\nu_\mu +^{56} Fe \longrightarrow \mu^- + X$, we also obtain appropriate spectrum averaged cross sections. For the former case we average over the Michel spectrum and for the latter case we average over the LSND[11] spectrum. These averaged cross sections are given by the formula:

$$< \sigma_\nu >= \frac{\int_{threshold}^{E_\nu^{max}} \sigma(E_\nu)\phi(E_\nu)\, dE_\nu}{\int_{threshold}^{E_\nu^{max}} \phi(E_\nu)\, dE_\nu} \qquad (26)$$

where E_ν^{max} is the maximal neutrino energy of the spectrum and $\phi(E_\nu)$ is the neutrino spectrum. For the case of the electron neutrino reaction, the

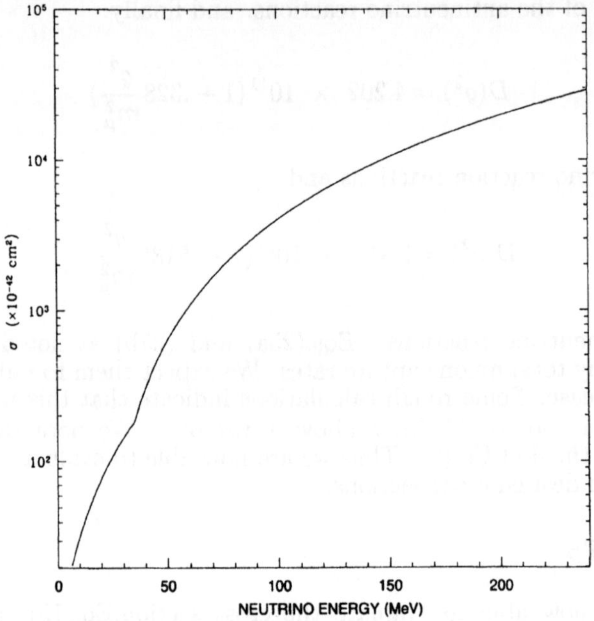

Fig. 1. Plot of the cross section for the reaction, $\nu_e + {}^{56}Fe \longrightarrow e^- + X$, as a function of incoming electron neutrino energy.

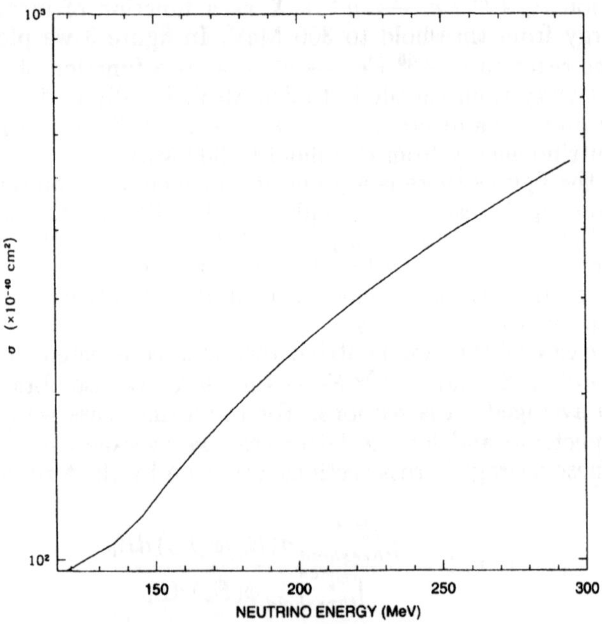

Fig 2. Plot of the cross section for the reaction, $\nu_\mu + {}^{56}Fe \longrightarrow \mu^- + X$, as a function of incoming muon neutrino energy.

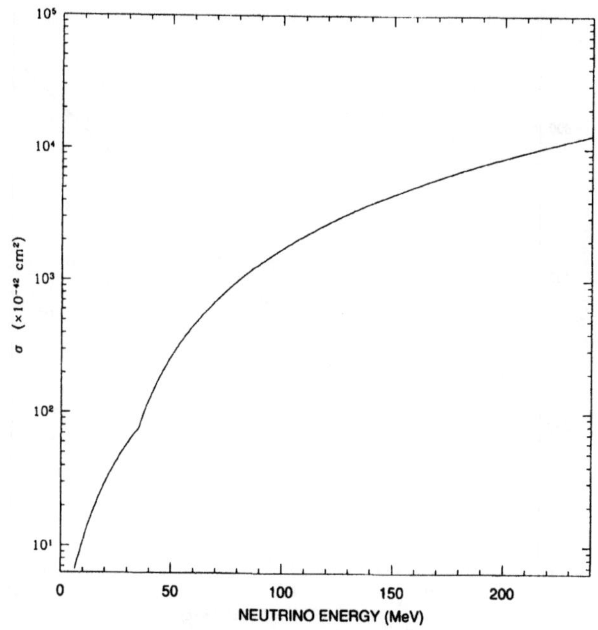

Fig 3. Plot of the cross section for the reaction,$\bar{\nu}_e +^{56} Fe \longrightarrow e^+ + X$, as a function of incoming electron antineutrino energy.

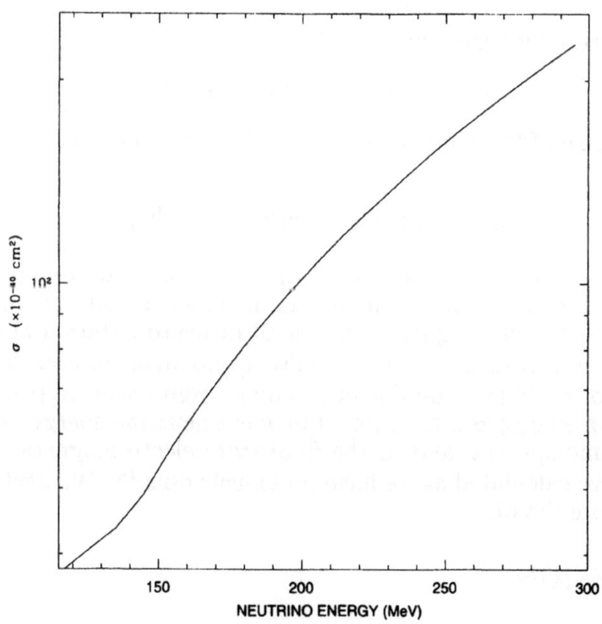

Fig 4. Plot of the cross section for the reaction,$\bar{\nu}_\mu +^{56} Fe \longrightarrow \mu^+ + X$, as a function of incoming muon antineutrino energy.

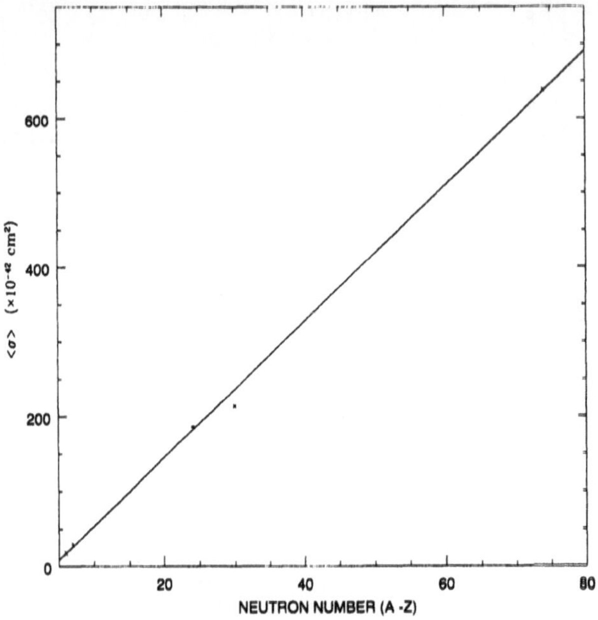

Fig 5. Plot of the Michel spectrum averaged cross sections for several reactions as a function of neutron number, $A-Z$. The actual points are indicated by X's and the solid straight line is our fit.

Michel spectrum averaged cross section is :

$$< \sigma_{\nu_e} >= 214 \ \pm 25\% \ \times \ 10^{-42} \ cm^2 \tag{27a}$$

and for the case of the muon neutrino the LAMPF spectrum averaged cross section yields:

$$< \sigma_{\nu_\mu} >= 134.8 \ \pm 25\% \ \times \ 10^{-40} \ cm^2. \tag{27b}$$

Finally we must note our sources of error. These are due to an approximately 5 percent error in D from errors in the muon capture results, and a 15 percent error stemming from our use of ratios to obtain a D appropriate to ^{56}Zn and uncertainties in the impulse approximation used to obtain the q^2 behavior of D. The remaining error comes from uncertainties in the average excitation energy as a function of incident neutrino energy, uncertainties in the neutrino spectra, and in the final state electromagnetic interactions which we have calculated as we have previously done[14]. We next discuss our results in more detail.

4. DISCUSSION

As we remarked near the beginning of this paper, we generally expect the neutrino cross sections to increase as the number of neutrons in the nucleus increases. We list below the results of three calculations for other nuclei plus that for ^{56}Fe. These are:

$$< \sigma_{^{12}C} >= 17.43 \ \times \ 10^{-42} \ cm^2 \tag{28a}$$

$$< \sigma_{^{13}C} >= 28.9 \times 10^{-42}\ cm^2 \qquad (28b)$$

$$< \sigma_{^{56}Fe} >= 214 \times 10^{-42}\ cm^2 \qquad (28c)$$

$$< \sigma_{^{127}I} >= 638 \times 10^{-42}\ cm^2 \qquad (28d)$$

where we have used cross sections for the electron neutrino reactions averaged over the Michel spectrum. It is immediately seen that indeed the cross sections do increase with the increasing number of neutrons in the nuclear target. What is less obvious is that this increase is essentially linear in the number of neutrons. In figure 5 we plot the cross sections versus the number of neutrons, $A - Z$. The actual points are indicated by X's. The data is very well fit by the straight line:

$$< \sigma >= (-37.23 + 9.11[A - Z]) \times 10^{-42}\ cm^2 \qquad (29).$$

We have used our calculated values[8,9,12] for the averaged cross sections but they are within 10 to 15 percent of the published experimental values for the ^{12}C cross section[1,13] and with published experimental results[3,4] for the ^{56}Fe and with the unpublished but circulated results[15] for the ^{127}I cross sections. The number of neutrons in these nuclei range from 6 to 74.

It is probably unwise to make particular claims for Eq.(29) as the number of points is small and the error bars are large. Hopefully the number of nuclei for which data is available will increase with time, particularly if the ORLaND project goes forward[21]. However the available data seems to imply that for low energy inclusive processes,the cross sections are dominated by the number of neutrons and that other effects such as screening are relatively small. We intend to investigate many of these questions in a future paper.

There has been a previous theoretical calculation using modern shell model techniques[22] of the inclusive electron neutrino reaction in ^{56}Fe. The result of this calculation is $< \sigma >= 229 \times 10^{-42}\ cm^2$. This is in strikingly good agreement with the value given by Eq. (27a) obtained in this paper by an entirely different method. Furthermore as we have noted,there are some experimental results from KARMEN[3] and some more recent results not yet published in a journal article but available in a published dissertation[4]. These values are $251 \pm 0.83\ (stat) \pm 0.42 (sys) \times 10^{-42}\ cm^2$ and $256 \pm 1.08\ (stat) \pm 0.43\ (sys) \times 10^{-42}$ respectively. Although the total uncertainties are at the 40 to 50 percent level in these measurements, the general agreement of calculations and experiment is quite heartening. We particularly wish to note that that our use of Eqs.(19) and (20) to produce an appropriate muon capture rate for the calculation of neutrino reactions in ^{56}Fe given here has produced very reasonable results, not only for iron, but over a wide range of nuclei from $A - Z = 7$ to $A - Z = 74$. This gives us additional confidence in this method.

To conclude we mentioned earlier that our values for D given by Eqs. (25a) and (25b) which are largely based on the total muon capture rate must begin to fail with increasing neutrino energy. Some rough calculations place this at 150 to 200 MeV above the reaction threshold. However it may be possible to extend the range of our calculation for the ^{56}Fe case. There exists higher energy neutrino reaction data for neutrino reactions in iron[5] for which q^2 sufficiently small may be available to extend the range of D. We do note that our hadron tensor in terms of structure functions may be written as:

$$Q_{\mu\nu} = -W_1(g_{\mu\nu} + \frac{q_\mu q_\nu}{Q^2}) + \frac{W_2}{M^2}(p_\mu + \rho q_\mu)(p_\nu + \rho q_\nu) - \frac{i}{M} W_3 \epsilon_{\mu\nu\lambda\sigma} q^\lambda p^\rho \quad (30)$$

where $\rho = \frac{p \cdot q}{Q^2}$. The structure functions W_1, W_2, and W_3 are known over a range of Q^2. Thus an extension of our present work to higher energy is certainly possible and we leave it to a future time.

5.REFERENCES

1 B. Zeitnitz,Prog. Part. Nucl. Phys. **13**,445(1995).

2 G. Drexlin,private communication.

3 R. Maschuw,Prog. Part. Nucl. Phys. **40**,183(1998)

4 C. Ruf,Dissertation Universitat Bonn,Mensch und Buch Verlag, ISBN 3-933346-59-2.

5 W. G. Seligman et al.,Phys. Rev. Lett. **79**,1213(1997).

6 C. James,Proceedings of the Intersections of Particle and Nuclear Physics,CIPANP2000,Editors: Z. Parsa and W.J. Marciano,AIP Conference Proceedings, Vol. 549,Melville,NY,777(2000).

7 S.L. Mintz and D.F. King, Phys. Rev. **C30**,1585(1984).

8 S.L. Mintz and M. Pourkaviani, Nucl. Phys. **A573**501(1994).

9 S.L. Mintz and M. Pourkaviani,Nucl. Phys. **A584**,665(1995).

10 S.L. Mintz and M. Pourkaviani,Nucl. Phys. **A589**,724(1995).

11 S.L. Mintz and M. Pourkaviani,Nucl. Phys. **A594**,346(1996).

12 S.L. Mintz,Nucl. Phys. **A672**,503(2000).

13 R.C. Allen et al.,Phys. Rev. Lett. **64**,1871(1990).

14 C. Athanassopoulos et al.,Phys. Rev. **C54**,2717(1996).

15 K. Lande,private communication.

16 Y. Torizuka et al.,Proceedings of the International Conference on Nuclear Structure Studies Using Electron Scattering and Photoreactions, Sendai, Japan,Editors:K. Shoda and H. Ui,Tohoku University,Tomizawa ,Sendai,Japan,171(1972).

17 Y. Torizuka et al.,Proceedings of the International Conference on Photonuclear Reactions and Applications,Editor: Barry L. Berman,Ernest O. Lawrence Livermore Laboratory,University of California, Vol. 1,Livermore California,675(1973).

18 B.Goulard and H.Primakoff, Phys. Rev.**135**,**B**1139(1964).

19 B. Goulard and H. Primakoff,Phys. Rev. **C10**,2034(1974).

20 T. Suzuki,D.F. Measday, and J. P. Roalsvig,Phys.Rev.**C35**, 2212(1987).

21 ORLaND Proposal Draft,1999.

22 E. Kolbe,K. Langanke,and G. Martinez-Pinedo,Phys. Rev. **C60**, 052801(1999).

NATURE'S HIGHEST ENERGY PARTICLE ACCELERATORS

Nature's Highest Energy Particle Accelerators

Elliott D. Bloom*

1. ACTIVE GALACTIC NUCLEI (AGN) AN OVERVIEW.

Active galaxies are the largest class of high-energy gamma ray emitters and identify sights where extreme particle acceleration is taking place. In this short paper I will review some of the overall features of these astronomical sources and indicate how their study can give important experimental information on fundamental physics, including relativistic gravity and high-energy particle production and acceleration. There is an excellent overview of AGN from this perspective in a recent review article by Madejski (1999). The slides, containing many illuminating color figures, from this talk are available on the web as a PDF file (Bloom, 2000). The two other papers in this session, presented by S. Ritz and S. Colgate, indicate dramatic developments in observational capability for AGN in the high-energy gamma ray regime with the Gamma Ray Large Area Telescope now in construction, and important theoretical developments in understanding how jets are produced in AGN sources, respectively.

Active galaxies produce vast amounts of energy from a very compact central volume (for example, the scale of the diameter of the volume is about 3 light hours in NGC5506 based on X-ray variability measurements, see Krolik, Done, & Madejski (1993)). The prevailing explanation for this phenomenon is that AGN are powered by accretion onto super-massive black holes (BH) having masses of $10^6 - 10^{10}$ solar masses (with Schwarzschild radius $\equiv R_s = 3\times10^6 - 3\times10^{10}$ km). This model of AGN includes energetic (multi-TeV), highly collimated, relativistic jets emerging from the core of the AGN near the BH for some classes of sources. Such jets are seen in high detail via radio maps of some of these sources. These models imply an economy of interpretation in observational astronomy that has resulted in a collapse of categories of AGN. This is due to the realization that the orientation of the AGN jet, or the AGN plane of symmetry, relative to the viewing angle from the earth can dramatically change what is observed. Thus source classes that were previously thought to have intrinsic differences are now unified.

---

*Stanford Linear Accelerator Center, MS-98, 2575 Sand Hill Road, Menlo Park, CA 94025.

The Role of Neutrinos, Strings, Gravity, and Variable Cosmological Constant in Elementary Particle Physics, Edited by Kursunoglu *et al.*, Kluwer Academic/Plenum Publishers, New York, 2001

2. A PHENOMENOLOGICAL VIEW OF AGN

In this section I will give a brief overview of the phenomenology of AGN trying to emphasize the salient features of these sources that make them of interest not only to high-energy astrophysicists, but also to high-energy physicists like myself.

Blazars are flat-radio-spectrum (radio loud) AGN, a class whose members include BL Lac and highly polarized and optically violently variable (OVV) quasars. Quasars are among the most distant objects in the universe with the most distant observed having a redshift of 5.8 (Fan, et al., 2000). Their associated jets commonly exhibit superluminal motion (Vermeulen & Cohen 1994). Blazars are incredibly powerful and highly variable. Currently favored models of blazars posit that we are viewing these objects almost down the throat of a highly collimated jet of relativistic particles (Blandford & Königl, 1979) emanating from an accreting supermassive BH (Blandford & Znajek, 1977).

The wideband spectra of blazars have two pronounced components: one broad peak at low energies (LE) extending from IR to soft x-ray, and a second broad peak at high energies (HE) in the gamma-ray band. Blazars that are hosted in quasars (QHBs in, for example, OVV quasars) and radio-selected BL Lac objects (RBLs or LBLs) typically have the LE peak in the IR band. On the other hand, the majority of BL Lac objects and X-ray selected BL Lacs (XBLs), have the LE peak in the UV or soft x-ray bands. The high level of polarization observed from radio to UV implies that the LE component is most likely produced via synchrotron radiation in a macroscopic scale magnetic field. The luminosity ratio of the HE to LE components is systematically larger for QHBs than for XBLs, and the Lorentz factors of electrons contributing to the peak in the $\nu F(\nu)$ spectrum (ergs/sec/cm^2 at the instrument) is much lower for QHB than for XBL, despite the fact that magnetic field is comparable in the QHB and XBL (Kubo et al., 1998).

Blazar AGN photon emission spans an enormous frequency range, from the radio to multi-TeV. In a number of blazars that have been observed at high-energy, the majority of the total source power is at > 100 MeV, and extends to 10 TeV or higher. Blazars have been observed to emit as much as ~ $f\times10^{49}$ ergs/s in the energy range of 100 MeV to 5 GeV (von Montigny et al. 1995; Mukherjee et al. 1997), where f is the (1/beaming factor) ~ 10^{-3}, but is somewhat uncertain. The beaming of photons is required if they are not to be absorbed via $\gamma\gamma \rightarrow e^+e^-$ on their way out from the source (Maraschi et al. 1992; Mattox et al. 1993; McNaron-Brown et al. 1995; Dermer & Gehrels 1995).

The very high-energy emission can only be observed for "close by" AGN ($z << 1$) due to absorption by the extra galactic background light (EBL) for energies > 20 GeV or so (Macminn & Primack, 1995; Madau & Phinney 1996; also see Salamon & Stecker 1997). Galaxies undergoing starbursts during the epoch of galaxy formation produced the EBL. The absorption of high-energy gamma rays occurs over cosmological distance via $\gamma\gamma \rightarrow e^+e^-$ interactions with near-ultraviolet, optical and near-infrared photons that make up the EBL. The cross section for this interaction is maximized when $\varepsilon \approx (1/3) \times (1\ TeV/E)$ [eV], where ε is the EBL photon energy in eV, and E is the gamma-ray energy in TeV. Measuring the characteristics of the EBL as a function of intensity and z can provide unique information on the formation of galaxies at early epochs, and will test models for structure formation in the universe, such as those in which a neutrino mass of 5 to 10 eV plays an important role (see Macminn & Primack 1996).

Another major class of AGN is Seyfert galaxies. Type 1 galaxies are characterized by the presence of both narrow and broad optical emission lines, superposed onto a relatively featureless blue continuum, and type 2 that are characterized by having only narrow optical emission lines. Seyferts tend to be closer than blazars, and so are more evolved. They show weak or no radio signals. In addition, the unified model of AGN indicates that we are viewing Seyferts 2s more from the side, while we view Seyfert 1s more aligned to the (relatively weak) "jet" direction. Seyfert 1 galaxies show lines that are emitted from the accretion disk. The huge line widths from the Doppler broadening (1000's km/sec) imply that the line forming regions are moving at high velocities around a compact, central, massive object. The narrow lines of Seyfert 2s come from absorption of photons from the central region in the torus of gas and dust in the plane of the galaxy and re-emission in longer wavelengths. These narrow lines are usually forbidden lines, but sometimes allowed narrow lines as well (Madejski, 2000). These lines are typically Doppler broadened to ~100's km/sec widths.

AGN show rapid time variability. The type 2 Seyfert galaxy NGC5506 shows X-ray intensity doubling times of about 10^4 seconds in 2-10 keV X-rays from a chaotic X-ray time series (see Krolik, Done, & Madejski 1993). This time scale is indirectly repeated in Mkn 501 that shows a doubling of the >300 GeV photon rate in about 600 seconds (D'Vali, M. et al., 1999). Mkn 501 is a blazar and the Wipple air Cherenkov telescope is observing the very high energy gamma rays shooting from the superluminal jet with bulk $\Gamma \sim 10$. Thus in the rest frame of the jet material the intensity doubling time is ~ 6×10^3 seconds, or about the same as for NGC5506 in the X-rays. Thus for these sources we believe that the X-ray or gamma radiation originates in a compact volume ~ 10^{-4} pc in size. A BH horizon of this size corresponds to a BH mass ~ 10^9 solar masses.

Perhaps the best evidence that a strong gravitational field is present at the center of AGN comes from measurements of the shape of the Fe K_α fluorescence line from Seyfert 1 galaxies. Observations of this line in MCG:-6-30-15 (Tanaka, et al., 1995), show a relativistically redshifted and distorted 6.4 keV Fe line from a geometrically thin, but optically thick accretion disk that is being illuminated by X-rays. The two-peaked shape of the spectrum - more intense blue Doppler peak at 6.4 keV, and weaker red Doppler peak at 5.1 keV - is characteristic of matter flowing in a disk-like structure close to a BH (George & Fabian, 1991). To fit the shape of the spectrum, both the relativistic Doppler shift coming from the rotation of the disk, as well as an overall gravitational red shift are needed. Recent results from XMM (Hasinger, 2000) show a few examples of fits to such spectra from Seyfert 1 galaxies, including one that is indicative of a rapidly spinning Kerr hole at the center of the AGN. The XMM data constrain the source of the Fe line to be inside $3R_S$ with > 3 σ significance. According to Hasinger, on the order of 100 distorted line measurements should be completed within the XMM mission lifetime.

3. IS THE ENGINE OF AGN A BLACK HOLE?

The basic technique currently available for searching for the existence of a BH in a galaxy is measuring a mass concentration its center. This is a classical mechanical method using Kepler's laws. If the mass density is high enough, as it is in the center of our galaxy where there is a mass density of at least 2×10^{11} solar masses/pc^3 inside a radius of 0.015 pc (Ghez et al. 2000), the case seems very strong as no other mechanism is known that can produce such a mass density. If there is a BH in our galaxy it is dormant, as the Milky Way does not have an AGN. Trying to prove this case for AGN is more difficult as they are much further away than the center of our galaxy. However, given some exotic phenomena that are actually observable, one can do surprisingly well.

The NASA Hubble Space Telescope has imaged a spiral-shaped disk of hot gas in the core of M87. The image also shows a bright streak moving from the center of the AGN at ~ 42° to the earth viewing angel, which is the optical jet (synchrotron radiation) powered by the central engine. Hubble has the resolution to enlarge the central accretion disk feature, and has made 5 spectrographic observations measuring the radial velocity of the visible matter in the neighborhood of the nucleus, within a radius of about 1arcsecond of the center of the disk (Harms et al. 1994). Two of the measurements were made at diametrically opposite locations, centered 0.25 arcseconds to either side of the nucleus along the major axis that is perpendicular to the jet, and one was made on the center of the disk. The resulting spectroscopic Doppler shifts from the last three measurements imply, through Kepler's laws, a central mass of $2.4 \pm 0.7\times10^9$ solar masses. The HST resolution value of ~0.5 arcseconds at 15 Mpc corresponds to a containment radius of about 18 pc, or an average density of 1×10^5 (solar masses)/pc^3. A BH of this mass has a horizon radius ~ 2.5×10^{-4} pc. The much lower mass density obtained makes for a relatively weak case for a central BH as compared to the Milky Way example.

Perhaps the most exotic observation showing the presence of a Keplerian disk around a putative BH includes measurement of not only the velocity of the disk, but also its acceleration. This dual measurement is capable of measuring the mass of the putative BH independently of the otherwise uncertain estimates of its distance. These results were obtained using Very Large Baseline Interferometry observations of megamasers in the vicinity of the nucleus of the Seyfert 2 galaxy NGC 4258, and were first reported by Miyoshi et al. (1995). See Bragg et al. (2000) for a recent review. The H_2O masing activity can only be observed where the velocity gradient along the line of sight is close to zero. This condition implies that activity can be seen at locations where the disk motion is approximately tangent to, or perpendicular to, the line of sight. The radio measurements at 22.23... GHz, the frequency of the H_2O masers in the disk, are sufficiently numerous and accurate that they allow a reconstruction of the disk geometry in great detail. The resulting model of the disk implies the existence of a mass concentration of 3.9×10^7 solar masses inside a radius of ~ 0.2 pc, or an average density of ~1×10^{10} (solar masses)/pc^3, only a factor of 20 or so below the Milky Way nucleus density. A BH of this mass has a horizon radius of 3.8×10^{-6} pc. The velocity centroid of the disk is consistent with the optical systemic velocity of the galaxy, and the derived rotation axis of the disk is well aligned with a 100 pc-scale radio jet. Measurements also yield an upper limit on the magnetic field of < 300 mG (1σ limit on the parallel, or toroidal, component of the field.)

Thus the existence of a massive BH at the core of at least two galaxies, the Milky Way and NGC 4258 seems very likely in that no other explanation reasonably presents itself. If there are two, there must be many.

4. THE JET MODEL OF AGN EMISSION.

The jet model of AGN emission is well established observationally, as compared to gamma ray burst models where no direct observations of jets emanating from the burst have been directly observed. This is well illustrated in multi-wavelength images of the active galaxy NGC 4261 (see web figure, 1994) that have been superimposed. The galaxy is an elliptical (imaged in the optical) with two enormous jets of material emanating from its core region (imaged in the radio).

AGN jets have been observed in radio and X-ray imaging of the same object. Recent Chandra imaging of the jet in PKS 0637-752, in 0.3 – 10 keV X-rays, have been compared to previous radio images of the same jet structure and find a striking congruence in morphology in multi-wavelengths (Chartas et al., 2000). The jet structure is over a Mpc long, and has multi-"knots" along its length. The structural similarity is qualitatively striking, but also quantitatively confirmed in the paper.

What is the relationship of the central BH and the powerful jets? Models have posited that the BH powers the jets. There are two general models that have been proposed. One proposes that the source of energy for the jet is the spin of a Kerr BH (Blandford and Znajek, 1977), the other proposes that the BH gravity drives the accretion disk and the jet is formed in the disk (see the paper of S. Colgate in this proceedings). If the spin of the Kerr hole is in the same direction as the rotation of the accretion disk, the disk inner edge can in principle come within a fraction of R_s of the horizon of the spinning hole. The resulting geometry enhances enormously the gravitational effects on the disk as compared to a Schwarzschild hole.

Recent results have indicated that it is likely that jets are formed by an accretion disk around the central black hole, which is threaded by a magnetic field (Junor, et al, 1999) supporting the model described by S. Colgate in his paper at this conference. In the Junor, et al. paper the jet's full opening angle is calculated from radio observations as a function of distance from the core of M87. They find that at 0.004 pc the opening angle of the jet, estimated from a Gaussian full-width at quarter max fit, is about 60°. The jet angle slowly decreases to about 20° at 1 pc and reaches an asymptotic value of about 6°-7° at 150 pc from the core at which value it remains to 10^3 pc where the measurements stop. As discussed above, the horizon radius for the BH candidate in M87 is about 2.5×10^{-4} pc. If we assume that the shrinking of the jet angle is due to the increasing jet velocity along the jet axis, relative to its transverse velocity (initial random velocity of material of the jet), these observations seem to favor an MHD model in which the jet is accelerated via a disk outflow, rather than emanating already collimated (at high velocity) from the vicinity of the black hole horizon.

In electron-positron models of AGN jets the main source of the high-energy gamma rays is the Compton up scattering of low-energy photons by the high-energy, very non-thermal electrons and positrons in the jet (compared to the radio through X-ray where the direct Synchrotron radiation tends to dominate). There are two classes of models for the source of soft photons. In the synchrotron self Compton (SSC) models the source is synchrotron radiation produced by the high-energy electrons themselves. In the external radiation Compton (ECR) models, either the accretion disk (~10 eV) (Dermer, Schlickeiser, & Mastichiadis, 1992), or the emission line clouds and/or intercloud medium provides the photons (Sikora, Begelman, & Rees, 1994; Blandford & Levinson, 1995). For an example of a recent comparison of this type of model to multiwavelength data, see Dermer, Sturner, & Schlickeiser (1997).

Note that it is still controversial whether the jets are primarily made of electron-positron plasma or electron-proton plasma (Celotti & Fabian, 1993). While electron-positron models for the blazars reasonably represent the observed multi-wavelength spectra, there is a very interesting possibility that proton-initiated cascades contribute substantially in some blazars (Mannheim & Biermann, 1992). In the electron-proton version, π^0 production and decay play important parts in the very high-energy gamma-ray spectrum, and ν emission should eventually be observed from the jets. Future observations e.g., GLAST at high energy (see paper of S. Ritz in these proceedings), should have the sensitivity to probe further the incredible engine that drives the AGN acceleration process by observing AGNs whose jets are not pointing at the earth to within a few degrees. Dermer, Sturner, & Schlickeiser (1997); Sambruna et al. (1996) have estimated the off-axis emission of photons from AGN jets. Their calculation range over 20 orders of magnitude in energy, from 10^{-12} MeV, or radio wavelengths, to 10^8 MeV (= 100 TeV). To achieve this range of energies for the critical testing of AGN models necessitate multiwavelength observations of AGN by radio, optical, and X-ray, and gamma ray observatories.

To give the reader a feeling for this AGN jet model, a brief description of the Dermer, Sturner, & Schlickeiser (1997) model follows. The calculation gives the instantaneous spectral power flux and the time-integrated spectral power fluence produced as a function of energy of the observed gammas and the angle of these gammas to the jet direction.

Relativistic electrons are injected into an on-axis jet blob moving away from a central black hole of mass M (solar masses), with bulk Lorentz factor $\Gamma \sim 10$ (for gamma ray bursts $\Gamma_{grb} \sim 100$ is needed). Addition assumptions have the electrons instantaneously injected into the co-moving blob frame with an isotropic angular distribution and a $1/E_e^2$ electron energy distribution, with $1<E_e/m_e<10^5$ (m_e the mass of the electron). The jet blob radius is r_b, assumed to be spherical in the co-moving frame, and the blob consists of a thermal electron-positron plasma with density $n_{th} = (\sigma_T \times r_b)^{-1}$, where σ_T is the Thomson (inverse Compton scattering) cross section. (Note that in the calculations the blob length is scaled to 1000 gravitational radii $\equiv 10^3 \times 1.5 \times 10^5$ [cm] ×M.) The derivation of the high-energy photon fluences is restricted to the regime where either synchrotron radiation losses or inverse Compton scattering losses with photons from the external radiation field dominate the electron energy loss rate and hence the photon production from the jet.

The magnetic field permeating the region of photon production is estimated at 1 Gauss, which drives the synchrotron radiation photon production from the electron beam. Also, the electron jet is moving through a radiation field, which drives the inverse Compton scattering, that is, up-scattering photon production. The radiation field is assumed in the calculation to be a 10 eV monochromatic, spherical radiation field (Sikora et al., 1994), with total inverse Compton scattering depth, τ^s ($\tau^s_{-2} \equiv \tau^s/100$), surrounding the supermassive black hole ($M_8 \equiv M$ [solar masses]/10^8 [solar masses]). A description of the actual central engine energy source mechanism is not included. There is also a scale distance for the radiation field from the black hole, R_{sc} ($R_{0.1} \equiv R_{sc}$ [pc]/0.1[pc]). The intensity of the external photon field is given by equation (33) from Dermer, Sturner, & Schlickeiser (1997) as $4.3 \times 10^3 \times (l_{Edd} \times M_8 \times \tau^s_{-2}/R_{0.1}^2)$, where the luminosity of the disk accretion photon field is written in terms of the normalized Eddington limit luminosity, $l_{Edd} \equiv L_{disk}$ [ergs/s]/($1.26 \times 10^{46} \times M_8$)[ergs/s]. The model calculations presented in Dermer, Sturner, & Schlickeiser (1997) take $l_{Edd} = M_8 = \tau^s_{-2} = R_{0.1} = 1$, and the AGN is placed at $z=1$ (with no EBL absorption correction). The power going into the jet is set equal to the Eddington luminosity.

Using this model, Bloom & Wells (1997) have estimated the off-axis luminosity for two very nearby AGNs, Centaurus A at 2 Mpc and M87 at 12 Mpc (NB: $z = 0.03$ is ~120 Mpc). They find that M87 with a jet axis angle of 42±5° with respect to the earth direction may show an observable signal in GLAST above ~10 GeV. At lower energies the diffuse background will dominate. Centaurus A has a jet axis of about 68° with respect to the earth direction, and according to the above jet model should show no observable signal. However, EGRET has observed a tentative signal in a large error box centered on Centaurus-A (Hartman, et al. 1999), which is unexpected in current off-axis AGN jet models.

5. SUMMARY AND CONCLUSIONS

- Multiwavelength measurements in radio, IR, X-ray, and γ-rays have yielded an explosion of information about AGN over the past 10 years.
- All of this data has sharpened our picture of the structure of AGN, with a unified model emerging, and the central engine most likely a super massive BH. The jets of the AGN, for those AGN that show jets, are powered buy the gravity of the BH. However, an electro-magnetic dynamo developed from the rotating magnetic field threaded in the Keplerian disk feeding the BH converts the BH gravitation energy to EM energy, and is the direct driver of the jet acceleration (cf. S. Colgate paper in this proceedings).
- Over the next 10 years new missions– Chandra, XMM, the reincarnation of Astro-E, and GLAST, combined with dramatic improvements in ground based facilities promise a further explosion of data that is bound to dramatically impact our current ideas, and our knowledge of strong gravity (cf. S. Ritz paper in this proceedings).

ACKNOWLEDGEMENTS

I thank Behram Kursunoglu for inviting me to organize the session on Nature's Highest Energy Accelerators. I found the Conference to be very informative, and look forward to future Conferences. I also thank Greg Madejski of SLAC for his help in preparing this talk, and his participation in the session as our annotator. Department of energy contract DE-AC03-76SF00515 supported this work.

REFERENCES

Blandford, R. D., and Königl, A., 1979, Relativistic jets as compact radio sources, *ApJ*, **232**:34.

Blandford, R. D., and A. Levinson, 1995, Pair cascades in extragalactic jets. 1: Gamma rays, *ApJ*, **441**:79.

Blandford, R. D., and Znajek, R. L., 1977, Electromagnetic extraction of energy from Kerr black holes, *MNRAS*, **179**:433.

Bloom, E. D., 2000, *ftp://ftp.slac.stanford.edu/groups/ek/elliott/EDBCoralGables12-00.pdf*

Bloom, E. D., and Wells, J. D., 1998, Multi – GeV photons from electron - dark matter scattering near active galactic nuclei, Phys. Rev., **D57**:1299.

Bragg, A. E., et al., 2000, Accelerations of water masers in NGC 4258, *ApJ*, **535**:73.

Celotti, A., and A. C. Fabian, 1993, The kinetic power and luminosity of parsec-scale radio jets - an argument for heavy jets, *MNRAS*, **264**:228.

Chartas, G., et al., 2000, The Chandra x-ray observatory resolves the x-ray morphology and spectra of a jet in PKS 0637-572, *ApJ*, **542**:655.

Dermer, C. D., and N. Gehrels, 1995, Two classes of gamma-ray-emitting active galactic nuclei, ApJ, **447**:103.

Dermer, C. D., R. Schlickeiser, and A. Mastichiadis, 1992, High-energy gamma radiation from extragalactic radio sources, *A&A*, **256**:L27.

Dermer, C. D., S. J. Sturner, and R. Schlickeiser, 1997, Nonthermal Compton and synchrotron processes in the jets of active galactic nuclei, *ApJS*, **109**:103.

D'Vali, D., et al. 1999, Whipple observations of BL Lac objects, in: Proceedings of the *26th International Cosmic Ray Conference*, Salt Lake City, **3**:422, *http://www.icrc1999.utah.edu/~icrc1999/root/icrc.html* .

Fan, X, et al., 2000, The discovery of a luminous z =5.80 quasar from the Sloan Digital Sky Survey, *AJ*, **120**:1167.

George, I. M., & Fabian, A. C. 1991, X-ray reflection from cold matter in active galactic nuclei and X-ray binaries, *MNRAS*, **249**:352.

Ghez, A. M., et al. 2000, The accelerations of stars orbiting the Milky Way's central black hole, *Nature*, **407**:349.

Harms, R., et al. 1994, HST FOS spectroscopy of M87: Evidence for a disk of ionized gas around a massive black hole, *ApJ*, **435**:L35.

Hasinger, G., 2000, Recent relativistic results from Newton and Chandra, Invited Talk, *20th Texas Symposium on Relativistic Astrophysics*, 10-15 December 2000 - Austin, Texas.

Junor, W., et al, 1999, Formation of the radio jet in M87 at 100 Schwarzschild radii from the central black hole, *Nature*, **401**:28.

Krolik, J., Done, C., & Madejski, G. 1993, X-ray light curves of active galactic nuclei are phase incoherent, *ApJ*, **402**:432.

Kubo, H., et al., 1998, ASCA Observations of Blazars and Multiband Analysis, ApJ, **504**:693.

Macminn, D., and J. R. Primack, 1996, Probing the ERA of Galaxy Formation via TeV Gamma-Ray Absorption by the Near Infrared Extragalactic Background, *SSRv*, **75**:413.

Madau, P., and E. S. Phinney, 1996, Constraints on the extragalactic background light from gamma-ray observations of high-redshift quasars, *ApJ*, **456**:124.

Madejski, G.M. 1998, Black holes in active galactic nuclei: observations, in: *Theory of Black Hole Accretion Disks*, M. Abramowicz, G. Bjornsson, and J. Pringle, ed., Cambridge University Press, p. 21, and references therein.

Mannheim, K., and P. L., Biermann, 1992, Gamma-ray flaring of 3C 279 - A proton-initiated cascade in the jet?, *A&A*, **253**:L21.

Maraschi, L., G. Ghisellini, and A. Celotti, 1992, A jet model for the gamma-ray emitting blazar 3C 279, *ApJ*, **397**:L5.
Mattox, J. R., et al., 1993, The EGRET detection of quasar 1633 + 382, *ApJ*, **410**:609.
McNaron-Brown, K., et al., 1994, in *AIP Conf. Proc. No. 304, Proc. 2nd Compton Symposium*, C. Fichtel, N. Gehrels, and J. P. Norris ed., AIP, New York, 587.
Miyoshi, M., et al., 1995, Evidence for a black hole from high rotation velocities in a sub-parsec region of NGC4258, *Nature*, **373**:127.
Mukherjee, R., et al., 1997, EGRET observations of high-energy gamma-ray emission from blazars: an update, *ApJ*, **490**:116.
Salamon, M. H., and F. W. Stecker, 1998, Absorption of high-energy gamma rays by interactions with extragalactic starlight photons at high redshifts and the high-energy gamma-ray background, *ApJ*, **493**:547.
Sambruna, R. M., et al., 1996, On the Spectral Energy Distributions of Blazars, *ApJ*, **463**:444.
Sikora, M., M. C. Begelman, and M. J. Rees, 1994, Comptonization of diffuse ambient radiation by a relativistic jet: The source of gamma rays from blazars?, *ApJ*, **421**:153.
Tanaka, Y., et al., 1995, Gravitationally redshifted emission implying an accretion disk and massive black hole in the active galaxy MCG:-6-30-15, *Nature*, 375:659.
Vermeulen, R. C., and Cohen, M. H., 1994, Superluminal motion statistics and cosmology, *ApJ*, **430**:467.
von Montigny, C., et al., 1995, High-energy gamma-ray emission from active galaxies: EGRET observations and their implications, *ApJ*, **440**:525.
Web figure, 1994, *http://www.seds.org/hst/ngc4261.html* .

THE ORIGIN OF ALL COSMIC RAYS: A SPACE-FILLING MECHANISM

Stirling A. Colgate, and Hui Li*

Abstract:

There is a need for one mechanism to accelerate cosmic rays universally over the full energy spectrum , isotropically, and space filling. The current view is a theory based upon a series of mechanisms, patched to fit various spectral regions with a mechanism for the origin of the UHCRs still in doubt. We suggest that the reconnection of force-free magnetic fields produced by the twisting of all imbedded magnetic flux by the vorticity motion of all accretion or condensations both within the Galaxy as well as the metagalaxy is the universal mechanism. This leads to the acceleration of all cosmic rays with both total energy and individual energies up to the highest observed of 3 x 10^{20} ev and predicting an upper limit of 10^{23} ev. There are three primary, and we believe compelling reasons for adopting this different view of the origin of CRs. (1) The energy source is space filling and isotropic, thereby avoiding any anisotropy's due to single sources , e.g., supernovae remnants and AGN. (2) The galactic and particularly the extragalactic energy source is sufficient to supply the full energy of a universal galactic and extragalactic spectrum of 10^{60} to 10^{61} ergs sufficient to avoid the GZK cut-off. (3) Efficient $E_{parallel}$ acceleration from reconnection of force-free fields is well observed in the laboratory whereas collisionless shock acceleration still eludes laboratory confirmation.

1. Introduction

The widely accepted mechanism of collisionless shock acceleration is most frequently substantiated by the prediction of a power-law spectrum not much different from that observed. Instead we observe that any non-thermal acceleration mechanism that results in a fractional loss in number of particles for a given fractional gain in energy results in a power -law spectrum . Whatever mechanism of acceleration that results in the minimum fractional loss of particles per fractional gain in particle energy results in that particular spectral index that is a minimum index value. The number of particles at higher energy then will exceed those of a steeper (more negative index) spectrum. This will then will be selectively chosen as the preferred mechanism for producing the higher energy particles.

*T-6, MSB-288, LANL, Los Alamos, NM87545 colgate@lanl.gov

The Role of Neutrinos, Strings, Gravity, and Variable Cosmological Constant in Elementary Particle Physics, Edited by Kursunoglu *et al.*, Kluwer Academic/Plenum Publishers, New York, 2001

Reconnection is the dissipation of the parallel currents that maintain the force-free fields, resulting in a parallel electric field and hence parallel acceleration. Magnetic energy is stored primarily in the tension of force-free fields. It is released or converted only by the dissipation, $E_{parallel} \times J_{parallel}$, of the currents that maintain the magnetic stressed (force-free) state. The sources of such twisted force-free field energy and flux are the many mass condensations, which always occur with finite angular momentum, e.g., accretion disks or rotating stars. The large winding number during the dissipation of this angular momentum assures that all imbedded flux will necessarily undergo continuous winding until either the angular momentum is dissipated or reconnection takes place. This assures that at least several orders of magnitude of twist i.e., a winding number of at least several hundred or greater will take place.

The transition between this picture of transient, multiple, highly twisted, flux-tubes within our galaxy (and between galaxies) and the observed state of relatively ordered flux must occur by reconnection. The acceleration of cosmic rays in this picture is then a sum of both individual particles that are accelerated in single coherent regions of high winding number and random diffusion in energy space of particles that stochastically traverse many such coherent regions of E-parallel acceleration (and deceleration.) The advantage of reconnection acceleration is that it is well grounded in laboratory measurements e.g. tokamak interruptions, whereas shock acceleration is still an unfounded speculation.

2. Why do we need an alternate Theory of the Origin of Cosmic Rays?

1) The observed isotropy of CRs requires a space-filling mechanism both for galactic CRs as well as for the UHCRs of extra galactic origin. If supernova shocks are the sources within our galaxy, then the diffusion mean-free path for galactic CRs after acceleration must be very long, >1 kpc, and yet not escape the galaxy. The spacing between the few near-by supernovae in the CR life-time is large enough to lead to an observed anisotropy. In order that the spatial fluctuations and therefore anisotropy due to a small number of contributing supernova be small enough. and furthermore that the surrounding "leaky box" be extraordinarily isotropic. The same argument applies to the UHCRs if they are accelerated locally within AGN.

2) The plasma physics of the required scattering mechanism necessary to produce shock acceleration is unlikely because of individual particle damping within the large non-linear amplitude waves of the turbulence necessary for the scattering itself. The average number of scatterings per particle required for shock acceleration of one e-fold in relativistic energy is approximately $(v_{shock}/c)^2 \sim 10^5$ to 10^6 scatterings. These scatterings must occur from strong, self-excited, hydromagnetic waves with a loss per scattering of less than 10^{-5} to $^{-6}$ per doubling in energy. Such a small loss or high reversibility is very unlikely unless proven in the laboratory. The hydromagnetic turbulence must be "strong" in order that a particle be scattered within a few Larmor orbits, ~10, in dimension, which are then the mean free path of each scattering event.

3) The evidence for acceleration due to quasi-parallel heliosphere shocks from satellite measurements near the earth is weak, in some cases with small or no evidence for the expected strong hydromagnetic turbulence, and where acceleration is demonstrated only after a few shock traversals by the accelerated particles. Thus no more than 10 kT acceleration has been observed whereas 10^{20} kT must be achieved.

4) Laboratory experiments necessary to demonstrate shock acceleration have not yet succeeded and space observations have failed to observe the necessary turbulence of the collisionless shock.

5) The argument that a unique power-law spectra resulting from shock acceleration is not compelling.

6) The above argument of number of scatterings per shock crossing and a non-linear wave amplitude of B/delta B ~ 10, implies a shock thickness of $n R_{Larmor} = 10\ (c/v_{shock})\ R_{Larmor} = 1$ kpc at 10^{15} ev, the "knee" of the spectrum, and therefore a thickness significantly greater than the thickness of the galaxy.

3. Power-law Spectrum

The accepted theory of cosmic ray acceleration is shock wave acceleration in the ISM driven by supernova (Axford, Leer, and Skadron, 1977; Bell, 1978; Blanford and Ostriker, 1988). "This acceptance has been largely based upon the good agreement between the "universal" power-law spectrum predicted by shock acceleration i.e., the power-law index becomes:

$$s=(d \ln N)/(d \ln E) \sim -(2+e)$$

depending only on the Mach number and the observed or inferred particle spectra." (see Blanford and Eichler, 1987 for a review, and many papers by P. Biermann for a more accurate comparison.) This belief that a nearly correct power-law spectral index alone is unique is instead, a less restrictive condition than commonly believed. Any accelerator for which a fractional gain in energy, (d ln E), by a few particles is accompanied by a fractional loss, -(d lnN), in number of the remainder will give a power-law:

$$dN/N = -s(dE/E)\,.$$

The fractional loss for a fractional gain in energy is what would be expected for a rigidity dependent loss mechanism where the probability of a relativistic particle being scattered out of an acceleration region is inversely proportional to its energy or rigidity.

For values of s <~2, i.e. a smaller fractional loss, the integral energy becomes asymptotically large, $\int_0^E N dE \approx E^{s-2}$ and at some energy will truncate or limit the acceleration mechanism, destroying the confinement and hence the accelerating mechanism itself. Hence, it is not likely that at any one time we should see many such accelerators occurring naturally in the Galaxy. On the other hand accelerators with $s >> 2$ will produce a steep spectrum that, relative to another less steep one, i.e. s closer to 2, will be lost relative to the less steep mechanisms above some critical energy. Hence it is likely that the spectrum of any observed mechanism should be close to $s=2+e$.

4. Isotropy

Any mechanism for accelerating cosmic rays that depends primarily upon supernova will have a problem producing the necessary isotropy presently observed (Richardson and Osborn, 1974). From a Tev to 10 to 100 times greater, where shower arrays are most sensitive to anisotropy, this anisotropy is 10^{-3} or less. Within the lifetime, t, of cosmic rays of 10^7 years and a diffusion coefficient of D, there is a characteristic distance, d, within which particles can diffuse to the observer. The anisotropy of the diffusive flux will be of the order of the maximum fractional differences of the source distribution of $N_{supernova}$ number of events averaged over the time, t. This fractional difference is of order $(N_{sn})^{1/2}$. If all the supernova within the galaxy at a typical rate of 1/30 years in the life time of 10^7 years, or $3x10^5$ supernova were all equally contributing to the local flux, then the anisotropy would be of the order $(3x10^5)^{-1/2} = 1.8x10^{-3}$ a value somewhat larger than observed. The diffusion distance, d, must then be significantly greater than the dimension of the galaxy, 30 kpc. Under these circumstances the diffusion coefficient must be greater than $3x10^{31}$ cm^2 /s or a mean free path for scattering greater than 1 kpc. This is $3x10^4$ greater than what is assumed for the necessary scattering for the diffusive shock mechanism where the mean free path is frequently assumed to be roughly 10 Larmor radii at 10 Tev. If this difference in scattering length indeed occurs due to a decrease in hydromagnetic turbulence after the passage of the shock, then one is faced with the further difficulty that the cosmic rays would escape from the galaxy unless the whole galaxy were surrounded by a near perfect "leaky box" with an anisotropic imperfection of diffusive loss of less than 5×10^{-4}. This is hard to imagine in view of current galactic dynamo theory where a dipole field surrounds the Galaxy yet at the same time inside the box cosmic rays diffuse relatively easily. If a smaller diffusion coefficient, D, is imagined, then in a flat Galaxy, N_{sn} is proportional to D and the probable anisotropy due to fluctuations becomes large, proportional to $D^{-1/2}$. Hence an unlikely circumstance must exist of first a very small diffusion coefficient, then a very large one, and finally a near perfectly symmetric surrounding leaky diffusive box.

5. Particle Scattering Losses

It is well recognized that in diffusive shock acceleration a particle must traverse the shock many times, $c/v_{shock} \sim 300$ in order to double in energy. However, in order to return to the shock this number of times by random scattering, the mean number of scattering events becomes the square of this number of crossings, or 10^5 scatterings. One recognizes that the accelerated particles are likely to be the exceptional ones that cross the shock the requisite number of times with fewer scatterings, but nevertheless the fractional loss per scattering must be very small for the average particle, which is the particle "lost" from acceleration by diffusion to convection at each energy. Hence, recognizing the many orders of magnitude of acceleration required, in order not to modify the power-law exponent below the good agreement with theory, the fractional irreversible loss per scattering must be less than $\sim 10^{-5 \text{ to } -6}$. This is an extraordinarily small loss in the presence of what must be strong turbulence in order to have the short scattering lengths required (~10 Larmor orbits) to reach the highest energies (Lagage and Caesarsky). One suspects the excitation of at least one mode of non-linear damping of the hydromagnetic waves at the level of $10^{-5 \text{ to } -6}$. Such non-linear damping induces heating and hence damping of those particles furthest ahead of the shock that excite the waves in the first place. The particles furthest ahead are those of greatest rigidity and therefore most sensitive to modifying the spectrum.

6. The Reconnection Theory of Acceleration

The specific angular momentum of matter in the galaxy rotating at an w= 10^{-8} per year and a local radius corresponding to roughly one solar mass, (~3pc) is 3×10^{22} cm^2/s. This is some 2×10^4 larger than the Keplearian orbit at a solar radius. This factor of angular momentum must be exchanged with the Galaxy as a whole in order to form a star, and still more angular momentum must be removed in order to collapse to a neutron star. The most likely mechanism for the transport of this angular momentum is by a turbulent, a-viscosity, the Rossby vortex mechanism, (Li et al. 2001) in the disk-like accretion of a star through the T-tauri stage. (If the torque is transmitted by magnetic stress, this makes the following effects even larger.). Similarly the magnetic flux threading the original matter at the density of the ISM is also orders of magnitude (300) larger than the flux that would allow compression to a solar radius. Hence magnetic flux must also be expelled from the condensing matter as well. Therefore this flux, in order to escape the matter, must undergo reconnection during the process of condensation. However the subsequent winding or twisting of this residual flux requires a far larger reconnection process since the winding produces a far larger total flux than the original.

The picture at every stage of condensation is that of a helical, twisted flux tube that extends some 50 to 100 turns beyond the source. The number of turns is interpreted observationally from the topology of the "jets" or collimated radio sources which are presumed to be just these helical flux tubes from accretion onto the black holes of AGN. They become illuminated in the radio by the accelerated electrons due to the reconnection of the force-free flux of the twisted flux tube. The same picture applies to the bi-polar flows observed in star formation. The relation between the number of turns and the angle of the "jet" is that the minimum energy force-free field of the helix is close to a 45 degrees where the f and z components of the field are equal. Hence the number of turns, n, will be of the order (1/jet angle) x $\ln(L_{max}/R_{min})$, or several hundred.

One possible location to accelerate the UHECRs, E> 10^{18} ev, is in the force-free field helix next to the black hole of nearly every galaxy. Here it is assumed that the reconnection dissipates the total current associated with the B_f component of the field. If this current is interrupted as a sequence of multiple reconnection events, but phased along a local flux surface at c by the accelerated particles themselves, the total potential becomes of the order:

$$V = n\, B_{max}\, R_{min}\, 300\, (2\text{pln}\, n) = 2\times10^{23} \text{ ev.}$$

where $B_{max} = 10^4$ gauss, and $R_{min} = 1$ au.
Individual near-by (100Mpc) AGN sources should then be seen in the UHECR distributions, contrary to observations.

7. Extra Galactic Acceleration

Our current view (Colgate and Li 2000, 2001) is that this force-free helical field fills the intergalactic space with magnetic flux, $\sim10^{-6}$ G per galaxy that decays by reconnection over a Hubble time. The total energy of $\sim3\times10^{61}$ ergs per galaxy spacing volume ($\sim10^{74}$ cm^3) is close to the limiting energy that can be supplied by the black hole and is observed as the minimum energy necessary to supply the synchrotron radiation of many radio lobes. This energy is also only slightly greater than the energy of the extra-galactic spectrum itself, assuming the UHECRs are extrapolated back to a Gev with the typical galactic index of -2.7. If one uses the slightly flatter slope of -2.5 implied by the spectrum $E > 10^{18}$ ev, then there is enough energy in the intergalactic fields on this model to re accelerate the CRs ~100 times during a Hubble time. These particles should then be confined primarily to the galaxy walls with a loss to the voids in 1/100 of the Hubble time. Because of this rapid loss to the voids of CRs accelerated in the walls, the remaining young population of UHCRs should then not be subject to the GZK cut off in the spectrum. This lost to the voids occurs by random walk in the coherent helical fields produced by each galaxy. The same upper energy limit as above, close to the black hole should apply in the IGM because the magnetic flux is presumed to be the same.

8. Summary

Accretion onto galactic black holes, neutron stars, cataclysmic variables, and T-tauri stars will all lead to tearing mode reconnection associated with twisted fields by all condensations. The energy of these condensations is far more than required to maintain the cosmic rays of our Galaxy.

Laboratory experiments can be performed to simulate both magneto hydrodynamics as well as the tearing mode reconnection and the associated $E_{parallel}$ acceleration of the "run-away" particles. The spheromak experiments are a step in this direction. Interruptions in tokomaks are already laboratory proof of this acceleration. We need to perform more laboratory plasma experiments to observe reconnection in the near vacuum limit with the resulting parallel electric fields. Without laboratory experiments, as for example shock acceleration, we are still uncertain about the origin of cosmic rays.

9. REFERENCES

Axford, W.I., Leer, E., and Skadron, G., 1977, ICRC.. 15th ICRC, 11, 132.

Bell, A.R., 1978, MNRAS, 182,147.

Blanford, R.D. and Ostriker, J.P., 1988, ApJ, 221, L29.

Blanford, R.D. and Eichler, D., 1987, *Physics Reports,* 154,1,1987.

Colgate, S.A. and Li, Hui, 1999, Astrophys. Space Sci., 264. 357

Colgate, S.A. and Li, Hui, 2000, "The Magnetic Fields of the Universe and Their Origin" IAU Symp 195, ASP Conf. Series 334, eds. P.C.H. Martens and S. Tsurta 1999

Colgate, S.A. and Li, H. & Pariev, V., 2001, "The Origin of the Magnetic Fields of the Universe: The Plasma Astrophysics of the Free Energy of the Universe" Physics of Plasmas, accepted

Jokipii, J.R., Morfill, G., 1987, Astrophys. J. **312**, 170

Lagage, P.O., Cesarsky, C.J., 1983, Astron. & Astroph. **118**, 223

PRIMORDIAL BLACK HOLES AND THE ASYMMETRICAL DISTRIBUTION OF SHORT GRB EVENTS

David B. Cline *

1. INTRODUCTION

The search for evidence for primordial black holes (PBHs) has continued since the first discussion by Hawking [1]. In fact, this was about the time that gamma-ray bursts (GRBs) were first discovered, making a natural association with PBHs. However, in the intervening years it has become clear that the time history of the typical GRB is not consistent with the expectations of PBH evaporation [2].

While the theory of PBH evaporation has been refined, there are still no exact predictions of the GRB spectrum, time history, etc. However, reasonable phenomenological models have been made, and the results again indicate that most GRBs could not come from PBHs [3]. In addition, there are new constraints on the production of PBHs in the Early Universe that indicate that the density of PBHs in the Universe should be very small, but not necessarily zero [4].

After the initial discovery of GRBs, it took many years to uncover the general properties. Around 1984, several GRBs were detected, indicating that there was a class of short bursts with dime duration of ~ 100 ms and a very short rise time [5]. A separate class of GRBs was declared. This classification seems to have been forgotten and then rediscovered by some of us.

It is still possible that there is a sizable density of PBHs in our Galaxy and that some of the GRBs could be due to PBH evaporation. Recently we showed that the BATSE 1B data have a few events that are consistent with some expectation of PBH evaporation (short bursts with time duration <200 ms and that are consistent with $V/V_{max} \sim ½$.

2. PRIMORDIAL BLACK HOLE EVAPORATION AND FIREBALL

Ever since the theoretical discovery of the quantum-gravitational particle emissions from black holes by Hawking, there have been many experimental searches (see

* Department of Physics and Astronomy, Box 951647 University of California Los Angeles Los Angeles, CA 90095-1647, USA

The Role of Neutrinos, Strings, Gravity, and Variable Cosmological Constant in Elementary Particle Physics, Edited by Kursunoglu *et al.*, Kluwer Academic/Plenum Publishers, New York, 2001

Ref. [6] for details) for high-energy gamma-ray radiation from PBHs. They would have been formed in the Early Universe and would now be entering their final stages of extinction. The violent final-stage evaporation or explosion is the striking result of the expectation that the PBH temperature is inversely proportional to the PBH mass, e.g., $T_{PBH} \approx 100$ MeV (10^{15} g/m_{PBH}), since the black hole becomes hotter as it radiates more particles and can eventually attain extremely high temperatures.

In 1974, S. Hawking showed in a seminal paper that an uncharged, non-rotating black hole emits particles with energy between E and $E + dE$ at a rate per spin of helicity state of

$$\frac{d^2N}{dtdE} = \frac{\Gamma_s}{2\pi h}\left[\exp\left(\frac{8\pi GME}{hc^3}\right) - (-1)^{2s}\right]^{-1}, \quad (1)$$

where M is the PBH mass, s is the particle spin, and Γ_s is the absorption probability [1]. It can be considered that this particle emission comes from the spontaneous creation of particle–antiparticle pair escapes to infinity, while the other returns to the black hole. Thus, the PBH emits massless particles, photons, and light neutrinos, as if it were a hot black-body radiator with temperature $T \approx 10^{16}$ g/M MeV, where M is the black hole mass [3]. A black hole with one solar mass, $M_1 \approx 2 \times 10^{33}$ g, has an approximate temperature of 10^{-3} K, while a black hole with mass of 6×10^{14} g has a temperature of ~ 20 MeV. The temperature of a black hole increases as it loses mass during its lifetime. The loss of mass from a black hole occurs at a rate, in the context of the standard model of particle physics, of

$$\frac{dM}{d_{t(2)}} = -\frac{\alpha(M)}{M^2}, \quad (2)$$

where $\alpha(M)$, the *running constant*, counts the particle degree of freedom in the PBH evaporation. The value of $\alpha(M)$ is model dependent. In the standard model (SM), with a family of three quarks and three leptons, it is given [2,6,13] as $\alpha(M) = (0.045\ S_{j=+1/2} + 0.162\ S_{j=1}) \times 10^{-4}$, where $S_{j=+1/2}$ and $S_{j=1}$ are the spin and color degrees of freedom for the fermions and gauge particles, respectively. For the SM, $\alpha(M) = 4.1 \times 10^{-3}$.

A reasonable model of the running coupling constant is illustrated in Figure 1, where the regions of uncertainty are indicated [2][7]. These are the regions where there could be a rapid increase in the effective number of degrees of freedom due to the quark–gluon phase (QGP) transition. The phase transition would lead to a rapid burst in the PBH evaporation or, at high energy, there could be many new particle types that would also lead to an increase in the rate of evaporation. Also shown in Figure 1 are the regions in PBH temperature where short duration GRBs may occur when the PBH mass is either 10^{14} or 10^9 g.

Black holes at the evaporation stage at the present epoch can be calculated as having $M_* \approx [3\alpha(M_*)\tau_{evap}]^{1/3} \approx 7 \times 10^{14}$ g for $\alpha(M_*) \approx 1.4 \times 10^{-3}$. The bound on the number of black holes at their critical mass, constrained by the observed diffuse gamma-ray background, has been put in the 10–100 MeV energy region [4]:

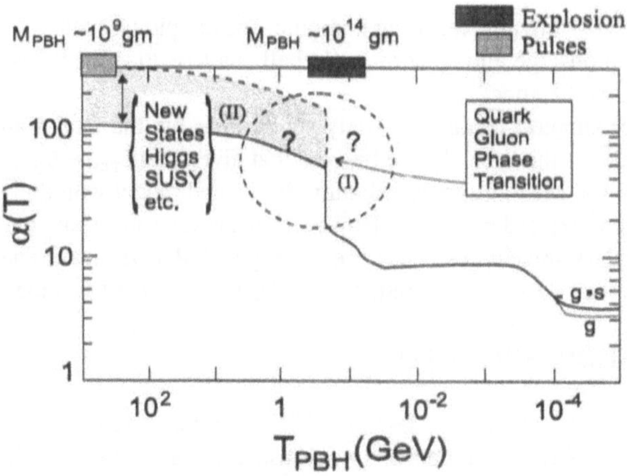

Figure 1 Running coupling or density of states factor α, showing regions of uncertainty due to the QGP transitions (I) or the increase in the number of new elementary particles (II). It is possible that intense short GRBs could occur at either of these temperatures (the decrease of mass of the black hole is given by $dm_{PBH}/dt = -\alpha/m^2_{PBH}$). A rapid mass decrease or burst can occur when $m_{PBH} \leq 10^{10}$ g or if α changes rapidly near the QGP transition.

$$N = \left.\frac{dn}{d(\ell n M)}\right|_{M=M_*} \leq 10^5\,\mathrm{pc}^{-3}. \qquad (3)$$

Thus, the number of black holes with critical mass M_* in their final state of evaporation is

$$\frac{dn}{dt} = -\frac{3\alpha(M_*)}{M_*^3} = N = 2.2\times 10^{-10}\,N\,\mathrm{pc}^{-3}\,\mathrm{yr}^{-1} \quad . \qquad (4)$$

Based on previous calculations and numerous direct observational searches for high-energy radiation from an evaporating PBH, we might conclude that it is not likely to single be able to single out such a monumental event. However, we pointed out a possible connection between very short GRBs observed by the BATSE team and PBH evaporation emitting very short energetic gamma rays [2][6]. If we want to accept this possibility, we may have to modify the method of calculating the particle emission spectra from an evaporating PBH, in particular, at or near the quark-gluon plasma], for a review, and phase transition temperature at which the T_{PBH} arrives eventually. We briefly discussed that inclusion of the QGP effect around the evaporating PBH at the critical temperature may drastically change the resulting gamma-ray spectrum [7][6]. The QGP interactions around the evaporating PBH form an expanding hadronic (mostly pions) matter fireball. Shortly after the decay of pions, the initial hadronic fireball converts to a fireball with mixtures of photons, leptons, and baryons.

Using the simplest picture, i.e., only π^0s produced in the QGP transition, we can obtain the properties of the fireball. We find out that given that $L_{PBH} \sim L_{QGP} \sim 5 \times 10^{34}$ ergs and $T_{PBH} \sim T_{QGP} \geq 160$ MeV, a simple radiation-dominated model would give $r_s \sim 10^9$ cm, which implies that τ is of order 100 ms. However, this is very uncertain and could even be the order of seconds in some cases. Thus one can expect GRBs from a fireball to have both a very short rise time (≤ 1 μs) and duration (~50–200 ms) also in this model.

3. STUDY OF SHORT GRBs [8] **

We have studied all events with T_{90} less than 200 ms and then refit the time profile using the TTE data with the BATSE 3B data set (we restrict this part of the analysis to BATSE 3B data. We found 12 events that have a fitted time duration of less than 100 ms. These events are listed in Table 1 with the fitted time duration. We then closely inspected the 12 events

TABLE 1
EVENT SELECTION AND PROPERTIES (BATSE 3B)

Trigger Number	Duration from TTE Fit (s)	Hardness Ratio	Comments	Used in This Analysis
01453......	0.006 ± 0.0002	6.68 ± 0.33	Poor event, precursor	No
0512......	0.014 ± 0.0006	6.07 ± 1.34	Good event	Yes
01649......	0.020 ± 0.0080	....	Fair event ‡	Yes
00207......	0.030 ± 0.0019	6.88 ± 1.93	Good event	
02615......	0.034 ± 0.0032	5.42 ± 1.15	Good event	Yes
03173......	0.041 ± 0.0020	5.35 ± 0.27	Poor event, precursor	No
02463......	0.049 ± 0.0045	1.60 ± 1.55	Good event	No
00432......	0.050 ± 0.0018	7.46 ± 1.17	Good event	Yes
00480......	0.062 ± 0.0020	7.14 ± 0.96	Good event	Yes
03037......	0.066 ± 0.0072	4.81 ± 0.98	Good event	Yes
02132......	0.090 ± 0.0081	3.64 ± 0.66	Good event	Yes
00799......	0.097 ± 0.0101	2.47 ± 0.39	Fair event ‡	No

‡ Fair event, small additive structure

and found that some have additional structure. Of the 10 good/fair events, 8 have a time duration of 66 ms or less. Our goal is to select a similar class of events, so we only study single-peak events which constitute the bulk of short bursts. As we will show, these events appear to be almost identical in all features, expect for BATSE trigger 2463,

** This section adapted from Reference 8.

which we delete in this analysis. This event has a clearly different energy spectrum from the bulk of the short GRB events. This sample constitutes the set of events studied here.

To obtain a better understanding of the short bursts, we discuss two individual bursts that have more detailed information than the bulk of the events. According to the arguments given above, we expect all the short burst to be very similar and, therefore, we assume that the behaviour of these special bursts is likely an example for all short bursts. If these events are typical of the short bursts, then we can see a clear behaviour in the fine time structure and the detected γ energy distribution. We start with the incredible GRB trigger 512. We note the detailed fine structure for BATSE trigger 512, which has the finest time structure of any GRB observed to date -- possibly down to 20 μs level. This was a very bright burst and allowed unpredicted time information.

The PHEBUS GRB detection (see PHEBUS Catalog in Terekov et al. 1994 [8]) has recorded two very interesting short time events shown in Figure 2. As far as can be

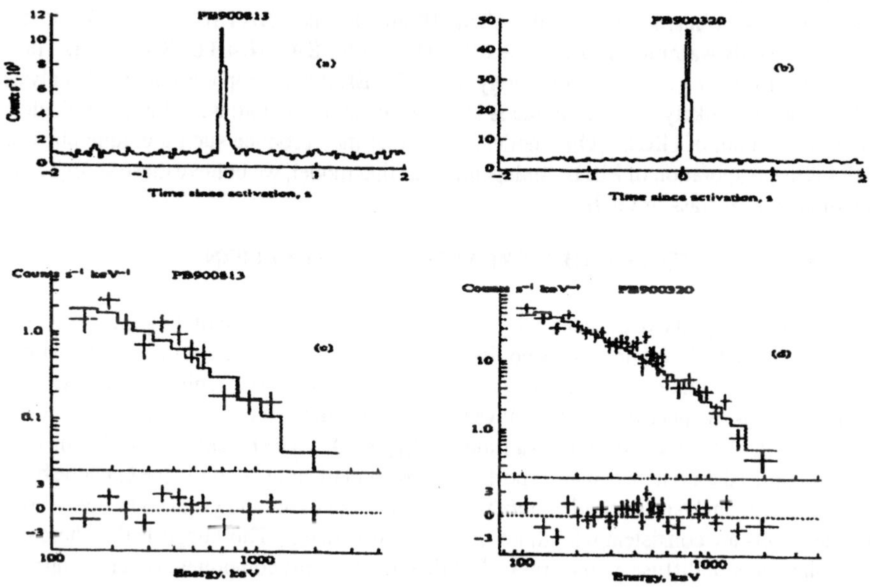

Figure 2. Two short bursts from the PHEBUS detector: Time profiles of (a) PB 900813 and (b) PB 900320, and energy spectrum of (c) PB 900813 and (d) PB 900320 [8].

determined, these events are identical (note that the energy distribution are fit to a synchrotron and are identical). Because of a thicker absorber, the PHEBUS detector has a larger energy-range capacity than that of the BATSE detector.

Thus, this detector can record photon energies up to 180 MeV in contrast to BATSE, which is only sensitive to about 580 keV due to the absorbers. Note that these two PHEBUS events have photon energies above 1 MeV. Thus, the short GRBs have energetic photons in the spectrum.

We classify all GRBs into three different categories: one with $\tau > 1s$ (long, L), one with $1s > \tau > 0.1s$ (medium, M), and one with $\tau \leq$ 100 ms (short, S), which is the focus of this investigation.

We note that the short bursts are strongly consistent with a $C_p^{=}3/2$ spectrum, indicating a Euclidean source distribution, as was shown previously in reference 6. In the medium (100 ms to 1 s) time duration, the ln N - ln S distribution seems to be non-Euclidean; in the long duration ($\tau >$ 1 s) bursts, the situation is more complicated as we have shown recently [8]. The $<V/V_{max}>$ for the S, M, and L class of events is, respectively, 0.52 ± 0.1, 0.36 ± 0.02, 0.31 ± 0.01. (In this case, we have used the BATSE 5B to obtain the best statistics.)

We have shown that the GRBs with $\tau <$ 100 ms likely are due to a separate class of sources and appear to be nearly identical in contrast to the bulk of GRBs. In this analysis, we have studied a small class of BATSE 3B TTE data in detail, and to improve the statistical power for some issues, we have used BATSE 4B and the latest BATSE 5B data. We do not believe this study warrants the use of the full TTE data for BATSE 4B or BATSE 5B, since the point is to show a general morphology of the GRBs, not a complete statistical analysis at this stage. It is likely that the source is local or Galactic, in contrast to the cosmological origin of the bulk of GRBs. One model source that may produce such a unique class of GRBs is the evaporation of PBHs. Independent of that model, we believe these short bursts constitute a third class of GRBs.

4. POSSIBLE RATE FOR GRB FROM PBH EVAPORATION

Over the past two decades, the reality of a diffuse component of the gamma-ray flux has been established. While there is no firm explanation of the source of these gamma rays, one possible candidate is due to the evaporation of PBHs in the Universe. In fact, the diffuse gamma-ray spectrum and flux have been used to put the only real limit on the density of PBHs in the Universe, leading to a limit of $\Omega_{PBH} < 10^{-7}$. The recent work of Dixon et al [9] has claimed the existence of an important component of gamma rays in the Galactic halo. This method of analysis is very different from previous analyses and gives some confidence that the results are consistent with those of other experiments. This suggests that there is a halo component of diffuse gamma rays. The flux level of this component is very similar to the extragalactic diffuse flux.

We now turn to a simple model that explains the current observations by assuming the existence of PBHs at the level of the relaxed Page-Hawking bound discussed previously and the Galactic clustering enhancement factors of 5×10^5 or greater. In Figure 3, we show the logic of the possible detection of individual PBHs by very short GRBs and the connection with the diffuse gamma-ray background. The rate is consistent with a few PBH evaporation for $(Pc)^3$ in our galaxy.

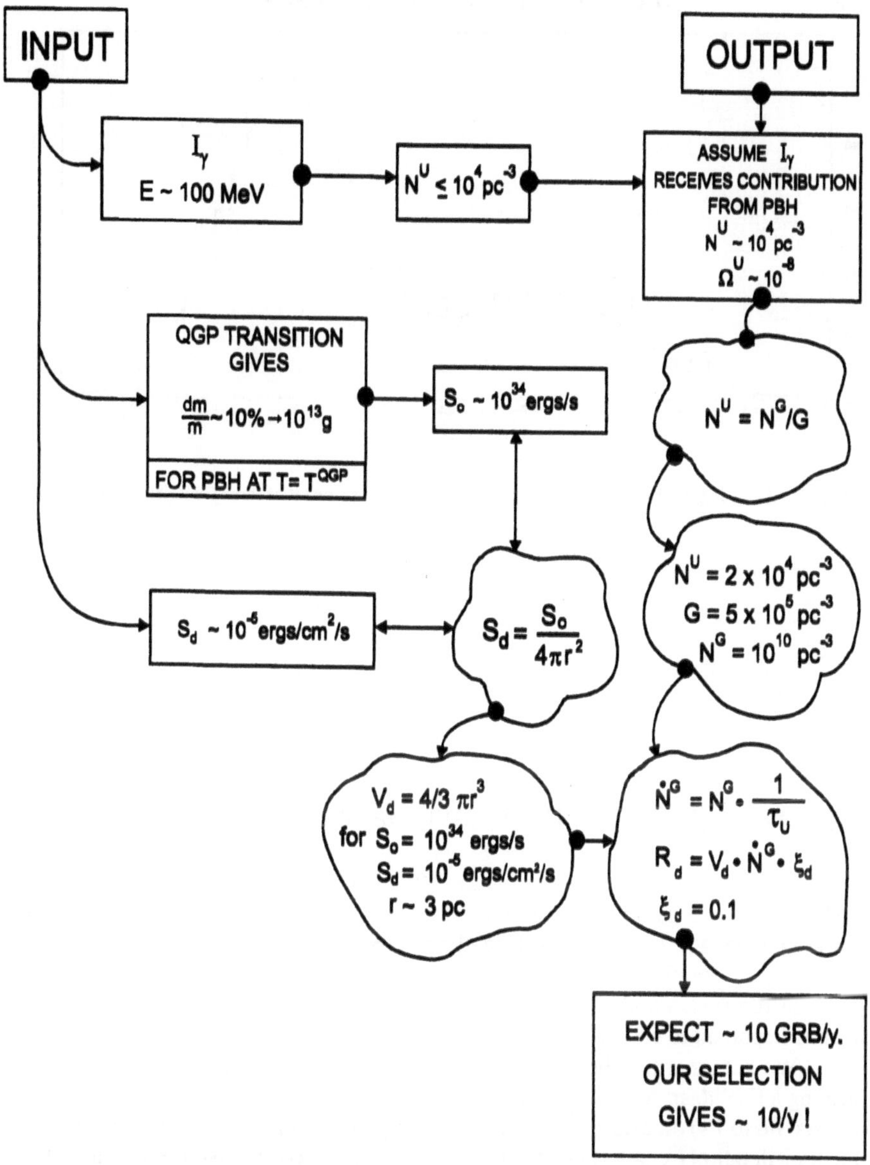

Figure 3. Schematic of the connection between GRB events and the diffuse γ background, I_γ, where N^U is density of PBHs in the Universe, Ω^U is ratio of density of the Universe, T^{QGP} is QGP transition temperature, S_o is GRB luminosity from PBH, N^G is number of PBHs in the Galaxy, G is galaxy clumping factor, S_d is fluence sensitivity of GRB detectors, r is distance to the source, $\dot{N}^G$ is rate of PBH decay in the Galaxy, τ_U is age of the Universe, R_d is number of GRBs detected per year from PBHs, ξ_d is GRB detector efficiency (including the fraction of energy detected).

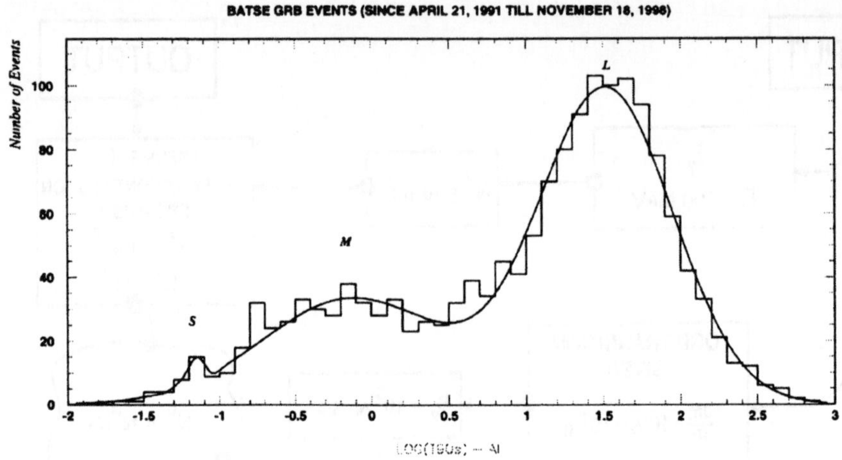

Figure 4. Time distribution T90 for all GRB events. The line is a result of three Gaussian curve fit.

5. DISCOVERY OF AN ANGULAR ASYMMETRY OF SHORT GRBs ***

Figure 4 shows the time distribution T90 for all GRBs from the BATSE detector up to Nov. 1998. Hereafter we will use this data. We divide the GRBs into three classes in time duration: long, L (τ >1 s); short, M (1 s > τ > 0.1 s); and very short, S (τ < 100 ms). We use the duration time of T90 for all of this analysis. Henceforth in this letter, we confine the discussion to the M and S classes of GRBs. Since these events are adjacent in time, it is important to contrast the behaviour.

We note that the short bursts are strongly consistent with a $C_p^{-3/2}$ spectrum, indicating a Euclidean source distribution, as was shown recently by ([8]). In the medium (from 100 ms to 1 s) time duration, the ln N - 1 - S distribution seems to be non-Euclidean; in the long duration (τ > 1 s) bursts, the situation is more complicated as we have shown recently ([8]). The $<V/V_{max}>$ for the S, M, and L class of events is, respectively, 0.52 ± 0.06, 0.36 ± 0.02, 0.31 ± 0.01.

One possibility for explaining these effects is that a big part of the short bursts may come from a local Galactic source. This explanation is likely not viable for the medium time bursts, since $<V/V_{max}>$ is 0.36 ± 0.02, indicating a likely cosmological source. The longer bursts are clearly mainly from cosmological distance.

*** Adapted from a UCLA report by D. Cline, C. Matthey, So. Otwinowski

We assume that the S GRBs constitute a separate class of GRBs and fit the time distribution in Figure 4 with a three-population model. The fit is excellent but does not in itself give significant evidence for a three-population model.

We now turn to the angular distributions of the S and M GRBs. In Figure 5A we show this distribution for the very short bursts. We can see directly that this is not an

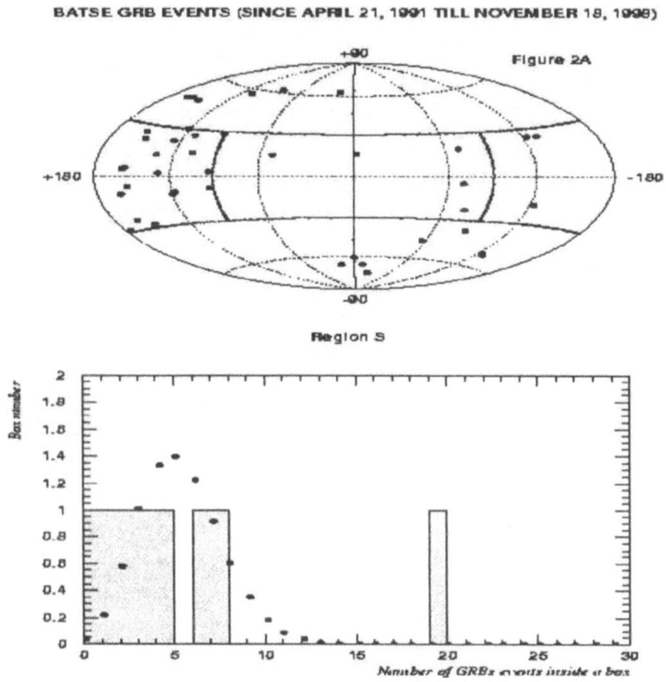

Figure 5 . (A) The Galactic coordinate angular distribution of the GRB events with very short time duration (S). We define eight regions that correspond to equal angular surface.

(B) The distribution of the number of GRB events in each of the eight regions. Points correspond to prediction of Poisson distribution for 42 events divided into eight equiprobable groups. Both distributions are normalized to the same surface.

isotropic distribution. To ascertain the significance of the anisotropy, we break up the Galactic map into eight equal probability regions. In Figure 2B, we show the distribution of events in the eight bins; clearly one bin has a large excess. To determine the statistical probability for such a deviation, we calculate the Poisson probability distribution for the eight bins with a total of 42 events. This distribution is also shown in Figure 2B.

The probability of observing 19 events in a single bin is 1.6×10^{-5}; we neglect the surface.

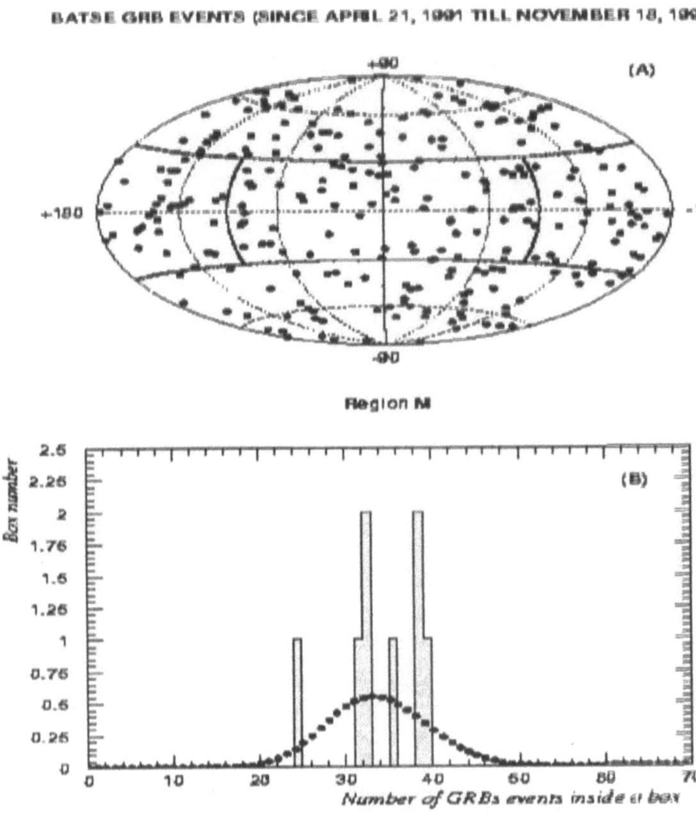

Figure 6. The Galactic coordinate angular distribution of the GRB events with short time duration (M). Eight regions correspond to equal angular surface (like those in Figure 2A). (B) The distribution of the number of GRB events in each of the eight regions. Points correspond to prediction in the Poisson (Gauss) distribution for 269 events divided into eight equiprobable

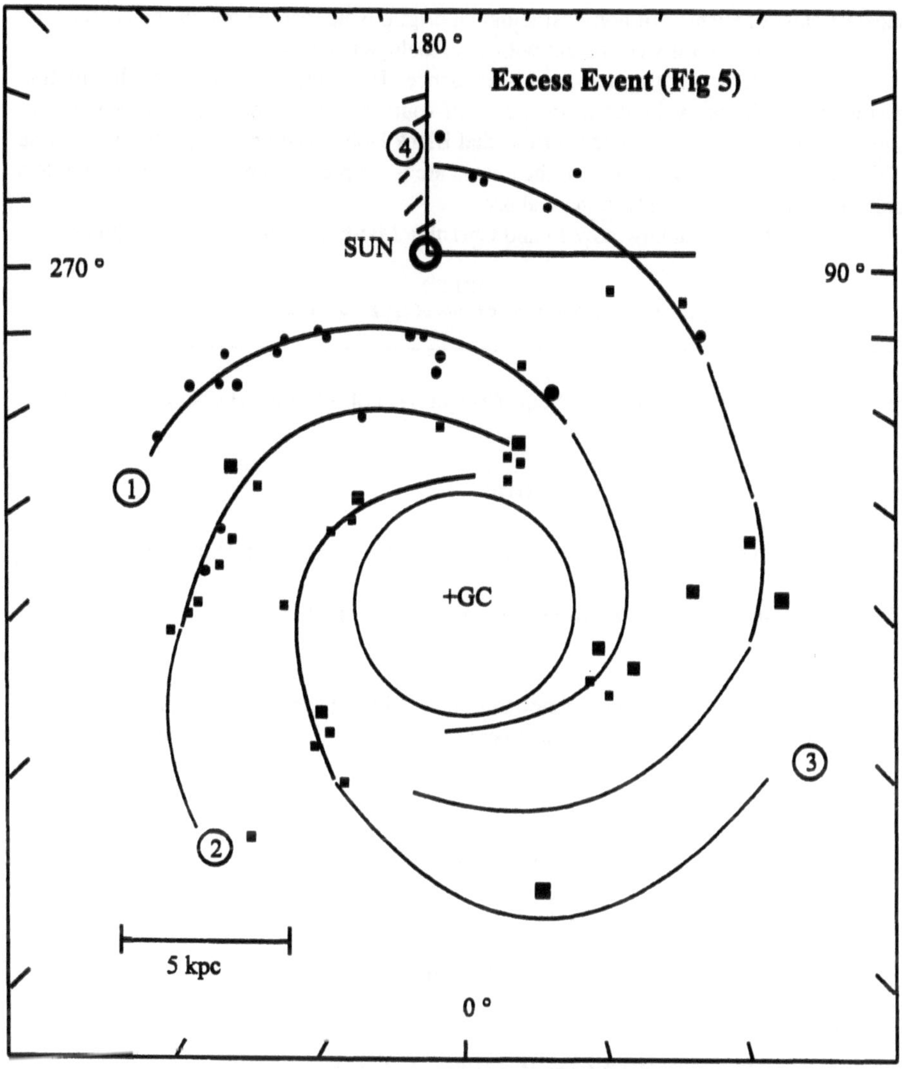

groups. Both distributions are normalized to the same surface.

Figure 7. The excess events could come from one of the Galactic Arms.

To contrast the distribution of the S GRBs and to test for possible errors in the analysis, we plot the same distributions for the M sample in Figures 6A and 6B. As can be seen from Fig. 6A, this distribution is consistent with isotropy. Figure 6B shows the same analysis as Figure 5B, indicating that there are no bins with a statistically significant deviation from the hypothesis of an isotropic distribution.

Of the 42 very short S GRBs, there are 19 in the excess region (Figure 5).

6. POSSIBLE ASSOCIATION WITH THE ORION ARM OF THE GALAXY [10]

We have presented evidence in this report that the very short time GRBs form a separate class of events - of non-cosmological origin. Note that the rest of the GRBs (class M and L) are fully consistent with a cosmological origin.

In table 2 we summarize this evidence. In addition the angular distribution asymmetry could be explained if an excess of sources were in the Orion Galaxy arm as shown in Fig. 7. It is possible that Primordial Black Holes could be concentrated in these Galactic arms. However, the rest of the events are isotropic, and are thus consistent with a uniform distribution of PBH in the Galaxy.

We thank Stan Otwinowski and Christina Matthey for their help with this work.

TABLE 2
Arguments for a New Exotic Class of GRBs in the Short Bursts

i) Time variations in some events (trigger 512) of ~ (10-20) μ s suggest the size of the source is

$$\Delta\gamma \leq \Delta t \times t \leq 6km$$

(much smaller than any astrophysical source except Black Holes) [8]

ii) The *ln N - ln S Distribution* suggests a local source ($<V/V_{max}> = 0.52 \pm 0.00$) [8]

iii) The *Angular Distribution* not consistent with a cosmologcal origin – and could be consistent with an enhancement of sources in the Orion Galactic Arm [10]

iv) The *Nearly Identical Nature* of the event suggests an exotic source [8]

REFERENCES

1. Hawking, S. W. 1974, Nature, 30, 248.
2. Cline, D. B. and Hong, W. P. 1992, ApJ, 401, L57.
3. Page, D. N. 1976, Phys. Rev. D, 13, 198.
4. MacGibbon, J. H., & Carr, B. J. 1991, ApJ, 371, 447.
5. Barat, C., Hayles, R. I., Hurley, K., Niel, M., Vedrenne, G., Estulin, I. V., & Zenchenko, M. 1984, ApJ, 285, 791.
6. Cline, D. B., Matthey, C., & Otwinowski, S. 1998b, in Proc. 14th IAP Colloq. (Paris: Editions Frontieres), 374.
7. Cline, D. B. 1996, Nucl. Phys. A, 610, 500.
8. Cline, D. B., Matthey, C., & Otwinowski, S. 1999, ApJ, 527, 825.
9. Dixon, D. 1997, quoted in "Halo Around Milky Way Is Reported" by Cole, K. C. Nov. 5, Los Angeles Times, 116 A1.
10. Report in Progress, D. Cline, C. Matthey & S. Otwinowski.

EXPLORING THE EXTREME UNIVERSE WITH THE GAMMA-RAY LARGE AREA SPACE TELESCOPE

S. Ritz*

1. INTRODUCTION AND MOTIVATION

The Gamma-ray Large Area Space Telescope (GLAST), a new satellite-based experiment to study high-energy cosmic gamma rays in the energy range 20 MeV - >300 GeV, is under development for launch in 2005. Much has been written about GLAST, and the most useful information can be found at the websites given in the references[1]. The slides from this talk are also available on the web[2]. GLAST will do fundamental science, with a very broad menu that includes systems with supermassive (10^6-10^{10} solar mass) black holes, gamma-ray bursts (GRBs), dark matter, solar physics, and probes of the early Universe. GLAST draws the interest of both the High Energy Particle Physics and High Energy Astrophysics communities. Indeed, the mission from its beginnings has been developed as a partnership between these communities.

Why study gamma rays? Gamma rays carry a wealth of information: unreprocessed photons offer a direct view into Nature's largest accelerators; similarly, the Universe is mainly transparent to gamma rays, so they probe cosmological volumes (any opacity is energy-dependent); conversely, gamma rays readily interact in detectors with a clear signature; and gamma rays are neutral, so there are no complications due to magnetic fields (galactic flux calculations don't have trapping time uncertainties, the photons point directly back to their sources, etc.).

Why this energy range? A glance at the "Grand Unified Photon Spectrum" of diffuse extragalactic photons[3] shows that this energy band is among the most poorly measured regions of the electromagnetic spectrum. History shows that opening new domains for exploration frequently leads to surprising discoveries. Although gamma rays in this energy range have no trouble reaching us from the edge of the visible Universe, they interact in the upper atmosphere and never reach the ground. For E>50-100 GeV, sufficient information from the air showers initiated by the primary gamma

*NASA Goddard Space Flight Center, Mail Code 661, Greenbelt, MD, 20771.

The Role of Neutrinos, Strings, Gravity, and Variable Cosmological Constant in Elementary Particle Physics, Edited by Kursunoglu *et al.*, Kluwer Academic/Plenum Publishers, New York, 2001

rays reaches the ground and can be measured with ground-based apparatus. However, measuring the flux of lower-energy gamma rays requires spaced-based apparatus. The difficulty here is that the cosmic gamma ray flux falls quickly with energy (typically a power law with index ~-2), so a large collecting area is required for a significant counting rate. Large area spaced-based detectors are technically quite challenging! The name of the game, therefore, is to push the high-energy frontier of space-based instruments upward and the low-energy threshold of ground-based instruments downward to have a substantial overlap in energy coverage. Ground-based and space-based gamma-ray detectors have capabilities that are complementary: ground-based detectors typically have good angular resolution, low duty cycle, huge collecting area, small field of view (FOV), and good energy resolution; space-based gamma ray detectors have good angular resolution, excellent duty cycle, relatively small collecting area, excellent FOV, and good energy resolution with relatively small systematic uncertainties. GLAST and the next generation ground-based instruments will, for the first time, have ~1 decade of energy overlap.

The GLAST energy range is also an important energy range for cosmology. Photons with E>10 GeV are attenuated by the diffuse field of UV-Optical-IR extragalactic background light (EBL). Unfortunately, the IR EBL limits the sources one can study with TeV gamma rays to a relatively local volume. However, this also presents an opportunity: the EBL over cosmological distances *can* be probed by gamma rays in the 10-100 GeV range. This sweet spot for cosmology is perfect for GLAST observations, and is a prime motivating factor for ground-based experiments such as STACEE and CELESTE that emphasize particularly low energy thresholds.

In many respects GLAST is the follow-on mission to EGRET, the very successful high-energy gamma ray instrument on the Compton Gamma Ray Observatory, which was launched in 1991 and de-orbited in 2000. Earlier instruments (SAS-2, COSB) had done the groundbreaking exploration of the gamma ray sky, establishing the galactic diffuse flux and identifying an extragalactic point source of gamma rays. EGRET then expanded the field dramatically by

- increasing the number of known point sources by a large factor (the final EGRET catalog contains 271 sources);
- providing broadband measurements covering the energy range ~20 MeV - ~30 GeV;
- discovering many yet-unidentified sources (the majority of sources in the 3rd catalog are unidentified);
- discovering a surprisingly large number of GeV-class Active Galactic Nuclei (AGN);
- discovering multi-GeV emissions from gamma ray bursts (GRBs); and
- discovering GeV emissions from the sun during solar flares.

EGRET provided a fine example of the great scientific yield often obtained when a new instrument probes previously unmeasured territory.

GLAST will explore the unexplored energy range above EGRET's reach, filling in the present gap in the photon spectrum, and will cover the very broad energy range 20 MeV - >300 GeV with superior acceptance and resolution. In particular, GLAST will have

- a huge FOV, covering ~20% of the sky at any instant;

- >four decades in energy coverage, including unattenuated acceptance in the largely unexplored region >10 GeV. By 10 GeV, EGRET lost half its acceptance relative to that at 1 GeV due to the "self-veto" effect;
- unprecedented point spread function (PSF), a factor >3 better than EGRET's for E>1 GeV;
- a large effective area, a factor >4 improvement over EGRET;
- a combined point source sensitivity >50 times better than EGRET's; and
- no expendables, resulting in a potentially long mission (5 years, with a 10 year lifetime goal) with little degradation.

2. A SHORT SUMMARY OF GLAST SCIENCE TOPICS

The basic features of the gamma-ray sky (E>100 MeV at EGRET sensitivity) are as follows: the diffuse extragalactic background, uniform across the sky with flux $\sim 1.5\times10^{-5}$ $cm^{-2}s^{-1}sr^{-1}$; the galactic diffuse flux in the plane of the galaxy, with flux approximately two orders of magnitude larger; high-latitude (extragalactic) point sources – typical fluxes measurable by EGRET are $O(10^{-7}\text{-}10^{-6})$ $cm^{-2}s^{-1}$; and galactic point sources. An essential characteristic of the gamma-ray sky is its extreme variability.

The origin of the diffuse extragalactic gamma-ray flux is a mystery. Is it really isotropic, produced at an early epoch in intergalactic space, or an integrated flux from a large number of yet-unresolved sources? GLAST, with a much larger effective area and much better angular resolution (PSF) will resolve a large fraction of the diffuse flux, if unresolved point sources are its origin. This constitutes a "no-lose" theorem for GLAST: either the diffuse flux will be resolved into ~10,000 high-energy gamma-ray sources (compared with EGRET's total catalog of 271 sources) to be studied in detail *or* GLAST will have discovered a truly diffuse flux of gamma rays from the early universe.

There is a very good candidate point source class to generate the apparent diffuse extragalactic flux: active galactic nuclei (AGN). Active galaxies produce vast amounts of energy (some flares reach 10^{45} ergs/s, with some assumptions, or about 10^{11} times solar luminosity – the total power output of ordinary stars in a typical galaxy -- in gamma rays) from a very compact central volume (roughly the size of our solar system). The prevailing idea[4] is that AGN are powered by accretion onto super-massive black holes (10^{6}-10^{10} solar masses). Models of AGN include energetic (multi-TeV), highly collimated, relativistic particle jets. The different emission characteristics of categories of AGN may be explained by the orientation of the jets with respect to us. In these models, high-energy gamma rays are emitted within a few degrees of the jet axis. Of course, these models must be carefully tested (see the very interesting talks in this session by Sterling Colgate and Elliott Bloom). Exactly how the huge particle kinetic energies are generated is an important topic of study during the coming decade.

AGN shine brightly in the GLAST energy range; in fact, for the "blazar" class of AGN, in which the jet axis points in our direction, the high energy gamma ray component can be dominant. A surprise from EGRET was the detection of dozens of AGN. This has two important implications: (1) AGN may be the key to solving the puzzle of the extragalactic diffuse emission, and (2) blazars provide a source of very high energy gamma rays at cosmological distances allowing us to probe huge volumes of space.

In particular, the systematic attenuation of AGN photons with both increasing energy and increasing redshift provides a probe of the density of the extragalactic background light (EBL), a bath of IR-optical-UV photons generated by shining stars. The EBL density is sensitive to the length of time the stars have been shining, so a measure of the EBL density gives valuable information about the era of galaxy formation[5]. When a high-energy gamma ray encounters the EBL bath of photons, if the center-of-mass energy is large enough for pair creation, the high energy flux can be attenuated. Kinematically, therefore, 10-100 GeV photons probe the optical-UV EBL and TeV photons probe the IR EBL. With large AGN statistics, GLAST will readily distinguish EBL models[6], provided the redshift determinations can be made for a sizable number of GLAST AGN.

Of the 271 sources in the EGRET 3rd Catalog, fully 172 are "unidentified", meaning that within the error circle provided by EGRET there are several counterparts known from other wavelengths. There is insufficient information to make a positive identification. GLAST will provide much more accurate point source localizations (typically 30 arcseconds to 5 arcminutes, depending on the source brightness) allowing unique identifications to be made, and solving another puzzle in present-day gamma-ray astrophysics.

GLAST will also address the question of the origin of the cosmic rays. In 1949, Fermi did the seminal work on shock acceleration leading to the hypothesis that all cosmic rays with $E<\sim 10^{14}$eV are produced in supernova remnants. This idea has yet to be proven unambiguously for hadron acceleration. However, it is expected that cosmic ray interactions with the gas swept up by the blast should produce π^0's that decay to gammas. Being able to resolve a substantial number of supernova remnant shells both spatially and spectrally, GLAST should easily confirm or refute the hypothesis.

Pulsars are very bright gamma ray sources, but here again the acceleration mechanism is not well understood. GLAST will uniquely distinguish models by probing the high-energy roll-offs in pulsar spectra (spatial volumes with 10^{12} gauss magnetic fields are opaque to gamma rays due to pair creation attenuation!) and by very different predicted population statistics of radio-loud and radio-quiet pulsars[7].

Gamma ray bursts (GRBs), also discussed at this conference, were detected by EGRET. However, due to the large event deadtime the information from EGRET on the time profile of high-energy emission from GRBs is extremely limited. What *is* known from EGRET on GRBs is quite puzzling: in one burst, very high-energy emission (~20 GeV) was detected ~75 minutes after the onset of the burst. GLAST, with an event dead time that is more than 3 orders of magnitude smaller than EGRET's, will open a wide window on the study of the high energy behavior of bursts. The very large FOV will also allow monitoring large patches of the sky, greatly increasing GRB statistics. Because of these capabilities, GLAST will provide unique information about the high-energy behavior of bursts. Once the burst spectra at high energy are better understood, they might even provide a probe of photon velocity dispersion effects[8] with energy that are suggested by some current ideas about Quantum Gravity. The large range of energies covered by GLAST, and good sensitivity to GRBs, will allow us to search for these effects within a single experiment. Of course, it may be very difficult to untangle any perceived dispersion from the intrinsic energy and time profiles of bursts, but this example illustrates the discovery potential opened by GLAST.

If the galactic dark matter problem is explained by a halo of Weakly Interacting Massive Particles (WIMPs), then annihilations of these halo particles could produce a detectable flux of gamma rays. This possibility is especially intriguing since one signature would be mono-energetic "lines" of gamma rays, set by the WIMP mass, at somewhere between 10's and many 100's of GeV.

Finally, it must be noted that this list of physics possibilities only contains the phenomena we already know, or of which we have already conceived; however, the most likely and exciting outcome of GLAST will be the discovery of the unanticipated. It is quite reasonable to expect that entirely new astrophysical gamma ray source classes will be discovered. There is already some hint from the EGRET data of what is awaiting us.

3. INSTRUMENTS

The central instrument of the GLAST mission is the Large Area Telescope (LAT). The LAT Instrument Team[1] consists of approximately 100 collaborators from some 20 institutions from around the world. The team is an unusual and fruitful partnership between high-energy particle physicists and high-energy astrophysicists. The Spokesperson and Principal Investigator is Peter Michelson (Stanford), and the LAT is being managed by SLAC-Stanford University. DoE and NASA together fund the LAT development and construction, along with significant contributions from agencies in Italy, France, Japan, and Sweden.

The instrument must measure the direction, energy and arrival time of high energy photons with good efficiency, and reject the much higher flux (10^4-10^5 larger) of background charged cosmic rays. The interaction of photons in the GLAST energy range with matter is dominated by pair conversion. This process is exploited for two purposes: (1) measuring the trajectories of the conversion electron and positron, the incident photon direction can be reconstructed; and (2) recognizing the characteristic features of the pair conversion signature provides a powerful method for rejecting background events. The fundamental limitation on the precision of the reconstructed photon direction at low energy is the multiple scattering of the conversion products in the converter/tracker material. Thus, the converter material is arranged in a stack of many thin layers, with charged particle tracking detectors interleaved. Such an arrangement is called a pair conversion telescope.

The conversion electron and positron continue to encounter material in the detector and lose energy by bremsstrahlung. These high-energy photons in turn pair-convert in the detector, and so on, resulting in a shower of particles materialized from the kinetic energy of the incident photon. This shower develops, and is measured, in a calorimeter (CAL) beneath the converter/tracker (TKR). An anticoincidence detector (ACD) surrounds the converter/tracker. Its main function is to detect the charged cosmic ray particles when they enter the instrument, thereby identifying the background events.

The GLAST LAT uses silicon strip detectors in the TKR, hodoscopically arranged bars of CsI in the CAL, and segmented scintillator tiles in the ACD. The TKR and CAL are subdivided into an array of 4x4 identical modules, which has many advantages, but the 16 modules function together as a single unit. The silicon strip detectors provide extremely high tracking efficiency, low noise occupancy, and high measurement precision; the CsI bars provide a 3-dimensional image of the shower development, which is essential for shower leakage corrections at high energy and for pattern recognition for background rejection. The ACD is segmented to mitigate the self-veto effect that reduced the high-energy sensitivity of EGRET: a high energy shower in the CAL produces a flash of low-energy photons ("backsplash") that Compton scatter in the ACD scintillator, producing a false veto signal.

The LAT design is the result of the hard work of many people; more than any other single person, Bill Atwood provided the technical foundation of the LAT design. Information about the LAT can be obtained from the references[1].

The LAT design is based on detailed Monte Carlo simulations. Using the simulation, reconstruction algorithms were developed to demonstrate the required cosmic ray background rejection; make realistic calculation of the effective area, PSF, and energy resolution; develop the trigger; and optimize the design. These simulations were validated in detailed beam tests[9] at SLAC in 1997 and 1999. The 1999 test was carried out using a flight-scale TKR and CAL module. This same hardware will be used in a balloon test in 2001. Additional beam tests on other LAT development hardware were performed at CERN.

A small second instrument, the GLAST Burst Monitor (GBM), augments the GLAST capabilities for studying gamma-ray bursts by providing extended spectral coverage to low energy and by providing on-board locations of bursts. These localizations are particularly important for bursts that occur outside the LAT field of view. The GBM principal investigator is Charles Meegan (Marshall Space Flight Center). The collaborating institutions are Marshall Space Flight Center, Max Planck Institute for Extraterrestrial Physics, and the University of Alabama.

4. SUMMARY

GLAST will address a wide variety of questions:

- What is going on near supermassive black holes? How do Nature's most powerful accelerators work?
- What are the EGRET unidentified sources?
- What is the origin of the diffuse gamma ray extragalactic background?
- What is the high-energy behavior of gamma-ray bursts?
- Are there other sources of gamma rays too faint for EGRET to detect? Are there further surprises in the poorly measured energy range?
- When and how did galaxies form?

The large menu of "bread and butter" science is accompanied by an enormous discovery potential.

GLAST represents the blossoming of an interdisciplinary field. Given the huge increment in sensitivity, implying a two-order-of-magnitude jump in the number of detectable sources, we anticipate the community of investigators interested in gamma-ray data will grow enormously during the GLAST era.

ACKNOWLEDGEMENT

I thank the organizers for inviting me to participate in this interesting meeting, which has such a rich and colorful history.

REFERENCES

1. http://www-glast.stanford.edu/, http://glast.gsfc.nasa.gov/, http://gammaray.msfc.nasa.gov/~gbm, and links therein.
2. http://lheawww.gsfc.nasa.gov/~ritz/global2000.pdf.
3. T.Ressell and M.Turner, FERMILAB-PUB-89/214 (1989).
4. see, *e.g.*, M.C.Begelman, R.D.Blandford, and M.J.Rees, Reviews of Modern Physics, **56**(1984)255.
5. F.Stecker et al, Astrophysical Journal **390**(1992)L49;
   D.MacMinn and J.Primack, Space Sci. Rev. **75**(1996)413; and
   P.Madau and S.Phinney, Astrophysical Journal **456**(1996)124.
6. A.Chen and S.Ritz, Astrophysical Journal, in press.
7. P.Gonthier, J.Berrier, M.Ouellette, S.O'Brien, and A.Harding, astro/ph-0012268, submitted to Astrophysical Journal(2001).
8. G.Amelino-Camelia, John Ellis, N.E.Mavromatos, D.V.Nanopoulos, and Subir Sarkar, Nature **393**(1998)763.
9. W.Atwood et al., Nuclear Instruments and Methods **A446**(2000)444; and
   SLAC-Pub 8682(2000), submitted to Nuclear Instruments and Methods.

THE LATEST DEVELOPMENTS IN HIGH ENERGY PHYSICS AND COSMOLOGY

COMMENTS ON SUPERSYMMETRY EXTRA DIMENSIONS AND THE ACCELERATING UNIVERSE

D. B. Lichtenberg*

Philosophy is written in a great book which is always open in front of our eyes (I mean: the universe), but one cannot understand it without first applying himself to understanding its language and know the characters in which it is written. It is written in the mathematical language.—Galileo

For any speculation which does not at first glance look crazy, there is no hope.—Freeman Dyson

1. INTRODUCTION

In the quote by Dyson, I think that he is hinting that there *is* hope for seemingly crazy speculations. Unfortunately, most speculations which at first glance look crazy turn out to be in fact crazy, or at any rate, wrong.

I am going to comment on some of the apparently crazy speculations of the latter part of the twentieth century, such as supersymmetry, strings and membranes, and extra dimensions. You should be aware that I have not made significant contributions to these areas, but am taking the attitude that this lack does not disqualify me from making critical remarks.

This Orbis Scientiae began on December 14, 2000, one hundred years to the day after Max Planck[1] gave a talk at at a meeting of the German Physical Society on his quantum hypothesis. Because Planck's talk was so revolutionary and because it has lessons for today, I am going to discuss Planck's hypothesis in some detail before going on to comment on recent speculations. My talk will be a little strange in that all the equations will be in the introduction and no new equations elsewhere.

As you know, classical physics led to a crisis in in the theory of blackbody radiation. Wilhelm Wien showed from thermodynamics that in a cavity the amount of radiant energy $U(\nu)$ of frequency ν is of the form

$$U(\nu) = \nu^3 f(\nu/T), \tag{1}$$

* Department of Physics, Indiana University, Bloomington, Indiana 47405. E-mail lichten@indiana.edu

The Role of Neutrinos, Strings, Gravity, and Variable Cosmological Constant in Elementary Particle Physics, Edited by Kursunoglu *et al.*, Kluwer Academic/Plenum Publishers, New York, 2001

where f is an unknown function that cannot be determined from thermodynamics alone.

Classical electrodynamics leads to the Rayleigh-Jeans law

$$U(\nu) \propto \nu^3 (T/\nu), \tag{2}$$

a result which not only goes to infinity at high frequency, but leads to an infinite total energy, a nonsensical result. Clearly, an ultraviolet cutoff is needed. Wien suggested an exponential cutoff; his formula can be written

$$U(\nu) \propto \nu^3 e^{-h\nu/kT}, \tag{3}$$

where k is Boltzmann's constant and h is an empirically determined constant which turns out to be Planck's constant. (Wien actually introduced the constant h/k.)

Wien's formula agrees quite well with the observed radiant energy emerging from a hole in a cavity both at low and at high frequencies, but disagrees with experiment at intermediate frequencies.

The puzzle of blackbody radiation remained until Planck[1,2] got the idea that radiation from a cavity had to be emitted in quantized units of energy. Planck originally tried quantizing in units of a certain energy E_0, using the relation[3]

$$E_n = nE_0, \quad n = 0, 1, 2... \tag{4}$$

but then showed that in order to get agreement with experiment, E_n should be quantized as

$$E_n = nh\nu. \tag{5}$$

By summing over the possible quantized energies rather than integrating over all possible classical energies, Planck obtained his well-known formula

$$U(\nu) \propto \nu^3 \frac{e^{-h\nu/kT}}{1 - e^{-h\nu/kT}}, \tag{6}$$

which agrees with experiment at all frequencies.

With this introduction, I can go on to discuss what I believe is the primary motive for the development of superstrings and extra dimensions: the incompatibility of general relativity and quantum mechanics.

2. GENERAL RELATIVITY AND QUANTUM MECHANICS

It is widely believed that when a classical theory, such as general relativity, has interactions with a quantum theory, such as the standard model of elementary particles, inconsistencies arise. However, as is well known, nobody has succeeded in quantizing Einstein's theory of relativity. The big stumbling block is that the quantized version of general relativity is nonrenormalizable and so in perturbation theory it requires an infinite number of constants, leaving it without predictive power. (Nobody yet knows how to solve a quantized version of general relativity nonperturbatively or whether such a theory makes sense.) What is needed is some sort of a modification that makes the quantized perturbation theory either renormalizable or finite.

There are two natural cutoffs for quantized general relativity: the Planck mass provides an ultraviolet cutoff and the inverse of the size of the visible universe provides an infrared cutoff. These cutoffs render the quantized theory finite. It might be more in the spirit of Wien to take an ultraviolet cutoff that will lead to best agreement with experiment. Unfortunately for this idea, however, is the fact that no quantized effects of general relativity have yet been observed.

Another possibility is that space and time are not continuous variables but are confined to points on a lattice. I would be astonished if such a lattice would be cubical. More plausible is the idea of a random lattice pushed some years ago by T.D. Lee and collaboraters.[4,5] Unfortunately, if a lattice exists, we are ignorant of the lattice spacing or structure.

It is esthetically more satisfying to render general relativity finite by some new idea. The most popular new idea is that strings (and their generalization, membranes) exist. Because strings are extended objects they do away with the infinities that arise from a field theory with point particles. However, so far we have no clue to the size of the strings or membranes. However, we expect the strings to be at least the Planck length, 10^{-33} cm, but less than of order 10^{-17} cm, or we would expect already to have some evidence of their existence. That leaves 16 orders of magnitude to play with.

A consistent theory of strings apparently requires both supersymmetry and extra dimensions. So far, there is no experimental evidence for either of these speculations, but the idea may be crazy enough to be right. Several apparently different superstring theories have been postulated to be different limits of one all-encompassing theory called M-theory.[6]

There is one important difference between Planck's quantum hypothesis and the hypothesis of the existence of superstrings (and/or supermembranes) and extra dimensions. Using only one parameter, Planck was able to obtain good agreement with the observed blackbody spectrum. On the other hand, nobody has made any prediction from superstrings that can be compared quantitatively with experiments that have been done so far. Furthermore, I do not know of anybody who has proposed an experiment to test superstring theory quantitatively, as opposed to qualitatively. I wonder how seriously physicists would have taken Planck if he had said only that quantized emission of light would lead to a finite blackbody spectrum (a qualitative prediction) but had not been able to predict what the spectrum should be. Worse than that, Planck would not have been able to choose between $E_n = nE_0$, the hypothesis he rejected because of its disagreement with experiment, and $E_n = nh\nu$, the hypothesis he accepted because of its agreement with experiment. It is because there is at present so little predictive power in superstring theories that we find several of them vying for favor.

3. THE ACCELERATING UNIVERSE

Recent evidence from fluctuations in the cosmic microwave background radiation indicates that the visible universe is flat, or nearly so.[7] We have not identified enough matter in the universe, whether visible or dark, to lead to a flat universe. So if the observations and their interpetation are correct, something else is needed.

Furthermore, evidence from measurments of the speed and distance of far-off supernovae seems to indicate that the expansion rate of the universe is speeding up,[8] rather than slowing down, as required by gravity.

Both the flatness and acceleration, if real, seem to require something more than general relativity. The extra ingrdient might be Einstein's cosmological constant or something more complicated, often called quintessence, which has a similar effect.[9] A cosmological constant can be considered to be a special case of quintessence.

I have speculated (unpublished) that it might be possible for matter existing in extra dimensions to lead to to an accelerating universe, that is, a universe in which the expansion is speeding up. This idea, if it worked, would give extra dimensions an additional reason for existing. However, although I could envision that peculiarly distributed matter in extra dimensions might lead to some kind of repulsion, I was not able to think of a scenario in which such matter would lead to a repulsion agreeing with the data. Therefore, I now am of the opinion that repulsive gravity (cosmological constant, quintessence, or whatever) arises from something filling our three-dimensional space and has nothing to do with extra dimensions. Note that I am giving an opinion, not a conclusion, because although I have not found a plausible mechanism arising from other dimensions, neither have I been able to rule it out.

At this point in my talk, Philip Mannheim,[10] an Orbis participant, said that he has thought of three mechanisms by which extra dimensions could lead to a repulsion in our three spatial dimensions, but he admitted that he doesn't know whether any of his mechanisms gives agreement with the observations.

4. CONCLUSIONS

At the present time particle physicists and cosmologists are faced with an inconsistency and a mystery. The inconsistency is that general relativity seems incompatible with quantum mechanics, and the mystery is that we do not know the nature of dark matter or of quintessence. Speculations about supersymmetry, strings, membranes, and extradimensions have apparently led to a method for quantizing a modified form of general relativity, but so far the ideas have not led to any quantitative conclusions.

Since Galileo, we are used to a theory being expressed in mathematical equations. However, to my knowlege, nobody has even written down, much less solved, the definitive equations of M-theory (if such equations exist). I quote from Michael Duff[6]: "So what is M-theory? There is still no definitive answer to this question..." In the spirit of Galileo, perhaps we should use the name "M-speculation" until we actually see equations describing M, but I suppose the term "M-theory" will be with us for yet a while.

Although we are lacking in quantitative predictions from the new speculations, supersymmetry (and possibly superstrings and extra dimensions) might be tested qualitatively in the future. Because we cannot predict at what energy supersymmetric particles and effects of superstrings and extra-dimensional gravitons might show up, experimentalists are forced to look where they can. There is still a chance for some of these things to be seen in the future at Fermilab or in further analysis of the LEP data. If not, those physicists who are looking for evidence for supersymmetry, superstrings, and extra dimensions will have to wait for their next best chance for discovery, which I think will come after the completion of the Large Hadron Collider around 2006.

5. REFERENCES

1. M. Planck, meeting of the German Physical Society, Dec. 14, 1900. The Society was established in 1845 as the Physical Society of Berlin, and was renamed the German Physical Society (*Deutsche Physikalische Gesellschaft*) in 1899.
   See the website http://www.aip.org/history/fall99/german.html.
2. M. Planck, *Verh. d. Deutsch. Phys. Ges.* **2**, 202 (1900); *ibid.* **2**, 237 (1900); Both papers are reprinted in *Ann Physik* **4**, 553 (1901).
3. F.K. Richtmyer, E.H. Kennard, and T. Lauritsen, *Introduction to Modern Physics*, Mc-Graw Hill, New York (1955), p. 127.
4. N.H. Christ, R. Friedberg, and T.D. Lee, *Nucl. Phys.* **B202**, 89 (1982), **B210**, 310, 337 (1982).
5. T.D. Lee, in *How Far Are We from the Gauge Forces*, ed. by A. Zichichi, Plenum Press, New York (1985), p. 15.
6. M.J. Duff, ed. *The World in Eleven Dimensions: Supergravity, Supermembranes, and M-theory*, Institute of Physics Publishing, Bristol (1999).
7. P. de Bernardis et al., *Nature* **404**, 955 (2000).
8. G. Fraser and E. Sanders, reporters, *CERN Courier* **40**, No. 5 (June 2000), p. 11.
9. Lawrence Krauss, *Quintessence: the mystery of missing mass in the universe*, Basic Books, New York (2000).
10. Philip Mannheim, to be published.

FIELD THEORY CORRELATORS AND STRING THEORY

S.S. Pinsky and U. Trittmann
Department of Physics, Ohio State University
Columbus, OH 43210, USA
pinsky,trittman@mps.ohio-state.edu

J.R. Hiller
Department of Physics, University of Minnesota-Duluth
Duluth, MN 55812, USA
jhiller@d.umn.edu

Abstract It appears that string-M-theory is the only viable candidate for a complete theory of matter. It must therefore contain both gravity and QCD. What is particularly surprising is the recent conjecture that strongly coupled QCD matrix elements can be evaluated though a duality with weakly coupled gravity. To date there has been no direct verification of this conjecture by Maldacena because of the difficulty of direct strong coupling calculations in gauge theories. We report here on some progress in evaluating a gauge-invariant correlator in the non-perturbative regime in two and three dimensions in SYM theories. The calculations are made using supersymmetric discrete light-cone quantization (SDLCQ). We consider a Maldacena-type conjecture applied to the near horizon geometry of a D1-brane in the supergravity approximation, solve the corresponding $\mathcal{N} = (8,8)$ SYM theory in two dimensions, and evaluate the correlator of the stress-energy tensor. Our numerical results support the Maldacena conjecture and are within 10-15% of the predicted results. We also present a calculation of the stress-energy correlator in $\mathcal{N} = 1$ SYM theory in 2+1 dimensions. While there is no known duality relating this theory to supergravity, the theory does have massless BPS states, and the correlator gives important information about the BPS wave function in the non-perturbative regime.

The Role of Neutrinos, Strings, Gravity, and Variable Cosmological Constant in Elementary Particle Physics, Edited by Kursunoglu *et al.*, Kluwer Academic/Plenum Publishers, New York, 2001

Introduction

Recently the conjecture that certain field theories admit concrete realizations as string theories on particular backgrounds has caused a lot of excitement. The original Maldacena conjecture [1] asserts that the $\mathcal{N} = 4$ supersymmetric Yang-Mills (SYM) theory in 3+1 dimensions is equivalent to Type IIB string theory on an $AdS_5 \times S^5$ background. However, more recently, other string/field theory correspondences have been conjectured. Attempts to rigorously test these conjectures have met with only limited success, because our understanding of both sides of the correspondences is usually insufficient. The main obstacle is that at the point of correspondence we want the curvature of space-time to be small in order to use the supergravity approximation to string theory. This requires a *non-perturbative* calculation on the field theory side. We use the method, SDLCQ, in the corresponding non-perturbative regime. SDLCQ, or Supersymmetric Discretized Light-Cone Quantization, is a non-perturbative method for solving bound-state problems that has been shown to have excellent convergence properties[2].

Aside from our numerical solutions, there has been very little work on solving SYM theories using methods that might be described as being from first principles. While selected properties of these theories have been investigated, one needs the complete solution of the theory to calculate the correlators. By a "complete solution" we mean the spectrum and the wave functions of the theory in some well-defined basis. The SYM theories that are needed for the correspondence with supergravity and string theory have typically a high degree of supersymmetry and therefore a large number of fields. The number of fields significantly increases the size of the numerical problem. Therefore, when presenting the first calculation of correlators in 2+1 dimensions, we consider only $\mathcal{N} = 1$ SYM.

An important step in these considerations is to find an observable that can be computed relatively easily on both sides of a string/field theory correspondence. It turns out that the correlation function of a gauge invariant operator is a well-behaved object in this sense. We chose the stress-energy tensor $T^{\mu\nu}$ as this operator and will construct this observable in the supergravity approximation to string theory and perform a non-perturbative SDLCQ calculation of this correlator on the field theory side.

1. String Theory

String theory contains solitons, the so-called D-branes on which modes can propagate as well as in the bulk. While in general these modes

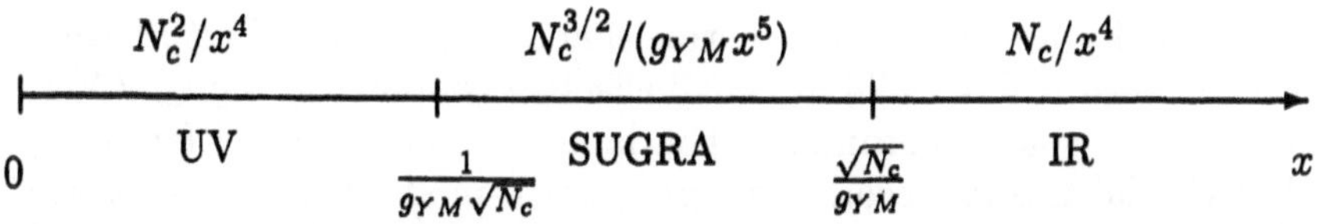

Figure 1. Phase diagram of two-dimensional $\mathcal{N} = (8,8)$ SYM: the theory flows from a CFT in the UV to a conformal σ-model in the IR. The SUGRA approximation is valid in the intermediate range of distances, $1/g_{YM}\sqrt{N_c} < x < \sqrt{N_c}/g_{YM}$.

couple, there exists a limit in which the bulk modes decouple from the modes on the D-brane; this is typically a low energy limit. In this limit the theory on N Dp-branes, separated by at most sub-stringy distances becomes a supersymmetric $SU(N)$ Yang-Mills theory. As the D-branes carry mass and charge, they can excite gravity modes in the bulk, for which in supergravity there exist equivalent solutions. One thus has a string/field theory correspondence. Naively, one would think that supergravity can only describe the large distance behavior of fields, but it turns out that one can trust these solutions as long as the curvature is small compared to the string scale. In this sense, the large N limit is a valid description[3].

The most prominent string/field theory correspondence is the so-called Maldacena conjecture[1], which assures that the conformal $\mathcal{N} = 4$ SYM in 3+1 dimensions, is equivalent to a type IIB string theory on a $AdS_5 \times S^5$ background. In the more general case of non-conformal theories, it turns out that a black p-brane solution, stretching in $p+1$ spacetime dimensions, of supergravity will correspond to a supersymmetric Yang-Mills theory in $p+1$ dimensions [3].

One can test these string/field theory correspondences, if one is able to construct and evaluate observables on both the string and the field theory regimes. Although this at first seems a hard task, because typically the small curvature regime of string theory, where the supergravity approximation allows quantitative calculations, falls into the strong coupling regime of the field theory side. We shall see that we can come up with scenarios where we can evaluate the field theory observable non-perturbatively.

1.1. Two-dimensional correlation functions from supergravity

It is instructive to take a closer look on the expected properties of $\mathcal{N} = (8,8)$ SYM in two dimensions, before we proceed to technical details on the string theory side. In the extreme ultra-violet (UV) this theory is conformally free and has a central charge $c_{UV} = N_c^2$. Perturbation theory in turn will be valid for small effective couplings $g = g_{YM}\sqrt{N_c}x$, where x is a space coordinate. For large distances, in the far infrared (IR), the theory becomes a conformal σ-model with target space $(R^8)^{N_c}/S_{N_c}$. The central charge is $c_{IR} = N_c$. It is a bit more involved to show that here perturbation theory breaks down when $x \sim \sqrt{N_c}/g_{YM}$, see *e.g.* Ref. [3].

The intermediate region, $1/g_{YM}\sqrt{N_c} < x < \sqrt{N_c}/g_{YM}$, where no perturbative field theoretical description is possible, is fortunately exactly the region which is accessible to string theory; or rather, to the supergravity (SUGRA) approximation to Type IIB string theory on a special background. It is that of the near horizon geometry of a D1-brane in the string frame, which has the metric

$$\begin{aligned} ds^2 &= \alpha' \hat{g}_{YM} \left(\frac{U^3}{g_s^2} dx_{\|}^2 + \frac{dU^2}{U^3} + U d\Omega_{8-p}^2 \right) \\ e^{\phi} &= \frac{2\pi g_{YM}^2}{U^3} \hat{g}_{YM}, \end{aligned} \tag{1}$$

where we defined $\hat{g}_{YM} \equiv 8\pi^{3/2} g_{YM}\sqrt{N_c}$. In the description of the computation of the two-point function we follow Ref. [4]. The correlator has been derived in Ref. [5], being itself a generalization of Refs. [6, 7].

First, we need to know the action of the diagonal fluctuations around this background to the quadratic order. We would like to use the analogue of Ref. [8] for our background, Eq. (1), which is not (yet) available in the literature. However, we can identify some diagonal fluctuating degrees of freedom by following the work on black hole absorption cross-sections [9, 10]. One can show that the fluctuations parameterized like

$$\begin{aligned} ds^2 &= \Big(1 + f(x^0,U) + g(x^0,U)\Big)\, g_{00}(dx^0)^2 \\ &\quad + \Big(1 + 5f(x^0,U) + g(x^0,U)\Big)\, g_{11}(dx^1)^2 \\ &\quad + \Big(1 + f(x^0,U) + g(x^0,U)\Big)\, g_{UU} dU^2 \\ &\quad + \Big(1 + f(x^0,U) - \frac{5}{7} g(x^0,U)\Big)\, g_{\Omega\Omega} d\Omega_7^2 \\ e^{\phi} &= \Big(1 + 3f(x^0,U) - g(x^0,U)\Big)\, e^{\phi_0}, \end{aligned} \tag{2}$$

satisfy the following equations of motion

$$\begin{aligned} f''(U) &= -\frac{7}{U} f'(U) + \frac{g_s^2 k^2}{U^6} f(U) \\ g''(U) &= -\frac{7}{U} g'(U) + \frac{72}{U^2} g(U) + \frac{g_s^2 k^2}{U^6} g(U). \end{aligned} \tag{3}$$

Without loss of generality we have assumed here that these fluctuations vary only along the x^0 direction of the world volume coordinates, and behave like a plane wave. One can interpret a D1-brane as a black hole in nine dimensions. The fields $f(U)$ and $g(U)$ are the the minimal set of fixed scalars in this black hole geometry. In ten dimensions, however, we see that they are really part of the gravitational fluctuation. Consequently, we expect that they are associated with the stress-energy tensor in the operator field correspondence of Refs. [6, 7]. In the case of the correspondence between $\mathcal{N}$=4 SYM field theory and string theory on an $AdS_5 \times S^5$ background, the superconformal symmetry allows for the identification of operators and fields in short multiplets [11]. In the present case of a D1-brane, we do not have superconformal invariance, and this technique is not applicable. Actually, we expect all fields of the theory consistent with the symmetry of a given operator to mix. The large distance behavior should then be dominated by the contribution with the longest range. The field $f(k^0, U)$ appears to be the one with the longest range since it is the lightest field.

Eq. (3) for $f(U)$ can be solved explicitly

$$f(U) = U^{-3} K_{3/2}\left(\frac{\hat{g}_{YM}}{2U^2}k\right), \tag{4}$$

where $K_{3/2}(x)$ is a modified Bessel function. If we take $f(U)$ to be the analogue of the minimally coupled scalar, we can construct the flux factor

$$\begin{aligned} \mathcal{F} &= \lim_{U_0\to\infty} \frac{1}{2\kappa_{10}^2}\sqrt{g}g^{UU}e^{-2(\phi-\phi_\infty)}\,\partial_U \log(f(U))|_{U=U_0} \\ &= \frac{NU_0^2k^2}{2g_{YM}^2} - \frac{N^{3/2}k^3}{4g_{YM}} + \ldots \end{aligned} \tag{5}$$

up to a numerical coefficient of order one which we have suppressed. We see that the leading non-analytic contribution in k^2 is due to the k^3 term. Fourier transforming the latter yields

$$\langle \mathcal{O}(x)\mathcal{O}(0)\rangle = \frac{N^{\frac{3}{2}}}{g_{YM}x^5}. \tag{6}$$

This is in line with the discussion at the beginning of this section. We expect to deviate from the trivial $(1/x^4)$ scaling behavior of the correlator at $x_1 = 1/g_{YM}\sqrt{N_c}$ and $x_2 = \sqrt{N_c}/g_{YM}$. This yields the phase diagram in Fig. 1. It is interesting to note that the entire N_c hierarchy is consistent in the sense of Zamolodchikov's c-theorem, which assures that the central charges obey $c(x) > c(y)$, whenever $x < y$ [12].

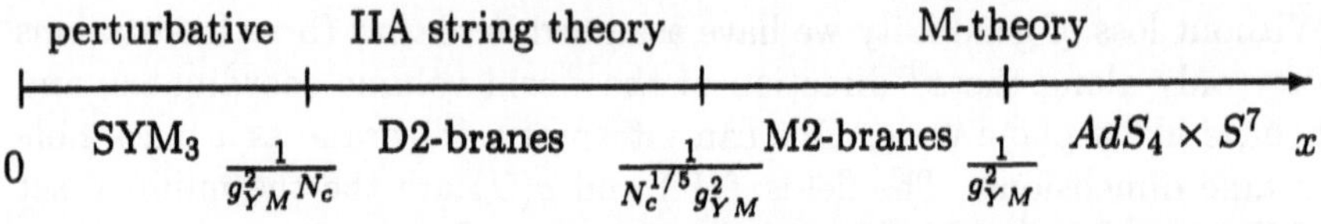

Figure 2. Phase diagram of three-dimensional $\mathcal{N} = (8,8)$ SYM: the theory flows from a perturbative SYM in the UV to a M-theory on $AdS_4 \times S^7$. The SUGRA approximation is valid in the intermediate range of distances, $1/g_{YM}^2 N_c < x < 1/g_{YM}^2 N_c^{1/5}$.

1.2. D2-branes and three-dimensional SYM

As stated in the introduction, in an analogous way one can show that a system of $D2$-branes corresponds in a certain limit to a Yang-Mills theory in three dimensions. It is again a $\mathcal{N} = (8,8)$ supersymmetric theory. Unfortunately, an observable like the correlation function of the stress-energy tensor has not yet been calculated for this theory. However, there are encouraging results both on the string and on the field theory side of the correspondence [3, 13].

We describe the phase-diagram of three-dimensional SYM with 16 supercharges here, following Ref. [3]. Later we will present a non-perturbative field theory calculation within the SDLCQ framework. The latter calculation is, however, of a theory with an $\mathcal{N} = 1$ supersymmetry. This theory might nevertheless share some features with the full $\mathcal{N} = (8,8)$ theory, *cf.* also the results of two-dimensional SYM with different supersymmetries in Sec. 2.

It can be argued that the theory has to be described by different degrees of freedom at different energy scales. At large N_c, one can use perturbation theory of SYM(2+1) in the far ultra-violet, *i.e.* at small distances. The supergravity solution, which in this regime is an approximation to type IIA string theory with D2-branes, can be trusted at intermediate distances r, $1/g_{YM}^2 N_c < r < 1/g_{YM}^2 N^{1/5}$. It has been conjectured that for large distances, $r > 1/g_{YM}^2 N^{1/5}$, an M-theory description is appropriate, while in the far infrared, $r \gg 1/g_{YM}^2$, this theory is equivalent to M-theory on an $AdS_4 \times S^7$ background, dual to a CFT with an $SO(8)$ R-symmetry. This picture is compiled in Fig. 2.

2. Field theory correlators and SDLCQ

Discretized Light-Cone Quantization (DLCQ) preserves supersymmetry at every stage of the calculation if the supercharge rather than the Hamiltonian is diagonalized [14, 15]. The framework of supersymmetric DLCQ (SDLCQ) allows one to use the advantages of light-cone quantization (*e.g.* a simpler vacuum) together with the excellent renormalization properties guaranteed by supersymmetry.

The technique of (S)DLCQ was reviewed in Ref. [16], so we can be brief here. The basic idea of light-cone quantization is to parameterize space-time using light-cone coordinates

$$x^{\pm} \equiv \frac{1}{\sqrt{2}}\left(x^{0} \pm x^{1}\right), \tag{7}$$

and to quantize the theory making x^+ play the role of time. In the discrete light-cone approach, we require the momentum $p_- = p^+$ along the x^- direction to take on discrete values in units of p^+/K where p^+ is the conserved total momentum of the system. The integer K is the so-called harmonic resolution, and plays the role of a discretization parameter. One can think of this discretization as a consequence of compactifying the x^- coordinate on a circle with a period $2L = 2\pi K/p^+$. The advantage of discretizing on the light cone is the fact that the dimension of the Hilbert space becomes finite. Therefore, the Hamiltonian is a finite-dimensional matrix, and its dynamics can be solved explicitly. In SDLCQ one makes the DLCQ approximation to the supercharges Q^i. Surprisingly, also the discrete representations of Q_i satisfy the supersymmetry algebra. Therefore SDLCQ enjoys the improved renormalization properties of supersymmetric theories. To recover the continuum result, K has to go to infinity. We finds is that SDLCQ usually converges much faster than the naive DLCQ.

In the three-dimensional case we also discretize the transverse momentum along the direction $x^{\perp}$; however, it is treated in a fundamentally different way. The transverse resolution is T, and we think of the theory as being compactified on a transverse circle of length l. Therefore, the transverse momentum is cut off at $\pm 2\pi T/l$ and discretized in units of $2\pi/l$. Removal of this transverse momentum cutoff therefore corresponds to taking the transverse resolution T to infinity.

Let us now review these ideas in the context of a specific super-Yang-Mills (SYM) theory. Actually, it turns out that the two-dimensional SYM is essentially 'included' in the three-dimensional case [17], in the sense that in the weak coupling limit the spectrum of the three-dimensional theory is that of the lower dimensional theory. We therefore describe here only the 'more general' three-dimensional theory and hint at the differences and changes that are to make to recover the two-dimensional theory. We start with $2+1$ dimensional $\mathcal{N}=1$ super-Yang-Mills theory [18] defined on a space-time with one transverse dimension compactified on a circle. The action is

$$S=\int d^2x\int_0^l dx_\perp \mathrm{tr}(-\frac{1}{4}F^{\mu\nu}F_{\mu\nu}+\mathrm{i}\bar{\Psi}\gamma^\mu D_\mu\Psi). \tag{8}$$

After introducing the light–cone coordinates $x^\pm=\frac{1}{\sqrt{2}}(x^0\pm x^1)$, decomposing the spinor Ψ in terms of chiral projections

$$\psi=\frac{1+\gamma^5}{2^{1/4}}\Psi,\qquad \chi=\frac{1-\gamma^5}{2^{1/4}}\Psi \tag{9}$$

and choosing the light-cone gauge $A^+=0$, we obtain the action in the form

$$\begin{aligned} S &= \int dx^+dx^-\int_0^l dx_\perp \mathrm{tr}\left[\frac{1}{2}(\partial_-A^-)^2+D_+\phi\partial_-\phi+\mathrm{i}\psi D_+\psi+\right.\\ &\left.+\mathrm{i}\chi\partial_-\chi+\frac{\mathrm{i}}{\sqrt{2}}\psi D_\perp\phi+\frac{\mathrm{i}}{\sqrt{2}}\phi D_\perp\psi\right]. \end{aligned}\tag{10}$$

A simplification of the light-cone gauge is that the non-dynamical fields A^- and χ may be explicitly solved from their Euler–Lagrange equations of motion

$$A^-=\frac{g_{\mathrm{YM}}}{\partial_-^2}J=\frac{g_{\mathrm{YM}}}{\partial_-^2}\left(i[\phi,\partial_-\phi]+2\psi\psi\right),\quad \chi=-\frac{1}{\sqrt{2}\partial_-}D_\perp\psi. \tag{11}$$

These expressions may be used to express any operator in terms of the physical degrees of freedom only. In particular, the light-cone energy, P^-, and momentum operators, P^+,$P^\perp$, corresponding to translation invariance in each of the coordinates $x^\pm$ and $x_\perp$ may be calculated explicitly as

$$P^+ = \int dx^-\int_0^l dx_\perp \mathrm{tr}\left[(\partial_-\phi)^2+\mathrm{i}\psi\partial_-\psi\right], \tag{12}$$

$$P^- = \int dx^-\int_0^l dx_\perp \mathrm{tr}\left[-\frac{g_{\mathrm{YM}}^2}{2}J\frac{1}{\partial_-^2}J-\frac{\mathrm{i}}{2}D_\perp\psi\frac{1}{\partial_-}D_\perp\psi\right], \tag{13}$$

$$P_\perp = \int dx^-\int_0^l dx_\perp \mathrm{tr}\left[\partial_-\phi\partial_\perp\phi+\mathrm{i}\psi\partial_\perp\psi\right]. \tag{14}$$

The light-cone supercharge in this theory is a two-component Majorana spinor, and may be conveniently decomposed in terms of its chiral projections

$$Q^+ = 2^{1/4}\int dx^-\int_0^l dx_\perp \mathrm{tr}\left[\phi\partial_-\psi-\psi\partial_-\phi\right], \tag{15}$$

$$Q^- = 2^{3/4}\int dx^-\int_0^l dx_\perp \mathrm{tr}\left[2\partial_\perp\phi\psi+g_{\mathrm{YM}}\left(\mathrm{i}[\phi,\partial_-\phi]+2\psi\psi\right)\frac{1}{\partial_-}\psi\right].$$

The action (9) gives the following canonical (anti-)commutation relations for propagating fields for large N_c at equal x^+:

$$\begin{aligned}[\phi_{ij}(x^-,x_\perp),\partial_-\phi_{kl}(y^-,y_\perp)] &= \{\psi_{ij}(x^-,x_\perp),\psi_{kl}(y^-,y_\perp)\} \qquad (16)\\ &= \frac{1}{2}\delta(x^- - y^-)\delta(x_\perp - y_\perp)\delta_{il}\delta_{jk}.\end{aligned}$$

Using these relations one can check the supersymmetry algebra

$$\{Q^\pm,Q^\pm+\} = 2\sqrt{2}P^\pm, \qquad \{Q^+,Q^-\} = -4P_\perp. \qquad (17)$$

In solving for mass eigenstates, we will consider only states which have vanishing transverse momentum, which is possible since the total transverse momentum operator is kinematical. Strictly speaking, on a transverse cylinder, there are separate sectors with total transverse momenta $2\pi N_\perp/L$; we consider only one of them, $N_\perp = 0$. On such states, the light-cone supercharges Q^+ and Q^- anti-commute with each other, and the supersymmetry algebra is equivalent to the $\mathcal{N} = (1,1)$ supersymmetry of the dimensionally reduced (*i.e.*, two-dimensional) theory [14]. Moreover, in the $P_\perp = 0$ sector, the mass squared operator M^2 is given by $M^2 = 2P^+P^-$.

As we mentioned earlier, in order to render the bound-state equations numerically tractable, the transverse momenta of partons must be truncated. First, we introduce the Fourier expansion for the fields ϕ and ψ, where the transverse space-time coordinate $x_\perp$ is periodically identified

$$\phi_{ij}(0,x^-,x_\perp) = \frac{1}{\sqrt{2\pi l}}\sum_{n^\perp=-\infty}^{\infty}\int_0^\infty \frac{dk^+}{\sqrt{2k^+}} \qquad (18)$$
$$\times\left[a_{ij}(k^+,n^\perp)e^{-ik^+x^- - i\frac{2\pi n^\perp}{l}x_\perp} + a^\dagger_{ji}(k^+,n^\perp)e^{ik^+x^- + i\frac{2\pi n^\perp}{l}x_\perp}\right],$$

$$\psi_{ij}(0,x^-,x_\perp) = \frac{1}{2\sqrt{\pi l}}\sum_{n^\perp=-\infty}^{\infty}\int_0^\infty dk^+ \qquad (19)$$
$$\times\left[b_{ij}(k^+,n^\perp)e^{-ik^+x^- - i\frac{2\pi n^\perp}{l}x_\perp} + b^\dagger_{ji}(k^+,n^\perp)e^{ik^+x^- + i\frac{2\pi n^\perp}{l}x_\perp}\right].$$

Substituting these into the (anti-)commutators (16), one finds

$$\begin{aligned}\left[a_{ij}(p^+,n_\perp),a^\dagger_{lk}(q^+,m_\perp)\right] &= \left\{b_{ij}(p^+,n_\perp),b^\dagger_{lk}(q^+,m_\perp)\right\}\\ &= \delta(p^+ - q^+)\delta_{n_\perp,m_\perp}\delta_{il}\delta_{jk}.\end{aligned}$$

The supercharges then take the following form:

$$Q^+ = \mathrm{i}2^{1/4}\sum_{n^\perp\in\mathbf{Z}}\int_0^\infty dk\sqrt{k} \qquad (20)$$

$$\times \left[b^\dagger_{ij}(k,n^\perp)a_{ij}(k,n^\perp) - a^\dagger_{ij}(k,n^\perp)b_{ij}(k,n^\perp)\right],$$

$$Q^- = \frac{2^{7/4}\pi \mathrm{i}}{l} \sum_{n^\perp \in \mathbf{Z}} \int_0^\infty dk \frac{n^\perp}{\sqrt{k}} \tag{21}$$

$$\times \left[a^\dagger_{ij}(k,n^\perp)b_{ij}(k,n^\perp) - b^\dagger_{ij}(k,n^\perp)a_{ij}(k,n^\perp)\right] +$$

$$+\frac{\mathrm{i}2^{-1/4}g_{\mathrm{YM}}}{\sqrt{l\pi}} \sum_{n_i^\perp \in \mathbf{Z}} \int_0^\infty dk_1 dk_2 dk_3 \delta(k_1+k_2-k_3)\delta_{n_1^\perp+n_2^\perp,n_3^\perp} \Big\{$$

$$\frac{1}{2\sqrt{k_1k_2}}\frac{k_2-k_1}{k_3}[a^\dagger_{ik}(k_1,n_1^\perp)a^\dagger_{kj}(k_2,n_2^\perp)b_{ij}(k_3,n_3^\perp)$$

$$-b^\dagger_{ij}(k_3,n_3^\perp)a_{ik}(k_1,n_1^\perp)a_{kj}(k_2,n_2^\perp)]$$

$$\frac{1}{2\sqrt{k_1k_3}}\frac{k_1+k_3}{k_2}[a^\dagger_{ik}(k_3,n_3^\perp)a_{kj}(k_1,n_1^\perp)b_{ij}(k_2,n_2^\perp)$$

$$-a^\dagger_{ik}(k_1,n_1^\perp)b^\dagger_{kj}(k_2,n_2^\perp)a_{ij}(k_3,n_3^\perp)]$$

$$\frac{1}{2\sqrt{k_2k_3}}\frac{k_2+k_3}{k_1}[b^\dagger_{ik}(k_1,n_1^\perp)a^\dagger_{kj}(k_2,n_2^\perp)a_{ij}(k_3,n_3^\perp)$$

$$-a^\dagger_{ij}(k_3,n_3^\perp)b_{ik}(k_1)a_{kj}(k_2,n_2^\perp)]$$

$$(\frac{1}{k_1}+\frac{1}{k_2}-\frac{1}{k_3})[b^\dagger_{ik}(k_1,n_1^\perp)b^\dagger_{kj}(k_2,n_2^\perp)b_{ij}(k_3,n_3^\perp)$$

$$+b^\dagger_{ij}(k_3,n_3^\perp)b_{ik}(k_1,n_1^\perp)b_{kj}(k_2,n_2^\perp)]\Big\}.$$

We now perform the truncation procedure; namely, in all sums over the transverse momenta $n_i^\perp$ appearing in the above expressions for the supercharges, we restrict summation to the following allowed momentum modes: $n_i^\perp = 0, \pm 1 \ldots \pm T$. Note that this prescription is symmetric, in the sense that there are as many positive modes as there are negative ones. In this way we retain parity symmetry in the transverse direction. The longitudinal momenta $k_i = n_i\pi/L$ are restricted by the longitudinal resolution according to $K = \sum_i n_i$.

The two-dimensional supercharges are essentially recovered, when we put $n_\perp$ to zero. In particular, the first term of the supercharge Q^-, Eq. (21), is absent in this case. Additionally, we have to adjust the normalization constants in front of the expressions for the supercharges.

2.1. Two dimensional correlators

Using SDLCQ, we can reproduce the SUGRA scaling relation, Eq. (6), fix the numerical coefficient, and calculate the cross-over behavior at $1/g_{YM}\sqrt{N_c} < x < \sqrt{N_c}/g_{YM}$. To exclude subtleties, *nota bene* issues of zero modes, we checked our results against the free fermion and the 't Hooft model and found consistent results.

Let us now focus on the theory in two dimensions. We would like to compute a general expression for the correlator of the form $F(x^-, x^+) = \langle \mathcal{O}(x^-, x^+)\mathcal{O}(0,0)\rangle$. In DLCQ one fixes the total momentum in the x^- direction, and it is natural to compute the Fourier transform and express it in a spectrally decomposed form

$$\begin{aligned} \tilde{F}(P_-, x^+) &= \frac{1}{2L}\langle \mathcal{O}(P_-, x^+)\mathcal{O}(-P_-, 0)\rangle \\ &= \sum_n \frac{1}{2L}\langle 0|\mathcal{O}(P_-)|n\rangle e^{-iP_+^n x^+}\langle n|\mathcal{O}(-P_-, 0)|0\rangle \ . \quad (22) \end{aligned}$$

The form of the correlation function in position space is then recovered by Fourier transforming with respect to $P_- = K\pi/L$. We can continue to Euclidean space by taking $r = \sqrt{2x^+x^-}$ to be real. The result for the correlator of the stress-energy tensor is

$$\begin{aligned} F(x^-, x^+) &= \sum_n \left|\frac{L}{\pi}\langle n|T^{++}(-K)|0\rangle\right|^2 \left(\frac{x^+}{x^-}\right)^2 \\ &\quad \times \frac{M_n^4}{8\pi^2 K^3} K_4\left(M_n\sqrt{2x^+x^-}\right), \quad (23) \end{aligned}$$

where M_i is a mass eigenvalue and $K_4(x)$ is the modified Bessel function of order 4. Note that this quantity depends on the harmonic resolution K, but involves no other unphysical quantities. In particular, the expression is independent of the box length L.

The matrix element $(L/\pi)\langle 0|T^{++}(K)|i\rangle$ can be substituted directly to give an explicit expression for the two-point function. We see immediately that the correlator has the correct small-r behavior, for in that limit, it asymptotes to

$$\left(\frac{x^-}{x^+}\right)^2 F(x^-, x^+) = \frac{N_c^2(2n_b + n_f)}{4\pi^2 r^4}\left(1 - \frac{1}{K}\right),$$

which we expect for the theory of $n_b(n_f)$ free bosons (fermions) at large K. On the other hand, the contribution to the correlator from strictly massless states is given by

$$\left(\frac{x^-}{x^+}\right)^2 F(x^-, x^+) = \frac{6}{K^3\pi^2 r^4}\sum_i \left|\frac{L}{\pi}\langle 0|T^{++}(K)|i\rangle\right|^2_{M_i=0}.$$

It is important to notice that this $1/r^4$ behavior at large r is *not* the one we are looking for at large r. First of all, we do not expect any massless physical bound state in this theory, and, additionally, it has the wrong N_c dependence. Relative to the $1/r^4$ behavior at small r, the $1/r^4$ behavior at large r that we expect is down by a factor of $1/N_c$. This behavior is suppressed because we are performing a large-N_c calculation. All we can hope is to see the transition from the $1/r^4$ behavior at small r to the region where the correlator behaves like $1/r^5$.

2.1.1 The $\mathcal{N} = (1,1)$ theory. Although it is the $\mathcal{N} = (8,8)$ theory in which we are ultimately interested in, we can, nevertheless, perform the computation of the correlation function in models with less supersymmetry. The evaluation of the correlator for the stress energy tensor in the $\mathcal{N} = (8,8)$ theory is especially hard because of the many degrees of freedom due to the large number of supercharges in that theory. We will show in Sec. 2.1.3 how to overcome this obstacle by exploiting a residual 'flavor' symmetry of the theory. To see how our numerical method works without these complications, it might be worthwhile to study the theory with supercharges (1,1). In the next section we will briefly cover the theory with a $\mathcal{N} = (2,2)$ supersymmetry.

It has been argued that the $\mathcal{N} = (1,1)$ SYM theory does not exhibit dynamical supersymmetry breaking. A physicist's proof that supersymmetry is not spontaneously broken in this theory was given in Ref. [4]. This theory is also believed not to be confining [19][20], and is therefore expected to exhibit non-trivial infra-red dynamics. The SDLCQ of the 1+1 dimensional model with $\mathcal{N} = (1,1)$ supersymmetry was solved in Refs. [14, 21], and we apply these results directly in order to compute (23). For simplicity, we work at large N_c. The spectrum of this theory at finite K consists of $2K-2$ exactly massless states, *i.e.* $K-1$ massless bosons, and their superpartners, accompanied by large numbers of massive states separated by a gap. There is numerical evidence that this gap closes in the continuum limit. At finite N_c, we expect the degeneracy of $2K-2$ exactly massless states to be broken, giving rise to precisely a continuum of states starting at $M = 0$ as expected.

The stress-energy correlator of this theory for various values of the harmonic resolution K, is shown in Fig. 3(a). We find the curious feature that it asymptotes to the inverse power law c/r^4 for large r. This behavior comes about due to the coupling $\langle 0|T^{++}|n\rangle$ with exactly massless states $|n\rangle$. The contribution to (23) from strictly massless states are given by

$$\left(\frac{x^-}{x^+}\right)^2 F(x^-,x^+) = \left|\frac{L}{\pi}\langle 0|T^{++}(k)|n\rangle\right|^2 \frac{M_n^4}{8\pi^2 k^3} K_4(M_n r)\bigg|_{M_n=0} \qquad (24)$$

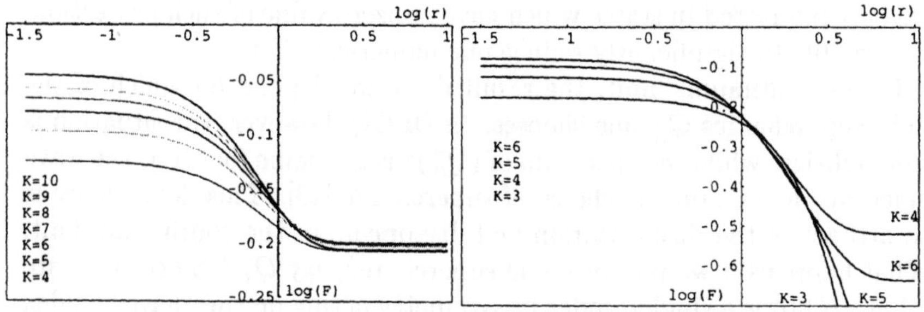

Figure 3. Log-Log plots of the two-dimensional correlation function $\langle T^{++}(x)T^{++}(0)\rangle \left(\frac{x^-}{x^+}\right)^2 \frac{4\pi^2 r^4}{N^2(2n_b+n_f)}$ v.s. r at $g_{YM}^2 N/\pi = 1$. Left: (a) $\mathcal{N} = (1,1)$ SYM for $K = 4$ to 10. Right: (b) $\mathcal{N} = (2,2)$ SYM for $K = 3, 4, 5, 6$.

$$= \left| \frac{L}{\pi} \langle 0|T^{++}(k)|n\rangle \right|^2_{M_n=0} \frac{6}{k^3\pi^2 r^4}. \tag{25}$$

We have computed this quantity as a function of the inverse harmonic resolution $1/K$ and extrapolated to the continuum limit. The data currently available suggests that the non-zero contribution from these massless states persists in this limit.

2.1.2 The $\mathcal{N} = (2,2)$ theory. Let us now turn to the model with $\mathcal{N} = (2,2)$ supersymmetry. The SDLCQ version of this model was solved in Ref. [22]. The result of this computation can be inserted into Eq. (23). The result is shown in Fig. 3(b). This model appears to exhibit the onset of a gapless continuum of states more rapidly than the $\mathcal{N} = (1,1)$ model as the harmonic resolution K is increased. Just as we found in the latter model, this theory contains exactly massless states in the spectrum. These massless states appear to couple to $T^{++}|0\rangle$ only for even K, and the overlap appears to be decreasing for growing K. It is believed that this model is likely to exhibit a power law behavior c/r^γ for $\gamma > 4$ for the T^{++} correlator for $r \gg g_{YM}\sqrt{N}$ in the large N_c limit [4].

2.1.3 The $\mathcal{N} = (8,8)$ theory. In principle, we can now calculate the correlator numerically by evaluating Eq. (23). However, it turns out that even for very modest harmonic resolutions, we face a tremendous numerical task. At $K = 2, 3, 4$, the dimension of the associated Fock space is 256, 1632, and 29056, respectively. The usual procedure is to diagonalize the Hamiltonian P^- and then to evaluate the projection of each eigenfunction on the fundamental state $T^{++}(-K)|0\rangle$. Since we are only interested in states which have nonzero value of such projection, we are able to significantly reduce our numerical efforts.

In the continuum limit, the result does not depend on which of the eight supercharges Q^-_α one chooses. In DLCQ, however, the situation is a bit subtler: while the spectrum of $(Q^-_\alpha)^2$ is the same for all α, the wave functions depend on the choice of supercharge [23]. This dependence is an artifact of the discretization and disappears in the continuum limit. What happens if we just pick one supercharge, say Q^-_1? Since the state $T^{++}(-K)|0\rangle$ is a singlet under R–symmetry acting on the "flavor" index of Q^-_α, the correlator (23) does not depend on the choice of α even at finite resolution!

We can exploit this fact to simplify our calculations. Consider an operator S commuting with both P^- and $T^{++}(-K)$, and such that $S|0\rangle = s_0|0\rangle$. Then the Hamiltonian and S can be diagonalized simultaneously. We assume in the sequel that the set of states $|i\rangle$ is a result of such a diagonalization. In this case, only states satisfying the condition $S|i\rangle = s_0|i\rangle$ contribute to the sum in (23), and we only need to diagonalize P^- in this sector, which reduces the size of the problem immensely. We can deduce from the structure of the state $T^{++}(-K)|0\rangle$ that any transformation of the form

$$\begin{aligned} a^I_{ij}(k) &\to f(I)a^{P[I]}_{ij}(k), \qquad f(I) = \pm 1 \\ b^\alpha_{ij}(k) &\to g(\alpha)b^{Q[\alpha]}_{ij}(k), \qquad g(\alpha) = \pm 1 \end{aligned} \tag{26}$$

given arbitrary permutations P and Q of the 8 flavor indices, commutes with $T^{++}(-K)$. The vacuum will then be an eigenstate of this transformation with eigenvalue 1. The requirement for $P^- = (Q^-_1)^2$ to be invariant under S imposes some restrictions on the permutations. In fact, we will require that Q^-_1 be invariant under S, in order to guarantee that P^- is invariant.

The form of the supercharge from [23] is

$$\begin{aligned} Q^-_\alpha &= \int_0^\infty [...]b^\dagger_\alpha(k_3)a_I(k_1)a_I(k_2) + ... \\ &+(\beta_I\beta^T_J - \beta_J\beta^T_I)_{\alpha\beta}[..]b^\dagger_\beta(k_3)a_I(k_1)a_J(k_2) + \dots . \end{aligned} \tag{27}$$

Here the β_I are 8×8 real matrices satisfying $\{\beta_I, \beta^T_J\} = 2\delta_{IJ}$.

Let us consider the expression for Q^-_1, Eq. (27). The first part of the supercharge does not include β matrices, and is therefore invariant under the transformation, Eq. (26), as long as $g(1) = 1$ and $Q[1] = 1$. We will consider only such transformations. The crucial observation for the analysis of the symmetries of the β terms is that in the representation of the β matrices we have chosen, the expression $B^\alpha_{IJ} = \left(\beta_I\beta^T_J - \beta_J\beta^T_I\right)_{1\alpha}$

may take only the values ± 2 or zero. Besides, for any pair (I, J) there is only one (or no) value of α corresponding to nonzero $\mathcal{B}$. Using this information, we may represent $\mathcal{B}$ in a compact form. With the definition [24]

$$\mu_{IJ} = \begin{cases} \alpha\,, & \mathcal{B}^{\alpha}_{IJ} = 2 \\ -\alpha\,, & \mathcal{B}^{\alpha}_{IJ} = -2 \\ 0\,, & \mathcal{B}^{\alpha}_{IJ} = 0 \text{ for all } \alpha \end{cases}\,, \tag{28}$$

together with the special choice of β matrices we get the following expression for μ

$$\mu = \begin{pmatrix} 0 & 5 & -7 & 2 & -6 & 3 & -4 & 8 \\ -5 & 0 & -3 & 6 & 2 & -7 & 8 & 4 \\ 7 & 3 & 0 & -8 & -4 & -5 & 6 & 2 \\ -2 & -6 & 8 & 0 & -5 & 4 & 3 & 7 \\ 6 & -2 & 4 & 5 & 0 & -8 & -7 & 3 \\ -3 & 7 & 5 & -4 & 8 & 0 & -2 & 6 \\ 4 & -8 & -6 & -3 & 7 & 2 & 0 & 5 \\ -8 & -4 & -2 & -7 & -3 & -6 & -5 & 0 \end{pmatrix}.$$

The next step is to look for a subset of the transformations, Eq. (26), which satisfy the conditions $g(1) = 1$ and $Q[1] = 1$ and leave the matrix μ invariant. This invariance implies that

$$Q[\mu_{P[I]P[J]}] = g(\mu_{IJ}) f(I) f(J) \mu_{IJ}\,. \tag{29}$$

The subset of transformations we are looking for forms a subgroup R of the permutation group $S_8 \times S_8$. Consequently, we will search for the elements of R that square to one. Products of such elements generate the whole group in the case of $S_8 \times S_8$. We will show later that this remains true for R. Not all of the Z_2 symmetries satisfying (29) are independent. In particular, if a and b are two such symmetries then aba is also a valid Z_2 symmetry. By going systematically through the different possibilities, we have found that there are 7 independent Z_2 symmetries in the group R. They are listed in Table 1. We explicitly constructed all the symmetries of the type, Eq. (26), which satisfy Eq. (29) using MATHEMATICA. It turns out that the group of such transformations has 168 elements, and we have shown that all of them can be generated from the seven Z_2 symmetries mentioned above.

In our numerical algorithm we implemented the Z_2 symmetries as follows. We can group the Fock states in classes and treat the whole class as a new state, because all states relevant for the correlator are singlets under the symmetry group R. As an example, consider the simplest non-trivial singlet

$$|1\rangle = \frac{1}{8} \sum_{I=1}^{8} \mathrm{tr}\left(a^{\dagger}(1, I) a^{\dagger}(K-1, I)\right) |0\rangle. \tag{30}$$

	a_1	a_2	a_3	a_4	a_5	a_6	a_7	a_8	b_2	b_3	b_4	b_5	b_6	b_7	b_8
1	a_7	a_3	a_2	a_6	a_8	a_4	a_1	a_5	b_2	$-b_3$	$-b_4$	$-b_6$	$-b_5$	b_8	b_7
2	a_3	a_6	a_1	a_5	a_4	a_2	a_8	a_7	$-b_4$	b_3	$-b_2$	$-b_5$	b_8	$-b_7$	b_6
3	a_8	a_7	a_6	a_5	a_4	a_3	a_2	a_1	$-b_3$	$-b_2$	b_4	$-b_5$	b_7	b_6	$-b_8$
4	a_5	a_4	a_8	a_2	a_1	a_7	a_6	a_3	$-b_2$	$-b_7$	b_8	b_5	$-b_6$	$-b_3$	b_4
5	a_8	a_3	a_2	a_7	a_6	a_5	a_4	a_1	$-b_5$	$-b_3$	b_7	$-b_2$	b_6	b_4	$-b_8$
6	a_5	a_8	a_7	a_6	a_1	a_4	a_3	a_2	$-b_8$	b_5	$-b_4$	b_3	$-b_6$	b_7	$-b_2$
7	a_4	a_6	a_8	a_1	a_7	a_2	a_5	a_3	$-b_2$	$-b_6$	b_5	b_4	$-b_3$	$-b_7$	b_8

Table 1. Seven independent Z_2 symmetries of the group R, which act on the 'flavor' quantum number of the different particles. Under the first of these symmetries, *e.g.*, the boson a_1 is transformed into a_7, etc.

Hence, if we encounter the state $a^\dagger(1,1)a^\dagger(K-1,1)|0\rangle$ while constructing the basis, we will replace it by the class representative; in this case, by the state $|1\rangle$. Such a procedure significantly decreases the size of the basis, while keeping all the information necessary for calculating the correlator. In summary, this use of the discrete flavor symmetry of the problem reduces the size of the Fock space by orders of magnitude.

In addition to these simplifications, one can further improve on the numerical efficiency by using Lanczos diagonalization techniques[25]. Namely, we substitute the explicit diagonalization with an efficient approximation. The idea is to use a symmetry preserving (Lanczos) algorithm. If we start with a normalized vector $|u_1\rangle$ proportional to the fundamental state $T^{++}(-K)|0\rangle$, the Lanczos recursion will produce a tridiagonal representation of the Hamiltonian $H_{LC} = 2P^+P^-$. Due to orthogonality of $\{|u_i\rangle\}$, only the (1,1) element of the exponential of the tridiagonal matrix $\hat{H}_{LC}$ will contribute to the correlator [26]. We exponentiate by diagonalizing $\hat{H}_{LC}\vec{v}_i = \lambda_i\vec{v}_i$ with eigenvalues λ_i and get

$$F(P^+,x^+) = \frac{|N_0|^{-2}}{2L}\left(\frac{\pi}{L}\right)^2 \sum_{j=1}^{N_L} |(v_j)_1|^2 e^{-i\frac{\lambda_j L}{2K\pi}x^+}.$$

Finally, we Fourier transform to obtain

$$F(x^-,x^+) \;=\; \frac{1}{8\pi^2K^3}\left(\frac{x^+}{x^-}\right)^2 \frac{1}{|N_0|^2}\sum_{j=1}^{N_L} |(v_j)_1|^2\lambda_j^2 K_4(\sqrt{2x^+x^-\lambda_i}),$$

which is equivalent to Eq. (23). This algorithm is correct only if the number of Lanczos iterations N_L runs up to the rank of the original matrix, but *in praxi* already a basis of about 20 vectors covers all leading contributions to the correlator.

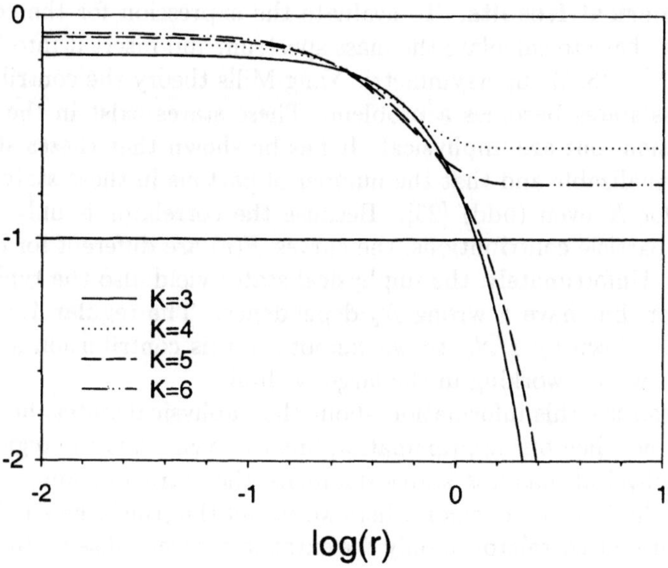

(a)

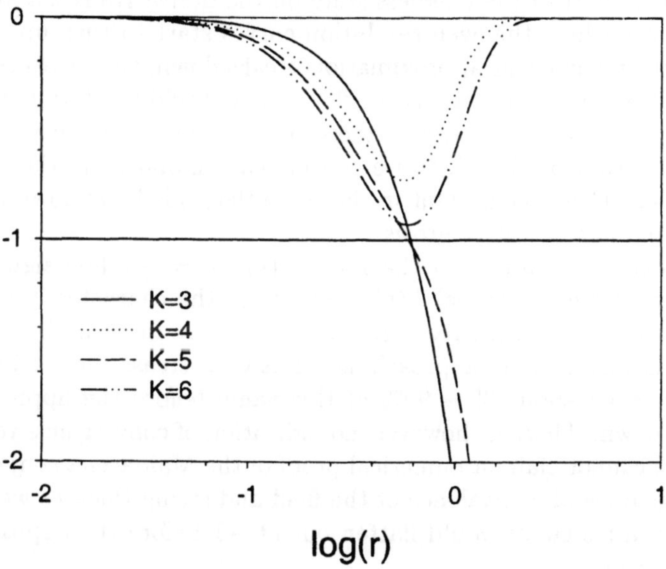

(b)

Figure 4. **Top: (a) Log-Log plot of $\mathcal{F}(r) = \langle T^{++}(x)T^{++}(0)\rangle \left(\frac{x^-}{x^+}\right)^2 \frac{4\pi^2 r^4}{N_c^2(2n_b+n_f)}$ vs. r for $g_{YM}^2 N_c/\pi = 1.0$, $K = 3,4,5$ and 6. Bottom: (b) the log-log derivative with respect to r of the correlation function in (a).**

Numerical Results. To evaluate the expression for the correlator $\mathcal{F}(r)$, we have to calculate the mass spectrum and insert it into Eq. (23). In the $\mathcal{N} = (8,8)$ supersymmetric Yang-Mills theory the contribution of massless states becomes a problem. These states exist in the SDLCQ calculation, but are unphysical. It has be shown that theses states are not normalizable and that the number of partons in these states is even (odd) for K even (odd) [23]. Because the correlator is only sensitive to two particle contributions, the curves $\mathcal{F}(r)$ are different for even and odd K. Unfortunately, the unphysical states yield also the typical $1/r^4$ behavior, but have a wrong N_c dependence. The regular $1/r^4$ contribution is down by $1/N_c$, so we cannot see this contribution at large r, because we are working in the large N_c limit.

We can use this information about the unphysical states, however, to determine when our approximation breaks down. It is the region where the unphysical massless states dominate the correlator sum. Unfortunately, this is also the region where we expect the true large-r behavior to dominate the correlator, if only the extra states were absent. In Fig. 4(a) for even resolution, the region where the correlator starts to behave like $1/r^4$ at large r is clearly visible. In Fig. 4(b) we see that for even resolution the effect of the massless state on the derivative is felt at smaller values of r where the even resolution curves start to turn up. Another estimate of where this approximation breaks down, that gives consistent values, is the set of points where the even and odd resolution derivative curves cross. We do not expect these curves to cross on general grounds, based on work in [4], where we considered a number of other theories. Our calculation is consistent in the sense that this breakdown occurs at larger and larger r as K grows.

We expect to approach the line $d\mathcal{F}(r)/dr = -1$ line signaling the cross-over from the trivial $1/r^4$ behavior to the characteristic $1/r^5$ behavior of the supergravity correlator, Eq. (6). Indeed, the derivative curves in Fig. 4(b) are approaching -1 as we increase the resolution and appear to be about $85 - 90\%$ of this value before the approximation breaks down. There is, however, no indication of convergence yet; therefore, we cannot claim a numerical proof of the Maldacena conjecture. A safe signature of equivalence of the field and string theories would be if the derivative curve would flatten out at -1 before the approximation breaks down.

2.2. Three-dimensional correlators

It remains a challenge to rigorously test the conjectured string/field theory correspondences. Although the so-called Maldacena conjecture maybe the most exciting one, because it promises insight into full four-dimensional Yang-Mills theories in the strong coupling regime, there are other interesting scenarios. For instance, it was conjectured that the supergravity solutions corresponding to $p+1$ SYM theories are black p-brane solutions, see *e.g.* Ref. [3]. Consequently, there are interesting testing scenarios also in three-dimensional spacetime. Numerically, of course, things get more difficult as the number of dimensions is increased. On the way to the full four-dimensional problem, it may be worthwhile to present our latest results on correlation functions in three dimensions; see also [13]. Fig. 7(b) shows the correlator for $\mathcal{N} = 1$ SYM(2+1) as a function of the distance r: it is converging well with the *transverse* cut-off T. To put things in perspective, we note that the construction of the largest Hamiltonian matrix involved in this calculation requires a Fock basis of approximately two million states. This is by a factor 100 more than we used in the test of the Maldacena conjecture described in this article, which itself was already substantially better than the first feasibility study [4].

The correlator of the energy momentum operator has been studied in conformal field theory in 2+1 dimensions [27], and this provides a reference point for our results. The structure of the correlators in conformal field theory is particularly simple in the collinear limit $x_\perp \to 0$, and we therefore find it convenient to work in this limit. From results in conformal field theory we expect that correlators behave as $1/r^6$ at small r, where we are probing deep inside the bound states. We have confirmed this $1/r^6$ behavior by an analytic calculation of the free-particle correlator in the DLCQ formalism [16].

The contributions of individual bound states have a characteristic length scale corresponding to the size of the bound states. On dimensional grounds one can show that the power behavior of the correlators are reduced by one power of r; so for individual bound states the correlator behaves like $1/r^5$ for small r. It then becomes a nontrivial check to see that at small r the contributions of the bound states add up to give the expected $1/r^6$ behavior. We find this expected result as well as the characteristic rapid convergence of SDLCQ at both small and intermediate values of r.

At large r the correlator is controlled by the massless states of the theory. In this theory there are two types of massless states. At zero coupling all the states of the $1+1$ dimensional theory are massless, and for non-vanishing coupling the massless states of the $1+1$ theory are promoted to massless states of the $2+1$ dimensional theory [17]. These states are BPS states and are exactly annihilated by one of the supercharges. This is perhaps the most interesting part of this calculation because the BPS masses are protected by the exact supersymmetry of the numerical approximation and remain exactly zero at all couplings. Commonly in modern field theory one uses the BPS states to extrapolate from weak coupling to strong coupling. While the masses of BPS states remain constant as functions of the coupling, their wave functions certainly do not. The calculation of the correlator at large r provides a window to the coupling dependence of the BPS wave functions. We find, however, that there is a critical coupling where the correlator goes to zero, which depends on the transverse resolution. A detailed study of this critical coupling shows that it goes to infinity linearly with the square root of the transverse resolution. Below the critical coupling the correlator converges rapidly at large r. One possible explanation is that this singular behavior signals the breakdown of the SDLCQ calculation for the BPS wave function at couplings larger than the critical coupling. If this is correct, calculation of the BPS wave function at stronger couplings requires higher transverse resolutions. We note that above the critical coupling (see Fig. 7 below) we do find convergence of the correlator at large r but at a significantly slower rate.

Let us now return to the details of the calculation. We would like to compute a general expression of the form

$$F(x^+,x^-,x^\perp) = \langle 0|T^{++}(x^+,x^-,x^\perp)T^{++}(0,0,0)|0\rangle. \qquad (31)$$

Here we will calculate the correlator in the collinear limit, that is, where $x^\perp = 0$. We know from conformal field theory [27] calculations that this will produce a much simpler structure.

The calculation is done by inserting a complete set of intermediate states $|\alpha\rangle$,

$$\begin{aligned} &F(x^+,x^-,x^\perp = 0) = \qquad (32)\\ &\sum_\alpha \langle 0|T^{++}(x^-,0,x^\perp = 0)|\alpha\rangle e^{-iP_\alpha^- x^+}\langle\alpha|T^{++}(0,0,0)|0\rangle. \end{aligned}$$

with energy eigenvalues P_α^-. The momentum operator $T^{++}(x)$ is given by

$$T^{++}(x) = \mathrm{tr}\left[(\partial_-\phi)^2 + \frac{1}{2}\left(i\psi\partial_-\psi - i(\partial_-\psi)\psi\right)\right] = T_B^{++}(x) + T_F^{++}(x)\,. \qquad (33)$$

In terms of the mode operators, we find

$$T^{++}(x^+,x^-,0)|0\rangle = \frac{1}{2Ll}\sum_{n,m}\sum_{n_\perp,m_\perp} T(n,m)e^{-i(P_n^+ + P_m^+)x^-}|0\rangle\,, \qquad (34)$$

where the boson and fermion contributions are given by

$$\frac{L}{\pi}T_B^{++}(n,m)|0\rangle = \frac{\sqrt{nm}}{2}\mathrm{tr}\left[a_{ij}^\dagger(n,n_\perp)a_{ji}^\dagger(m,m_\perp)\right]|0\rangle \tag{35}$$

and

$$\frac{L}{\pi}T_F^{++}(n,m)|0\rangle = \frac{(n-m)}{4}\mathrm{tr}\left[b_{ij}^\dagger(n,n_\perp)b_{ji}^\dagger(m,m_\perp)\right]|0\rangle\,. \tag{36}$$

Given each $|\alpha\rangle$, the matrix elements in (32) can then be evaluated, and the sum computed.

First, however, it is instructive to do the calculation where the states $|\alpha\rangle$ are a set of free particles with mass m. The boson contribution is

$$F(x^+,x^-,0)_B = \sum_{n,m,s,t}\left(\frac{\pi}{4L^2l}\right)^2 e^{-iP_n^-x^+-iP_n^+x^--iP_m^-x^+-iP_m^+x^-} \tag{37}$$
$$\times\sqrt{mnst}\,\langle 0|\mathrm{tr}[a(n,n_\perp)a(m,m_\perp)]\mathrm{tr}[a^\dagger(s,s_\perp)a^\dagger(t,t_\perp)]|0\rangle,$$

where the sum over n implies sums over both n and $n_\perp$, and

$$P_n^- = \frac{m^2+(2n_\perp\pi/l)^2}{2n\pi/L} \quad\text{and}\quad P_n^+ = \frac{n\pi}{L}\,. \tag{38}$$

The sums can be converted to integrals which can be explicitly evaluated, and we find

$$F(x^+,x^-,0)_B = \frac{i}{2(2\pi)^3}m^5\left(\frac{x^+}{x^-}\right)^2\frac{1}{x}K_{5/2}^2(mx)\,, \tag{39}$$

where $x^2 = 2x^-x^+$. Similarly for the fermions we find

$$F(x^+,x^-,0)_F = \sum_{n,m,s,t}\left(\frac{\pi}{8L^2l}\right)^2 e^{-iP_n^-x^+-iP_n^+x^--iP_m^-x^+-iP_m^+x^-} \tag{40}$$
$$\times(m-n)(s-t)\,\langle 0|\mathrm{tr}[b(n,n_\perp)b(m,m_\perp)]\mathrm{tr}[b^\dagger(s,s_\perp)b^\dagger(t,t_\perp)]|0\rangle.$$

After doing the integrals we obtain

$$F(x^+,x^-,0)_F = \frac{i}{4(2\pi)^3}\frac{m^5}{x}\left(\frac{x^+}{x^-}\right)^2\left[K_{7/2}(mx)K_{3/2}(mx)-K_{5/2}^2(mx)\right]\,. \tag{41}$$

We can continue to Euclidean space by taking $r=\sqrt{2x^+x^-}$ to be real, and, finally, in the small-r limit we find

$$\left(\frac{x^-}{x^+}\right)^2 F(x^+,x^-,0) = \frac{-3i}{8(2\pi)^2}\frac{1}{r^6}\,, \tag{42}$$

which exhibits the expected $1/r^6$ behavior.

Now let us return to the calculation using the bound-state solution obtained from SDLCQ. It is convenient to write

$$F(x^+,x^-,0)=\sum_{n,m,s,t}\left(\frac{\pi}{2L^2l}\right)^2\langle 0|\frac{L}{\pi}T(n,m)e^{-iP^-_{op}x^+-iP^+x^-}\frac{L}{\pi}T(s,t)|0\rangle\,, \tag{43}$$

where P^-_{op} is the Hamiltonian operator. We again insert a complete set of bound states $|\alpha\rangle$ with light-cone energies $P^-_\alpha=(M^2_\alpha+P^2_\perp)/P^+$ at resolution K (and therefore $P^+=\pi K/L$) and with total transverse momentum $P_\perp=2N_\perp\pi/l$. We also define

$$|u\rangle=N_u\frac{L}{\pi}\sum_{n,m}\delta_{n+m,K}\delta_{n_\perp+m_\perp,N_\perp}T(n,m)|0\rangle\,, \tag{44}$$

where N_u is a normalization factor such that $\langle u|u\rangle=1$. It is straightforward to calculate the normalization, and we find

$$\frac{1}{N_u^2}=\frac{K^3}{8}(1-\frac{1}{K})(2T+1)\,. \tag{45}$$

The correlator (43) becomes

$$F(x^+,x^-,0)=\sum_{K,N_\perp,\alpha}\left(\frac{\pi}{2L^2l}\right)^2e^{-iP^-_\alpha x^+-iP^+x^-}\frac{1}{N_u^2}|\langle u|\alpha\rangle|^2\,. \tag{46}$$

We will calculate the matrix element $\langle u|\alpha\rangle$ at fixed longitudinal resolution K and transverse momentum $N_\perp=0$. Because of transverse boost invariance the matrix element does not contain any explicit dependence on $N_\perp$. To leading order in $1/K$ the explicit dependence of the matrix element on K is K^3; it also contains a factor of l, the transverse length scale. To separate these dependencies, we write F as

$$F(x^+,x^-,0)=\frac{1}{2\pi}\sum_{K,N_\perp,\alpha}\frac{1}{2L}\frac{1}{l}\left(\frac{\pi K}{L}\right)^3e^{-iP^-_\alpha x^+-iP^+x^-}\frac{|\langle u|\alpha\rangle|^2}{lK^3|N_u|^2}\,. \tag{47}$$

We can now do the sums over K and $N_\perp$ as integrals over the longitudinal and transverse momentum components $P^+=\pi K/L$ and $P^\perp=2\pi N_\perp/l$. We obtain

$$\frac{1}{\sqrt{-i}}\left(\frac{x^-}{x^+}\right)^2F(x^+,x^-,0)=\sum_\alpha\frac{1}{2(2\pi)^{5/2}}\frac{M_\alpha^{9/2}}{\sqrt{r}}K_{9/2}(M_\alpha r)\frac{|\langle u|\alpha\rangle|^2}{lK^3|N_u|^2}\,. \tag{48}$$

In practice, the full sum over α is approximated by a Lanczos [25] iteration technique [24, 26] that eliminates the need for full diagonalization of the Hamiltonian matrix. For the present case, the number of iterations required was on the order of 1000.

Looking back at the calculation for the free particle, we see that there are two independent sums over transverse momentum, after the contractions are performed. One would expect that the transverse dimension is controlled by the dimensional scale of the bound state (R_B) and therefore the correlation should scale like $1/r^4 R_B^2$. However, because of transverse boost invariance, the matrix element must be independent of the difference of the transverse momenta and therefore must scale as $1/r^5 R_B$.

There are three commuting Z_2 symmetries. One of them is the parity in the transverse direction,

$$P : a_{ij}(k, n^\perp) \to -a_{ij}(k, -n^\perp), \qquad b_{ij}(k, n^\perp) \to b_{ij}(k, -n^\perp). \tag{49}$$

The second symmetry [28] is with respect to the operation

$$S : a_{ij}(k, n^\perp) \to -a_{ji}(k, n^\perp), \qquad b_{ij}(k, n^\perp) \to -b_{ji}(k, n^\perp). \tag{50}$$

Since P and S commute with each other, we need only one additional symmetry $R = PS$ to close the group. Since Q^-, P and S commute with each other, we can diagonalize them simultaneously. This allows us to diagonalize the supercharge separately in the sectors with fixed P and S parities and thus reduce the size of matrices. Doing this one finds that the roles of P and S are different. While all the eigenvalues are usually broken into non-overlapping S-odd and S-even sectors [29], the P symmetry leads to a double degeneracy of massive states (in addition to the usual boson-fermion degeneracy due to supersymmetry).

Numerical Results. The first important numerical test is the small-r behavior of the correlator. Physically we expect that at small r the bound states should behave as free particles, and therefore the correlator should have the behavior of the free particle correlator which goes like $1/r^6$. We see in (48) that the contributions of each of the bound states behaves like $1/r^5$. Therefore, to get the $1/r^6$ behavior of the free theory, the bound states must work in concert at small r. It is clear that this cannot work all the way down to $r = 0$ in the numerical calculation. At very small r the most massive state allowed by the numerical approximation will dominate, and the correlator must behave like $1/r^5$. To see what happens at slightly larger r it is useful to consider the behavior at small coupling. There, the larger masses go like

$$M_\alpha \simeq \sum_i \frac{(k_i^\perp)^2}{2P^+} \,. \tag{51}$$

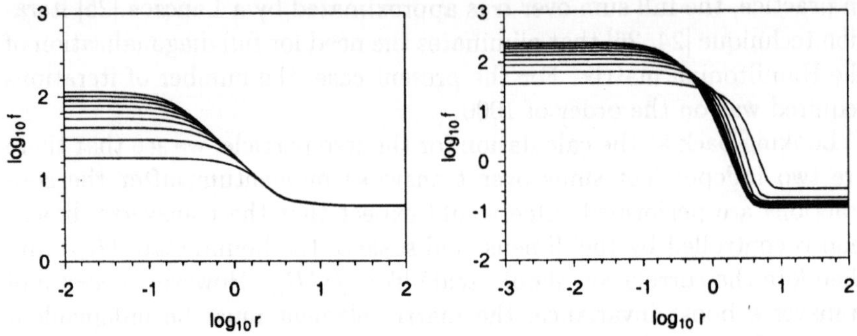

Figure 5. The log-log plot of the correlation function $f \equiv r^5\langle T^{++}(x)T^{++}(0)\rangle\left(\frac{x^-}{x^+}\right)^2 \frac{16\pi^3}{105}\frac{K^3 l}{\sqrt{-i}}$ vs. r. Left: (a) in units where $g = g_{\rm YM}\sqrt{N_c l}/2\pi^{3/2} = 0.10$ for $K = 4$ and $T = 1$ to 9; Right: (b) in units where $g = g_{\rm YM}\sqrt{N_c l}/2\pi^{3/2} = 1$ for $K = 5$ and $T = 1$ to 9.

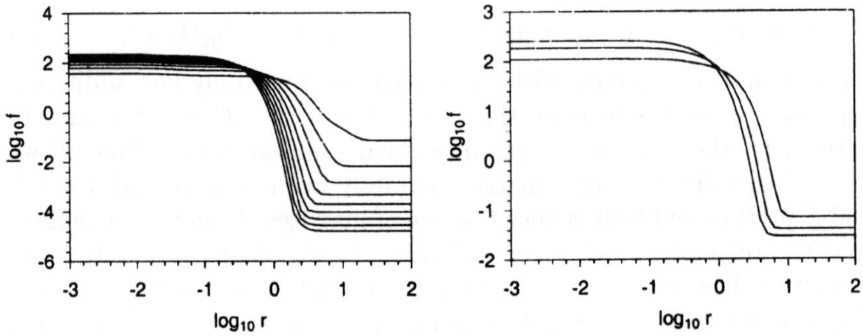

Figure 6. Left: (a) Same as Fig. 5(b), but for $g = g_{\rm YM}\sqrt{N_c l}/2\pi^{3/2} = 10$. Right: (b) Log-Log plot of the *three-dimensional* correlation function $f \equiv r^5\langle T^{++}(x)T^{++}(0)\rangle\left(\frac{x^-}{x^+}\right)^2 \frac{16\pi^3 K^3 l}{105\sqrt{-i}}$ vs. r for $g = g_{YM}^2\sqrt{N_c l}/2\pi^{3/2} = 1.0$ for $K = 6$ and $T = 1$ to 5.

Consequently, as we remove the $k^\perp$ cutoff, *i.e.* increase the transverse resolution T, more and more massive bound states will contribute, and the dominant one will take over at smaller and smaller r leading to the expected $1/r^6$. This is exactly what we see happening in Fig. 5 at weak coupling with longitudinal resolution $K = 4$ and 5.

The correlator converges from below at small r with increasing T, and in the region $-0.5 \le \log r \le 0.5$ the plot of r^5 times the correlator falls like $1/r$. In Fig. 7 at resolution $K = 5$ we see the same behavior for strong coupling ($g = g_{\rm YM}\sqrt{N_c l}/2\pi^{3/2} = 10$) but now at smaller r ($\log r \simeq -0.5$) as one would expect. Again at strong coupling we see that the correlator converges quickly and from below in T. All indications are that at small r the correlators are well approximated by SDLCQ, converge rapidly, and show the behavior that one would expect on general physical grounds. This gives us confidence to go on to investigate the behavior at large r.

The behavior for large r is governed by the massless states. From earlier work [30, 17] on the spectrum of this theory we know that there are two types of massless states. At $g = 0$ the massless states are a reflection of all the states of the dimensionally reduced theory in $1+1$. In 2+1 dimensions these states behave as $g^2 M_{1+1}^2$. We expect therefore that for $g \simeq 0$ there should be no dependence of the correlator on the transverse momentum cutoff T at large r. In Fig. 5(a) this behavior is clearly evident.

At all couplings there are exactly massless states which are the BPS states of this theory, which has zero central charge. These states are destroyed by one supercharge, Q^-, and not the other, Q^+. From earlier work [30] on the spectrum we saw that the number of BPS states is independent of the transverse resolution and equal to $2K-1$. Since these states are exactly massless at all resolutions, transverse and longitudinal convergence of these states cannot be investigated using the spectrum. These states do have a complicated dependence on the coupling g through their wave function, however. This is a feature so far not encountered in DLCQ [16]. In previous DLCQ calculations one always looked to the convergence of the spectrum as a measure of the convergence of the numerical calculation. Here we see that it is the correlator at large r that provides a window to study the convergence of the wave functions of the BPS states. In Fig. 7 we see that the correlator converges from above at large r as we increase T.

We also note that the correlator at large r is significantly smaller than at small r, particularly at strong coupling. In our initial study of the BPS states [17] we found that at strong coupling the average number of particles in these BPS states is large. Therefore the two particle components, which are the only components the T^{++} correlator sees, are small.

The coupling dependence of the large-r limit of the correlator is much more interesting than we would have expected based on our previous work on the spectrum. To see this behavior we study the large-r behavior of the correlator at fixed g as a function of the transverse resolution T and at fixed T as a function of the coupling g. We see a hint that something unusual is occurring in Fig. 7. For values of the coupling up to about $g = 1$ we see the typical rapid convergence in the transverse momentum cutoff; however, at larger coupling the convergence appears

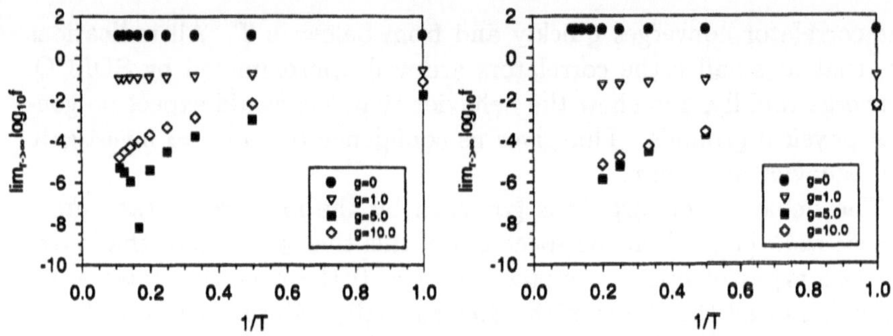

Figure 7. The large-r limit of the log of the correlation function $f \equiv r^5 \langle T^{++}(x) T^{++}(0)\rangle \left(\frac{x^-}{x^+}\right)^2 \frac{16\pi^3}{105}\frac{K^3 l}{\sqrt{-1}}$ vs. $1/T$ for [left](a) $K = 5$ and [right](b) $K = 6$ and for various values of the coupling $g = g_{\rm YM}\sqrt{N_c l}/2\pi^{3/2}$.

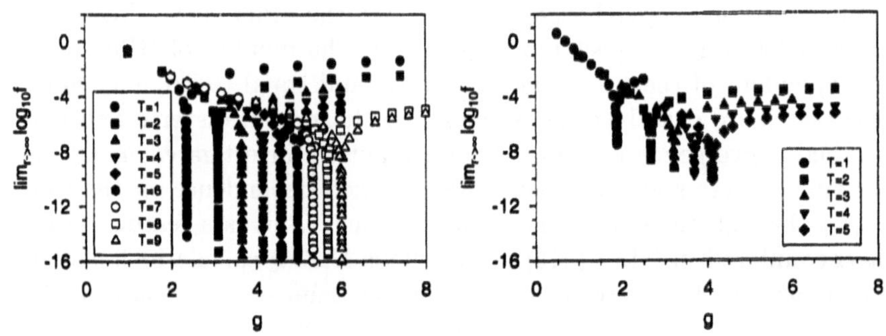

Figure 8. The large-r limit of the correlation function $f \equiv r^5 \langle T^{++}(x) T^{++}(0)\rangle \left(\frac{x^-}{x^+}\right)^2 \frac{16\pi^3}{105}\frac{K^3 l}{\sqrt{-1}}$ vs. $g = g_{\rm YM}\sqrt{N_c l}/2\pi^{3/2}$ for [left](a) $K = 5$ [right](b) $K = 6$ and for various values of the transverse resolution T.

to deteriorate, and we see that for $g = 5$ the correlator is smaller than at $g = 10$. We see this same behavior at both $K = 5$ and $K = 6$. We do not see this behavior at $K = 4$, but it is not unusual for effects to appear only at a large enough resolution in SDLCQ. In Fig. 8 we see that the correlator does not in fact decrease monotonically with g but rather has a singularity at a particular value of the coupling which is a function of K and T. Beyond the singularity the correlator again appears to behave well.

If we plot the 'critical' couplings, at which the correlator goes to zero, versus $\sqrt{T}$, as in Fig. 9, we see that they lie on a straight line, *i.e.* this coupling is a linear function of $\sqrt{T}$ in both cases, $K = 5$ and

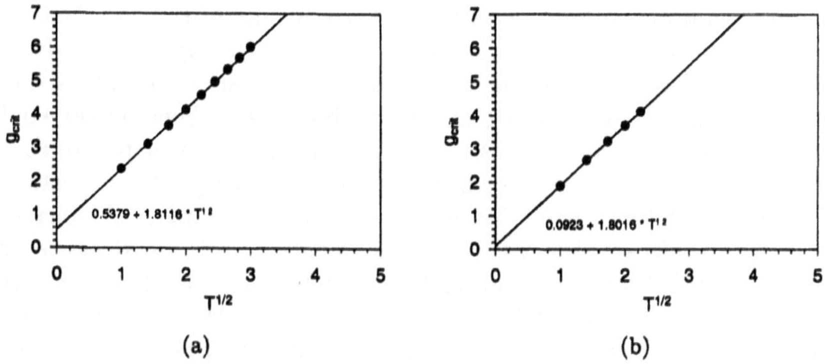

Figure 9. Critical coupling g_{crit} versus $\sqrt{T}$ for (a) $K = 5$ and (b) $K = 6$.

6. Consequently, the 'critical' coupling goes to infinity in the transverse continuum limit. It appears as though we have encountered a finite transverse cutoff effect. The most likely conclusion is that our numerical calculation of the BPS wave function is only valid for $g < g_{\text{crit}}(T)$. While the large-r correlator does converge above the critical coupling, it is unclear at this time if it has any significance. It might have been expected that one would need larger and larger transverse resolution to probe the strong coupling region, the occurrence of the singular behavior that we see is a surprise, and we have no detailed explanation for it at this time. We see no evidence of a singular behavior at small or intermediate r. This indicates, but does not prove, that our calculations of the massive bound states is valid at all g.

We do not seem to see a region dominated by the massive bound states, that is, a region where r is large enough that we see the structure of the bound states but small enough that the correlator is not dominated by the massless states of the theory. Such a region might give us other important information about this theory.

3. Conclusions

In this note we reported on progress in an attempt to rigorously test the conjectured equivalence of two-dimensional $\mathcal{N} = (8,8)$ supersymmetric Yang-Mills theory and a system of $D1$-branes in string theory. Within a well-defined non-perturbative calculation, we obtained results that are within 10-15% of results expected from the Maldacena conjecture. The results are still not conclusive, but they definitely point in right direction. Compared to previous work [4], we included orders of magnitude more states in our calculation and thus greatly improved the testing conditions. We remark that improvements of the code and the numerical method are possible and under way. During the calculation we noticed that contributions to the correlator come from only a small number of terms. An analytic understanding of this phenomenon would greatly accelerate calculations. We point out that in principle we could study the proper $1/r$ behavior at large r by computing $1/N_c$ corrections, but this interesting calculation would mean a huge numerical effort.

In this work we also discussed the calculation of the stress-tensor correlator $\langle 0|T^{++}(x)T^{++}(0)|0\rangle$ in $\mathcal{N}=1$ SYM in 2+1 dimension at large N_c in the collinear limit. We find that the free-particle correlator behaves like $1/r^6$, in agreement with results from conformal field theory. The contribution from an individual bound state is found to behave like $1/r^5$, and at small r such contributions conspire to reproduce the conformal field theory result $1/r^6$. We do not seem to find an intermediate region in r where the correlator behaves as $1/r^5$, reflecting the behavior of the individual massive bound states.

At large r the correlator is dominated by the massless BPS states of the theory. We find that as a function of g the large-r correlator has a critical value of g where it abruptly drops to zero. We have investigated this singular behavior and find that at fixed longitudinal resolution the critical coupling grows linearly with $\sqrt{T}$. We conjecture that this critical coupling signals the breakdown of SDLCQ at sufficiently strong coupling at fixed transverse resolution, T. While this might not be surprising in general, it is surprising that the behavior appears in the BPS wave functions and that we see no sign of this behavior in the massive states. We find that above the critical coupling the correlator still converges but significantly slower. It is unclear at this time if we should attach any significance to the correlator in this region.

This calculation emphasizes the importance of BPS wave functions which carry important coupling dependence, even though the mass eigenvalues are independent of the coupling. We will discuss the spectrum, the wavefunctions and associated properties of all the low energy bound states of $\mathcal{N}=1$ SYM in 2+1 dimensions in a subsequent paper [31].

A number of computational improvements have been implemented in our code to allow us to make these detailed calculations. The code now fully utilizes the three known discrete symmetries of the theory, namely supersymmetry, transverse parity P, Eq. (49), and the Z_2 symmetry S, Eq. (50). This reduces the dimension of the Hamiltonian matrix by a factor of 8. Other, more efficient storage techniques allow us to handle on the order of 2,000,000 states in this calculation, which has been performed on a single processor Linux workstation. Our improved storage techniques should allow us to expand this calculation to include higher supersymmetries without a significant expansion of the code or computational power. We remain hopeful that porting to a parallel machine will allow us to tackle problems in full 3+1 dimensions.

Acknowledgments

We would like to thank the organizers for the opportunity to speak at this workshop. This work was supported in part by the US Department of Energy.

References

[1] J. Maldacena, *Adv. Theor. Math. Phys.* **2** (1998) 231, hep-th/9711200.

[2] O. Lunin and S. Pinsky, in the proceedings of 11th International Light-Cone School and Workshop: New Directions in Quantum Chromodynamics and 12th Nuclear Physics Summer School and Symposium (NuSS 99), Seoul, Korea, 26 May - 26 Jun 1999 (New York, AIP, 1999), p. 140, hep-th/9910222.

[3] N. Itzhaki, J. Maldacena, J. Sonnenschein, and S. Yankielowicz, *Phys. Rev.* **D58** (1998) 046004, hep-th/9802042.

[4] F. Antonuccio, O. Lunin, S. Pinsky, and A. Hashimoto, JHEP **07** (1999) 029.

[5] A. Hashimoto and N. Itzhaki, *Phys. Lett.* **B465** (1999) 142–147, hep-th/9903067.

[6] S. S. Gubser, I. R. Klebanov, and A. M. Polyakov, *Phys. Lett.* **B428** (1998) 105, hep-th/9802109.

[7] E. Witten, *Adv. Theor. Math. Phys.* **2** (1998) 253, hep-th/9802150.

[8] H. J. Kim, L. J. Romans, and P. van Nieuwenhuizen, *Phys. Rev.* **D32** (1985) 389–399.

[9] M. Krasnitz and I. R. Klebanov, *Phys. Rev.* **D56** (1997) 2173–2179, hep-th/9703216.

[10] S. S. Gubser, A. Hashimoto, I. R. Klebanov, M. Krasnitz, *Nucl. Phys.* **B526** (1998) 393, hep-th/9803023.

[11] S. Ferrara, C. Fronsdal, and A. Zaffaroni, *Nucl. Phys.* **B532** (1998) 153, hep-th/9802203.

[12] A.B. Zamolodchikov, *JETP Lett.* **43** (1986) 730; *Sov.J.Nucl.Phys.* **46** (1987) 1090.

[13] J.R. Hiller, S. Pinsky, U. Trittmann, *Two-Point Stress-Tensor Correlator in $\mathcal{N} = 1$ SYM(2+1)*, hep-th/0101120.

[14] Y. Matsumura, N. Sakai, and T. Sakai, *Phys. Rev.* **D52** (1995) 2446, hep-th/9504150.

[15] A. Hashimoto and I.R. Klebanov, *Nucl. Phys.* **B434** (1995) 264.

[16] S.J. Brodsky, H.-C. Pauli, and S.S. Pinsky, *Phys. Rept.* **301** (1998) 299.

[17] F. Antonuccio, O. Lunin, and S. Pinsky, *Phys. Rev.* **D59** (1999) 085001.

[18] F. Antonuccio, O. Lunin, S. Pinsky, *Phys. Lett.* **B429** (1998) 327, hep-th/9803027.

[19] A. Armoni, Y. Frishman, and J. Sonnenschein, *Phys. Lett.* **B449** (1999) 76, hep-th/9807022.

[20] A. Armoni, Y. Frishman, and J. Sonnenschein, hep-th/9903153.

[21] F. Antonuccio, O. Lunin, and S. Pinsky, *Phys. Rev.* **D58** (1998) 085009.

[22] F. Antonuccio, H.-C. Pauli, S. Pinsky, S. Tsujimaru, *Phys. Rev.* **D58** (1998) 125006.

[23] F. Antonuccio, O. Lunin, H.-C. Pauli, S. Pinsky, S. Tsujimaru, *Phys. Rev.* **D58** (1998) 105024.

[24] J.R. Hiller, O. Lunin, S. Pinsky, U. Trittmann, *Phys.Lett.* **B482** (2000) 409–416.

[25] C. Lanczos, *J. Res. Nat. Bur. Stand.* **45**, 255 (1950); J. Cullum and R.A. Willoughby, *Lanczos Algorithms for Large Symmetric Eigenvalue Computations*, Vol. I and II, (Birkhauser, Boston, 1985).

[26] R. Haydock, in *Solid State Physics*, Vol. 35, H. Ehrenreich, F. Seitz, and D. Turnbull (eds.), (Academic, New York, 1980), p. 283.

[27] H. Osborn and A.C. Petkou, dimensions. *Ann. Phys.* **231** (1994) 311.

[28] D. Kutasov, *Phys. Rev.* **D48** (1993) 4980.

[29] G. Bhanot, K. Demeterfi, and I.R. Klebanov, *Phys. Rev.* **D48** (1993) 4980.

[30] P. Haney, J.R. Hiller, O. Lunin, S. Pinsky, and U. Trittmann, *Phys. Rev.* **D62** (2000) 075002, hep-th/9911243.

[31] J.R. Hiller, S. Pinsky, and U. Trittmann, in preparation.

A MICROSCOPIC BASIS FOR THE ENTROPY OF AdS_3 BLACK HOLE

S. Fernando and F. Mansouri[1]

1. INTRODUCTION

Since the discovery of the AdS_3 black hole [1, 2], the physical basis of its entropy has been investigated from a veriety of points of view. Some of the more prominent approaches to this problem have been compared and contrasted by Carlip [3]. Within the framework of pure gravity in $2+1$ dimensions, a direct method of obtaining the entropy of the BTZ black hole has been given by Strominger [4], in which use is made of the earlier work of Brown and Henneaux [5]. These authors demonstrated that the asymptotic symmetry group of any solution of the standard (metrical) Einstein theory with a negative cosmological constant, which satisfies certain boundary conditions, is generated by two copies of the Virasoro algebra with central charges

$$c_L = c_R = \frac{3l}{2G}, \tag{1}$$

where l is the radius of curvature of the AdS_3 space, and G is Newton's constant. Then, assuming that the ground state eigenvalue Δ_0 of the Virasoro generator L_0 vanishes, he obtained the Bekenstein-Hawking expression for the entropy. As pointed out by Strominger [4], in this derivation one must take for granted the existence of a quantum gravity theory with appropriate symmetries. In the absence of such a quantum theory, there will be no practical way of computing either Δ_0 or the value of the classical central charge given by Eq. (1) from first principles.

Other approaches to the entropy problem make use of the Chern-Simons theory representation of gravity in $2+1$ dimensions [6, 7]. Two common

[1]S. Fernando, Physics Dept., Northern Kentucky University, Highland Heights, KY 41099. F. Mansouri, Physics Dept., University of Cincinnati, Cincinnati, OH 45221.

The Role of Neutrinos, Strings, Gravity, and Variable Cosmological Constant in Elementary Particle Physics, Edited by Kursunoglu *et al.*, Kluwer Academic/Plenum Publishers, New York, 2001

features among them are that the free Chern-Simons theory is formulated on a manifold with boundary [8, 9, 10] and that they all lead to WZNW theories [11]. More recently, these scenarios have been further refined, improved, and extended [8, 13, 14, 15]. One important feature of a typical conformal field theory obtained in this way is that its central charge varies between the rank and the dimension of the gauge group. The relevant gauge groups for the AdS_3 black hole are two copies of the group $SL(2, R)$, so that in the corresponding Virasoro algebras the central charges vary in the range $1 \leq c \leq 3$. This is in sharp contrast to the values of the central charges given by Eq. (1), which are very large. Thus, it appears that the Chern-Simons approach fails to provide a quantum mechanical basis for the classical results of Brown and Henneaux.

In this work, we describe [16] a way to resolve this apparent contradiction by interpreting the classical asymptotic Virasoro algebra of Brown and Henneaux [5] as an "effective" symmetry characterized by an "effective central charge"in the sense defined by Carlip [3]. Then, rather than naively comparing central charges, we derive the consequences of this effective theory, including its "effective central charge" from yet another approach which makes use of Chern-Simons theory but which is physically very different from the ones mentioned above. For one thing, in our approach the Chern-Simons theory is coupled to a source. Then, since the BTZ black hole is a solution of source-free Einstein's equations [1, 2], it is clear that the manifold M on which the Chern-Simons theory is defined cannot be identified with space-time. Instead, as shown in previous work [17, 18], the classical black hole space-time can be constructed from the information encoded in the manifold M. In particular, this information supplied, mass, angular momentum, and the all important discrete identification group [1, 2] which distinguishes the black hole from anti-de Sitter space.

The coupling to a source also turns out to be essential in arriving at a microscopic description of the black hole entropy. This is because the corresponding conformal field theory on the boundary is specified with two copies of a *twisted* affine Kac-Moody algebra. They lead to two copies of a Virasoro algebra in which the value of the central charge remains the same as the theory without a source, but the eigenvalues of the operator L_0 are shifted and are non-vanishing. Using this result and the non-compactness of $SL(2, R)$ into account, we find that the asymptotic density of states for this microscopic theory agree with that given by Strominger [4] if we identify the Brown-Henneaux values for the central charge [5] with the effective central charge c_{eff} of our theory.

2. CHERN-SIMONS ACTION AND BOUNDARY EFFECTS

For a simple or a semi-simple Lie group, the Chern-Simons action has the form

$$I_{cs} = \frac{k}{4\pi} Tr \int_M A \wedge \left(dA + \frac{2}{3} A \wedge A \right) \tag{2}$$

where Tr stands for trace and

$$A = A_\mu dx^\mu \tag{3}$$

We require the 2+1 dimensional manifold M to have the topology $R \times \Sigma$, with Σ a two- manifold and R representing the time-like coordinate x^0. Moreover, we take the topology of Σ to be trivial in the absence of sources, with the possible exception of a boundary. Then, subject to the constraints

$$F^b[A] = \frac{1}{2}\epsilon^{ij}(\partial_i A_j^b - \partial_j A_i^b + \epsilon^b{}_{cd} A_i^c A_j^d) = 0 \tag{4}$$

the Chern-Simons action for a simple group G will take the form

$$I_{cs} = \frac{k}{2\pi} \int_R dx^0 \int_\Sigma d^2x \left(-\epsilon^{ij} \eta_{ab} A_i^a \partial_0 A_j^b + A_0^a F_a \right) \tag{5}$$

where $i, j = 1, 2$.

We want to explore the properties of the Chern-Simons theory coupled to a source for the group $SL(2,R)_L \times SL(2,R)_R$ on a manifold with boundary. Since the gauge group is semi-simple, the theory breaks up into two parts, one for each $SL(2,R)$, where by $SL(2,R)$ we mean its infinite cover. So, to simplify the presentation, we will study a single $SL(2,R)$. Much of what we discuss in this and the next section hold for any simple Lie group, G. Also, to establish our notation, we consider first the theory in the absence of the source.

The main features of a Chern-Simons theory on a manifold with boundary has been known for sometime [11, 19]. Here, with $M = R \times \Sigma$, we identify the two dimensional manifold Σ with a disk D. Then, the boundary of M will have the topology $R \times S^1$. We parametrize R with τ and S^1 with ϕ. In this parametrization, the Chern-Simons action on a manifold with boundary can be written as

$$S_{cs} = \frac{k}{4\pi} \int_M Tr(AdA + \frac{2}{3} A^3) + \frac{k}{4\pi} \int_{\partial M} A_\phi A_\tau. \tag{6}$$

The surface term vanishes in the gauge in which $A_\tau = 0$ on the boundary. In this action, let $A = \tilde{A} + A_\tau$ and $d = d\tau \frac{\partial}{\partial \tau} + \tilde{d}$. Then, the resulting constraint equations for the field strength take the form

$$\tilde{F} = 0. \tag{7}$$

They can be solved exactly by the ansatz [11, 19]

$$\tilde{A} = -\tilde{d}UU^{-1}, \tag{8}$$

where $U = U(\phi, \tau)$ is an element of the gauge group G. Using this solution, the Chern-Simons action given by Eq. (5) can be rewritten as

$$S_{WZNW} = \frac{k}{12\pi}\int_M Tr(U^{-1}dU)^3 + \frac{k}{4\pi}\int_{\partial M} Tr(U^{-1}\partial_\phi U)(U^{-1}\partial_\tau U)d\phi d\tau. \tag{9}$$

We thus arrive at a WZNW action and can take over many result already available in the literature for this model. As in any WZNW theory, the change in the integrand of this action under an infinitesimal variation δU of U is a derivative. We interpret this to mean that $U = U(\phi, \tau)$, i.e., it is independent of the third (radial) coordinate of the bulk. In other words, the information encoded in the disk depends only on its topology and is invariant under any scaling of the size of the disk.

The above Lagrangian is invariant under the following transformations of the U field [19]:

$$U(\phi, \tau) \to \overline{\Omega}(\phi)U\Omega(\tau), \tag{10}$$

where $\overline{\Omega}(\phi)$ and $\Omega(\tau)$ are any two elements of G. To obtain the conserved currents, let $U \to U + \delta U$. The corresponding variation of the action leads to $S_{WZNW} \to S_{WZNW} + \delta S_{WZNW}$, where

$$\delta S_{WZNW} = \frac{k}{2\pi}\int_{\partial M}(\partial_\tau(U^{-1}\partial_\phi U)\delta U. \tag{11}$$

This implies an infinite number of conserved currents:

$$J_\phi = -kU^{-1}\partial_\phi U = J_\phi^a T_a. \tag{12}$$

Here, T_a are the generators of the algebra g of the group G, and J_ϕ is a function of ϕ only because $\partial_\tau J_\phi = 0$.

If we expand J_ϕ in a Laurent series, we obtain

$$J_\phi = \Sigma J_n z^{-n-1}, \tag{13}$$

where $z = exp(i\phi)$. As usual, J_n satisfy the Kac-Moody algebra

$$[J_n^a, J_m^b] = f_c^{ab} J_{m+n}^c + kng^{ab}\delta_{m+n,0} \tag{14}$$

The corresponding energy momentum tensor for the action S_{WZNW} can be computed using the Sugawara-Sommerfield construction. For example, for the gauge group $SL(2, R)$,

$$T_{\phi\phi} = \frac{1}{(k-2)} g_{ab}{:}J_\phi^a(z)J_\phi^b(z){:} = \frac{1}{k-2}\Sigma{:}J_{n-m}^a J_m^a{:}z^{-n-2} = \Sigma L_n z^{-n-2}, \tag{15}$$

where

$$L_n = \frac{1}{k-2}\Sigma{:}J^a_{n-m}J^a_m{:}. \tag{16}$$

The L_n operators satisfy the following Virasoro algebra:

$$[L_n, L_m] = (n-m)L_{n+m} + \frac{c}{12}n(n^2-1)\delta_{n+m,0}, \tag{17}$$

with c the central charge. For $SL(2,R)$, it is given by $c = \frac{3k}{k-2}$. We note that for large negative values of k, the value of c approaches 3 which is the dimension of the group. We also note that this boundary WZNW theory has one, not the more usual two, Virasoro algebra. It will be shown below that when the Chern-Simons theory is coupled to a source on a manifold with the topology of a disk, the central charge of the Virasoro algebra of the corresponding modified WZNW theory remains the same as that in the source-free theory discussed above.

3. THE COUPLING OF THE SOURCE

Next, we couple a source to the Chern-Simons action on the manifold M with disk topology, which, as in the previous section, has the boundary $R \times S^1$. In general, we take the source to be a unitary representation of the group G. To be more specific, let us consider a source action given by [11, 19]

$$S_{source} = \int d\tau Tr[\lambda\omega(\tau)^{-1}(\partial_\tau + A_\tau)\omega(\tau)]. \tag{18}$$

Here $\lambda = \lambda^i H_i$ where H_i are elements of the Cartan subalgebra H of G. We will take λ^a to be appropriate weights. The quantity $\omega(\tau)$ is an arbitrary element of G. The above action is invariant under the transformation $\omega(\tau) \to \omega(\tau)h(\tau)$, where $h(\tau)$ commutes with λ.

Now the total action on M is,

$$S_{total} = \frac{k}{4\pi}\int Tr(AdA+\frac{2}{3}A^3)+\frac{k}{4\pi}\int A_\tau A_\phi+\int d\tau Tr[\lambda\omega(\tau)^{-1}(\partial_\tau+A_\tau)\omega(\tau)] \tag{19}$$

The new constraint equation takes the form,

$$\frac{k}{2\pi}\tilde{F}(x) + \omega(\tau)\lambda\omega^{-1}(\tau)\delta^2(x-x_p) = 0, \tag{20}$$

where x_p specifies the location of the source, heretofore taken to be at $x_p = 0$. The solution to the above equation is given by

$$\tilde{A} = -\tilde{d}\tilde{U}\tilde{U}^{-1}, \tag{21}$$

where [19]

$$\tilde{U} = U exp(\frac{1}{k}\omega(\tau)\lambda\omega^{-1}(\tau)\phi) \tag{22}$$

The new effective action on the boundary ∂M is then

$$S_{total} = S_{WZNW} + \frac{1}{2\pi}\int_{\partial M} Tr(\lambda U^{-1}\partial_\tau U). \tag{23}$$

This Lagrangian is also invariant under the following transformation:

$$U(\phi,\tau) \rightarrow \overline{\Omega}(\phi)U\Omega(\tau) \tag{24}$$

where $\Omega(\tau)$ commutes with λ. Varying the action under the above symmetry transformation, we get

$$\delta S_{total} = \delta S_{WZNW} + \delta S_{source}, \tag{25}$$

where

$$\delta S_{source} = \frac{1}{2\pi}\int Tr\left(-U^{-1}\delta U[U^{-1}\partial_\tau U, \lambda]\right). \tag{26}$$

Hence, the requirement that $\delta S_{total} = 0$ will give rise to the conservation equation [21]

$$\partial_\tau\left(-kU^{-1}\partial_\phi U\right) + [U^{-1}\partial_\tau U, \lambda] = 0. \tag{27}$$

The first term in this expression has the same structure as the current J_ϕ of the source free theory. Hence, requiring that $U(\phi,\tau) = U(\phi+\tau)$, we can write the new current $\hat{J}_\phi$ in terms of the current in the absence of the source as

$$\hat{J}_\phi = e^{\frac{\lambda}{k}(\phi+\tau)} J_\phi e^{-\frac{\lambda}{k}(\phi+\tau)} \tag{28}$$

It is easy to check that

$$\partial_\tau \hat{J}_\phi = 0 \tag{29}$$

With the new currents at our disposal, the next step is to see how this modification affects the properties of the corresponding conformal field theory. In this respect, we note from Eq. (28) that our new currents $\hat{J}_\phi$ are related to the currents J_ϕ in the absence of the source by a conjugation with respect to the elements of the Cartan subalgebra H of the group G. This kind of conjugation has been noted in the study of Kac-Moody algebras [20, 22, 23]: The algebra satisfied by the new currents fall in the category of *twisted* affine Kac-Moody algebras. So, to understand how the coupling to a source modifies the structure of the source-free conformal field theory, we follow the analysis of reference [20] and express the Lie algebra of the group G of rank r in the Cartan-Weyl basis. Let H^i be the elements of the Cartan subalgebra and denote the remaining generators by E^α. Then, with label $a = (i, \alpha)$,

$$[H^i, H^j] = 0\,; \qquad [H^i, E^\alpha] = \alpha^i E^\alpha \tag{30}$$

$$[E^\alpha, E^\alpha] = \left\{ \begin{array}{c} \epsilon(\alpha,\beta)E^{\alpha+\beta} \quad \text{if } \alpha+\beta \text{ is a root} \\ 2\alpha^{-2}(\alpha.H) \quad \text{if } \alpha = -\beta \\ 0 \quad \text{otherwise} \end{array} \right\} \tag{31}$$

In this expression, $1 \leq i, j \leq r$, and α, β are roots. Now we can rewrite the affine Kac-Moody algebra g of the source free theory of the last section in this basis as follows:

$$[H^i_m, H^j_n] = km\delta^{ij}\delta_{m,-n}\,; \qquad [H^i_m, E^\alpha_n] = \alpha^i E^\alpha_{m+n} \tag{32}$$

$$[E^\alpha_m, E^\alpha_n] = \left\{ \begin{array}{c} \epsilon(\alpha,\beta)E^{\alpha+\beta}_{m+n} \quad \text{if } \alpha+\beta \text{ is a root} \\ 2\alpha^{-2}(\alpha.H_{m+n} + km\delta_{m,-n}) \quad \text{if } \alpha = -\beta \\ 0 \quad \text{otherwise} \end{array} \right\} \tag{33}$$

We also note from the last section that in the absence of the source the element L_0 of the Virasoro algebra of the source-free theory and the currents J^a_n have the following commutation relations:

$$[L_0, J^a_n] = -nJ^a_n. \tag{34}$$

It can be seen from Eq. (28) that the new currents can be viewed as an inner automorphism of the algebra g in the form $\zeta(J_m) = \gamma J_m \gamma^{-1}$, where, suppressing the dependence on k, ϕ, τ,

$$\gamma = e^{i\chi \cdot H} \tag{35}$$

As a result of this inner automorphism on elements of the algebra g, we obtain a modified algebra $\hat{g}$ the elements of which in the Cartan-Weyl basis are given by [20]

$$\zeta(H^i) = H^i\,; \qquad \zeta(E^\alpha) = e^{i\chi\cdot\alpha}E^\alpha \tag{36}$$

If the map ζ is endowed with the property that $\zeta^N = 1$, then we must have $N\chi.\alpha = 2n\pi$, where n is an integer for all roots $\alpha \,\epsilon\, g$:

$$e^{i\chi.\alpha} = e^{\frac{2\pi i n}{N}}. \tag{37}$$

where n is a positive integer $\leq N-1$. As far as the currents obtained from the Chern-Simons theory coupled to a source are concerned, all possible values of N are allowed. As we will see in the next section, the choice of a particular value of N requires additional physical input. We also note that the automorphism ζ divides a suitable combination of the generators of $\hat{g}$ into eigenspaces $\hat{g}_{(m)}$.

Thus, the basis of $\hat{g}$ consists of the elements H^i_m and E^α_n where $m \epsilon Z$ and $n \epsilon (Z + \frac{\chi.\alpha}{2\pi})$. These operators satisfy a Kac-Moody algebra which has formally the same structure as that of g but with rearranged (fractional) values of the suffices. Hence the algebra $\hat{g}$ can be viewed as the "twisted" version of the algebra g.

Since the automorphism which relates the two algebras is of inner variety [20], we must look for features, if any, that distinguish the algebra $\hat{g}$ from its untwisted version g. These features depend on the extent to which we can undo the twisting. To this end, we introduce a new basis for $\hat{g}$

$$\hat{E}^\alpha_n = E^\alpha_{n+\frac{\chi.\alpha}{2\pi}}; \qquad \hat{H}^i_n = H^i_n + \frac{k}{2\pi}\chi^i\delta_{n,0}, \tag{38}$$

The new operators, $\hat{E}^\alpha_n$ and $\hat{H}^i_n$ satisfy the same commutation relations as the elements of the untwisted affine Kac-Moody algebra g. The corresponding conformal field theories are not identical, however. This can be seen most easily if we express the Virasoro generators $\hat{L}_n$ of the twisted theory in terms of untwisted generators:

$$\hat{L}_m = L_m - \frac{1}{2\pi}\chi^i H^i_n + \frac{k}{4\pi^2}\chi^i\chi^i\delta_{n,0}. \tag{39}$$

In particular, we get for $\hat{L}_0$,

$$\hat{L}_0 = L_0 - \frac{1}{2\pi}\chi^i H^i_0 + \frac{k}{4\pi^2}\chi^i\chi^i. \tag{40}$$

We will discuss the eigenvalues of these operators in the next section.

4. GROUND STATE EIGENVALUES OF L_0

In this section, we will briefly summarize some of the relevant information about the lowest (highest) weight representations of affine Kac-Moody algebras, which we need to construct the lowest weight representations of the twisted algebra obtained in the previous section. For compact Lie groups, the lowest weight unitary representations are also highest weight representations. However, for non-compact Lie groups such as $SL(2,R)$ there are discrete unitary representations which have either a lowest weight or a highest weight but not both. For the application to the black hole entropy, it is the former discrete series which turn out to be relevant. Moreover, we will see that some of the well known results which have been obtained for simple compact Lie groups will be altered when the gauge group is non-compact. This will prepare us for the application of this formalism to the entropy of the AdS_3 discussed in the next section. For a given Lie algebra, let the lowest weight state of an irreducible representation be $|\mu_0>$. Here, μ stands for the set μ^i in which the index i runs over the elements of the Cartan subalgebra. Then, in the Cartan-Weyl basis we have, by construction,

$$E^{\alpha}|\mu_0>=0, \qquad \alpha>0$$

Otherwise $\mu_0+\alpha$ will be higher than μ_0. Also

$$H^i|\mu_0>=\mu_0^i|\mu_0>$$

It follows that the other states in this representation can be constructed by applying the operator $E^{-\alpha}$ to the state $|\mu_0>$.

Since an affine Kac-Moody algebra has a simple Lie algebra as a subalgebra, then an irreducible representation of an affine Kac- Moody algebra also furnishes an irreducible representation of the corresponding Lie subalgebra. The states of this representation of the Lie sub-algebra are the ground states (vacuum states) upon which the lowest (highest) weight representation of the larger algebra can be built. These vacuum states are specified by the condition

$$J_n^a|\Psi>=0; \qquad n>0. \tag{41}$$

In other words, as we have seen above, an irreducible representation of an affine Kac-Moody algebra is characterized by the lowest weight μ_0 of its Lie subalgebra as well as the central term k. Thus, returning to the Cartan-Weyl basis, to extend this representation of the Lie subalgebra to a lowest weight representation of the corresponding affine Kac-Moody algebra, let, the lowest weight state of the latter be given by $|\Delta_0,\mu_0>$. Then,

$$L_0|\Delta_0,\mu_0>=\Delta_0|\Delta_0,\mu_0>; \qquad H_0^i|\Delta_0,\mu_0>=\mu_0^i|\Delta_0,\mu_0>$$

Now the operators E_n^α and L_n act on the lowest weight state as follows:

$$E_n^\alpha|\Delta_0,\mu>=0; \qquad n>0 \; or \; n=0 \; and \; \alpha>0.$$

$$L_n|\Delta,\mu_0>=0; \qquad n>0.$$

Hence, the other states of this representation can be obtained by the application of the operators E_{-n}^α, L_{-n}, H_{-n}^i, $n>0$, to the lowest weight state:

$$L_{-n}|\Delta_0,\mu_0>\sim|\Delta,\mu_0>; \qquad \Delta=\Delta_0+n.$$

$$E_{-n}^\alpha|\Delta_0,\mu_0>\sim|\Delta,\mu>; \quad \Delta=\Delta_0+n, \quad \mu=\mu_0+\alpha.$$

$$H_{-n}^i|\Delta,\mu>\sim|\Delta+n,\mu>.$$

In particular, let us obtain the eigenvalues of H_0^i and L_0 on the following states:

$$H_0^i(E_0^\alpha|\Delta,\mu>) = (\mu+\alpha)|\Delta,\mu+\alpha> .$$
$$L_0(L_{-n}|\Delta,\mu>) = (\Delta+n)|\Delta+n,\mu>$$

Let us now consider the lowest (highest) weight representations of the deformed algebra. In this case, the states are eigenstates of operators $\hat{L}_0$ and $\hat{H}_0$. In analogy with the representations of the untwisted algebra, we denote the states by $|\hat{\Delta},\mu>$, where $\hat{\Delta}$ is the eigenvalue of $\hat{L}_0$. Since this operator can be expressed in terms of L_0 as in Eq. (40), the lowest weight representations of the deformed algebra have the same structure as those of the untwisted algebra. Thus, we can write

$$\hat{L}_0|\hat{\Delta},\mu>= (L_0 - \frac{1}{2\pi}\chi^i H_0^i + \frac{k}{4\pi^2}\chi^i\chi^i)|\hat{\Delta},\mu> . \tag{42}$$

Or

$$\hat{L}_0|\hat{\Delta},\mu>= \hat{\Delta}|\hat{\Delta},\mu>, \tag{43}$$

where

$$\hat{\Delta} = \Delta - \frac{1}{2\pi}\chi^i\mu^i + \frac{k}{4\pi^2}\chi^i\chi^i. \tag{44}$$

In particular, for the lowest (highest) weight state we get

$$\hat{\Delta}_0 = \Delta_0 - \frac{1}{2\pi}\chi^i\mu_0^i + \frac{1}{4\pi^2}k\chi^2. \tag{45}$$

It is clear that the eigenvalues $\hat{\Delta}$ of the operator $\hat{\Delta}$ are shifted relative to the eigenvalues Δ of L_0. But, as can be verified directly, the value of the central charge c remains unchanged [20, 22, 23].

With minor exceptions, most of the derivation of our twisted Kac- Moody algebras from the Chern-Simons theory applies to any gauge group. But the relation between the irreducible representations of an affine Kac-Moody algebra and its Lie subalgebra imposes restrictions on the value of the central term k. For example, for $SU(2)$, the value of k is restricted to the non-negative values [20]. But for discrete unitary representations of $SL(2,R)$ with a lowest weight, the quantity k is restricted to [24]

$$k < -1. \tag{46}$$

It follows that in this case large negative values of k are allowed. We will take advantage of this feature in the application of this formalism to entropy of the AdS_3 black hole in section 5.

5. ENTROPY OF AdS_3 BLACK HOLE

As pointed out in the introduction, in the derivation of the entropy of the AdS_3 black hole by Strominger [4], use was made of the expression for the central charges of the two asymptotic Virasoro algebras obtained by Brown and Henneaux [5] using classical (non-quantum) arguments. They are given by

$$c_L = c_R = \frac{3l}{2G}. \tag{47}$$

where l is the radius of curvature of the AdS_3 space, and G is Newton's constant. The presence of such a symmetry indicates that there is a conformal field theory at the asymptotic boundary. It was shown by Strominger that the BTZ solution satisfies the Brown-Henneaux boundary conditions so that it possessed an asymptotic symmetry of this type. So, he identified the degrees of freedom of the black hole in the bulk with those of the conformal field theory at the infinite boundary. Then, using Cardy's formula [25] for the asymptotic density of states, he showed that for $l \gg G$ the entropy of this conformal field theory is given by

$$S = \frac{2\pi r_+}{4G}, \tag{48}$$

in agreement with Bekenstein-Hawking formula. Here, the quantity r_+ is the outer horizon radius of the black hole. An important assumption in this derivation was that the ground state eigenvalue Δ_0 of the operator L_0 vanishes.

The formula by Cardy [25] for the asymptotic density of states, leading to the above expression for entropy is given by

$$\rho(\Delta) \approx exp\{2\pi\sqrt{\frac{c\Delta}{6}}\}, \tag{49}$$

where $\rho(\Delta)$ is the number of states for which the eigenvalue of L_0 is Δ. The result holds when Δ is large and the lowest eigenvalue Δ_0 vanishes. From the analysis of the previous section, it is clear that in the conformal field theory arising from a Chern-Simons theory coupled to a source the eigenvalue $\hat{\Delta}_0$ does not vanish, so that the above Cardy formula must be appropriately modified. In such a case, the asymptotic density of states for large Δ is given by [3]:

$$\rho(\Delta) \approx exp\{2\pi\sqrt{\frac{(c-24\Delta_0)\Delta}{6}}\}\rho(\Delta_0) = exp\{2\pi\sqrt{\frac{c_{eff}\Delta}{6}}\}\rho(\Delta_0). \tag{50}$$

Thus, it is the latter formula which must be used in the application of our formalism to the black hole entropy. It is important to note that the expression for the asymptotic density of states given by Eq. (48) rests on the existence of a consistent conformal field theory with a well defined partition function. For Kac-Moody algebras based on compact Lie groups this can be established rigorously. But for Kac-Moody algebras based on non-compact groups such as $SL(2,R)$, no general proof exists. So, all the conformal field theories based on $SL(2,R)$, which have been made use of in connection with the AdS$_3$ black hole, including the present work, share this common weakness.

We want to show that the results obtained from such a microscopic analysis are in agreement with those given by Strominger [4]. In so doing, we will rely heavily on our previous results which dealt with the understanding of the macroscopic features of the BTZ black hole [17, 18]. We recall from these references that the unitary representations of $SL(2,R)$ which are relevant to the description of the macroscopic features of the black hole are the infinite dimensional discrete series which are bounded from below. These irreducible representations are characterized by a label F which can be identified with the lowest eigenvalue of the $SL(2,R)$ generator which is being diagonalized. In the literature of the $SL(2,R)$ Kac-Moody algebra [24], this (lowest weight) label, which is convenient for the description of this series, is often referred to as $-j$. Thus, the Casimir eigenvalues in the two notations are related according to

$$j(j+1) = F(F-1). \tag{51}$$

On the other hand, in the description of the black holes in terms of a Chern Simons theory with gauge group $SL(2,R)_L \times SL(2,R)_R$, the mass and the angular momentum of the black hole are related to the Casimir invariants of $j^2_\pm$ of this gauge group as follows [17, 18]:

$$j^2_\pm = \frac{1}{2}(lM \pm J). \tag{52}$$

So, for positive mass black holes we must have

$$F_\pm = \frac{1}{2}\left[1 + \sqrt{1 + 2(lM \pm J)}\right]. \tag{53}$$

In particular [17, 18], for the ground state $F_\pm \approx 1$, we must have $2(lM_0 + J_0) \ll 1$. We will identify this quantity with the ratio of the two scales of the theory:

$$2(lM_0 + J_0) \approx \frac{G}{l}. \tag{54}$$

Next, consider the determination of the Chern-Simons couplings $k_\pm$. These were referred to as $a_\pm^{-1}$ in references [17, 18]. Since the gauge group $SL(2,R)_L \times SL(2,R)_R$ is semi-simple, the couplings k_+ and k_- are independent. Also, since the two $SL(2,R)$ groups play a parallel role in our approach, we will focus on determining one of them, say, k_+. The other one can be obtained in a similar way. To this end, we recall from references [17, 18] that in our approach, the manifold M on which the Chern-Simons theory is defined is not space-time. This means that from the data encoded in M we must be able to obtain all the features of the classical black hole space-time. One of these is the discrete identification group of the black hole [1, 2]. This was obtained within our framework by considering the holonomies around the source in M. In this respect, we note that the Cartan sub-algebra of $SL(2,R)$ is one dimensional, and the corresponding weight is F as discussed above. Then, using non-abelian Stokes theorem [26] and equation (20), the holonomies can be evaluated. To get the correct discrete identification group, this implies that [17, 18]

$$\frac{2F_+}{k_+} = \pm\frac{1}{l}(r_+ + r_-) = \pm 2\sqrt{2\frac{G}{l}(lM_0 \pm J_0)}, \tag{55}$$

where r_+ and r_- are, respectively, the outer and the inner horizon radii of the black hole. A similar analysis can be carried out for k_-. Using the values of F and $2(lM_0 + J_0)$ for the ground state given above, we find for both couplings

$$k_+ = k_- = \pm\frac{l}{G}. \tag{56}$$

The sign of $k_\pm$ is not fixed by holonomy considerations alone. But from the discussion of the last section it is clear that we must choose the negative sign for both since $SL(2,R)$ is non-compact.

Our determination of $k_\pm$ is to be compared with other approaches in one or both k's are taken to be positive. In one approach [14], e.g., the manifold M was taken to be space-time, and the free Chern-Simons action led to the classical black hole solution in M. In that case, the signs of k_+ and k_- are opposite each other, so that one of them would have to be positive. This appears to be inconsistent with what we know about $SL(2,R)$ Kac- Moody algebras.

Next, consider the ground state eigenvalue $\hat{\Delta}_0$ for one of the two $SL(2,R)$ Virasoro algebras. Since the Cartan subalgebra is one dimensional, the sums in Eq. (43) consist of one term each, and, from Eq. (37), the quantity χ is given by

$$\chi = \frac{2\pi n}{N\alpha}. \tag{57}$$

The quantity μ_0 in Eq. (43) is clearly the weight of the ground state, i.e., the weight $F \approx 1$ described above. We also note that the root α is the weight of the adjoint representation of $SL(2,R)$, so that $\alpha = 1$. Then, Eq. (44) specialized to the case at hand will take the form

$$\hat{\Delta}_0 = \Delta_0 - \frac{n}{N} + k(\frac{n}{N})^2. \tag{58}$$

Also substituting for k from Eq. (54), we get

$$\hat{\Delta}_0 = \Delta_0 - \frac{n}{N} - \frac{l}{G}(\frac{n}{N})^2. \tag{59}$$

In this expression, the quantity Δ_0 is the ground state eigenvalue of L_0, and its value is not known but is often taken to be zero without á priori justification. For our purposes, it is only necessary that it be small compared to the last term.

Let us now compute the effective central charge c_{eff}, defined via Eq. (48), for our theory. It is given by

$$c_{eff} = c - 24\hat{\Delta}_0 = c - 24\Delta_0 - 24\frac{n}{N} + 24\frac{l}{G}(\frac{n}{N})^2. \qquad (60)$$

In this expression, $\frac{l}{G} \gg 1$ whereas $1 \leq c \leq 3$, and the $\frac{n}{N}$ is a fraction less than one. Assuming that Δ_0 is also relatively small, we get

$$c_{eff} \approx 24\frac{l}{G}(\frac{n}{N})^2. \qquad (61)$$

As for the ratio (n/N), we pointed out that its determination more requires physical input. The key point to keep in mind is that its value does not affect the qualitative features of the answer to the AdS_3 black hole entropy. The situation here is somewhat similar to the fixing of the ground state eigenvalue Δ_0 by Strominger [4]. In that work, the conformal field theory alone does not limit the continuous infinity of the possible values of Δ_0, and the requirement that it vanish has only *a posteriori* justification. Compared to this, we can make a stronger case for a particular choice of the ratio (n/N). This is because our starting point, i.e., the Chern-Simons theory coupled to a source leads, on the one hand, to the twisted Kac-Moody algebras described above and, on the other hand, to [17, 18] the classical BTZ solution [1, 2] for which the central charge for the asymptotic Virasoro algebra was given by Brown and Henneaux [5]. It is then necessary that the classical and quantum results which follow from the same Chern-Simons theory be consistent with each other. So, it is reasonable to require that the asymptotic density of states of the above quantum theory, as computed from Cardy- Carlip [3], agree, up to logarithmic terms, with the asymptotic density of states obtained by Strominger [4] using the traditional Cardy formula and the classical Brown- Henneaux value of central charge (45). The simplest way to satisfy this requirement is to set our effective central charge c_{eff} equal to the Brown-Henneaux central charge. This fixes $\frac{n}{N} = \frac{1}{4}$. That this requirement makes sense can also be seen by noting that for the Virasoro algebra obtained by Brown and Henneaux, the underlying Kac-Moody algebra is not known, so that there is no direct way of calculating its central charge c or its ground state eigenvalue from a fundamental Kac-Moody algebra. Then, taking the classical theory to be an "effective theory", we see that we can compute its "effective central charge" and its "effective ground state eigenvalue" from the above quantum theory for a particular choice of inner automorphism.

It follows that for this conformal field theory, the expression for the asymptotic density of states given by Eq. (48) reduces to

$$\rho(\hat{\Delta}) \approx exp\{2\pi\sqrt{\frac{c_{eff}\,\hat{\Delta}}{6}}\}\rho(\hat{\Delta}_0) = exp\{2\pi\sqrt{\frac{l\,\hat{\Delta}}{4G}}\}\,\rho(\hat{\Delta}_0). \qquad (62)$$

Modulo a logarithmic correction, this expression is identical with that used by Strominger [4]. This resolves the longstanding controversy in the traditional method of comparing the central charge of a conformal field theory obtained from the Chern-Simons approach with the (effective) classical results of Brown and Henneaux. One of the crucial features of our work which led to this resolution was the recognition that for $SL(2,R)$ Kac-Moody algebras, large negative values of k are allowed.

So far, we have dealt with the density of states for one $SL(2,R)$, say, $SL(2,R)_L$. This will contribute an amount S_L to the black hole entropy. Clearly, we can repeat this computation for the density of states of $SL(2,R)_R$. Then, to the extent that the logarithmic corrections can be neglected, the black hole entropy $S = S_L + S_R$ is in agreement with that given by Strominger [4]. More recently, the logarithmic contributions to the black hole entropy have been discussed in the literature [27, 28]. One would then have to assess the relative size of our logarithmic term compared to those given in these works.

6. ACKNOWLEDGEMENTS

This work was supported, in part by the Department of Energy under the contract number DOE-FG02-84ER40153. We would like to thank Philip Argyres and Alex Lewis for reading the manuscript and suggesting improvements.

7. REFERENCES

opt

1. M. Bañados, C. Teitelboim and J. Zanelli, Phys. Rev. Lett. **69** (1992) 1849
2. M. Bañados, M. Henneaux, C. Teitelboim and J. Zanelli, Phys. Rev. D **48** (1993) 1506.
3. S. Carlip, [hep-th/9806026].
4. A. Strominger, [hep-th/9712252], J. High Energy Physics **02** (1998) 009.
5. J. D. Brown and M. Henneaux, Comm. Math. Phys. **104** (1986) 207.

6. A. Achucarro, P. Townsend, Phys. Lett.B **180** (1986) 35.
7. E. Witten, Nucl. Phys. **B311** (1988) 46; **B323** (1989) 113.
8. S. Carlip, Phys. Rev. **D51** (1995) 632
9. A.P. Balachandran, L. Chandar, A. Momen, Nuc. Phys. **B461** (1996) 581; eprint archive [gr-qc/9506006].
10. M. Bañados, Phys. Rev **D52** (1995) 5815.
11. E. Witten, Comm. Math. Phys. **121** (1989) 351.
12. S. Carlip, Phys. Rev. **D55** (1997) 878.
13. M. Bañados, T. Brotz, M.E. Ortiz, [hep-th/9802076].
14. M. Bañados, [hep-th/9901148].
15. M. Bañados, [hep-th/9903178].
16. S. Fernando, F. Mansouri, [hep-th/0010153] and Cincinnati preprint UCTP107.00, to appear in the Proceedings of Johns Hopkins Workshop, Budapest, Hungary, August 19-22, 2000.
17. S. Fernando, F. Mansouri, Int. Jour. Mod. Phys. **A 14** (1999) 505; [hep-th/9804147].
18. S. Fernando, F. Mansouri, Proceedings of XXII Johns Hopkins Workshop on *Novelties in String Theory*, ed. L. Brink and R. Marnelius, Goteborg, Sweden, August 20-22, 1998; [hep-th/9901163].
19. S. Elitzur, G. Moore, A. Schwimmer and N. Seiberg, Nucl. Phys **B326** (1989) 108.
20. P. Goddard and D. Olive, Int. Jour. Mod. Phys. **A1** (1986) 303.
21. F. Ardalan, S. Fernando, F. Mansouri, Cincinnati preprint UCTP102.00, to appear in the Proceedings of the Sixth International Wigner Symposium, Istanbul, Turkey, August, 1999.
22. N. Sakai and P. Suranyi, Nucl. Phys. **B362**, 655 (1989).
23. J.K. Freericks and M.B. Halpern, Ann. Phys. (N.Y.) **188**, 258 (1988).
24. J. Balog, L. O'Raifeartaigh, P. Forgác, A. Wipf, Nuc. Phys. **B325** (1989) 225; S. Hwang, Nuc. Phys. **B354** (1991) 100; P.M.S. Petropoulos, Phys. Lett. **B 236** (1990) 151.
25. J.A. Cardy, Nucl. Phys. **B270** (1986) 186.
26. R.L. Karp, F. Mansouri, J.S. Rno, [hep-th/9903221]; Jour. Math. Phys. **40** (1999) 6033, [hep-th/9910173]
27. S. Carlip, [gr-qc/0005017]
28. J. Jing and M.L. Yan, [gr-qc/0005105]

CP–Violation in B–Decays Status and Prospects of BA$\overline{B}$AR [1]

Carsten Hast for the BA$\overline{B}$AR Collaboration [2]

1. INTRODUCTION

In this paper we report on our preliminary measurements of the CP-violating parameter sin2β which we performed during the summer of 2000. In last years Orbis Scientiae Giampiero Mancinelli gave an overview of the BA$\overline{B}$AR experiment and the PEP II accelerator at the Stanford Linear Accelerator Center, SLAC in California. Here only updates on the machine and detector performances are presented.

Several experiments are probing the standard model in these days by trying to measure CP-violation in the B-system. The most easy decay mode to analyze is $B^0 \rightarrow J/\psi K^0_s$. Here theretical uncertainties in calculating corrections to this decay mode are very small. The relatively large branching ratio of $\mathcal{O}(3 \times 10^4)$ and the experimental cleaness of this channel make it especially attractive. The decay channels wich were used in our analysis are $B^0 \rightarrow J/\psi K^0_s$ and the decay $B^0 \rightarrow \psi(2S) K^0_s$ with the K^0_s either decaying into $\pi^+\pi^-$ or $\pi^0\pi^0$.

The CP-violating asymmetry in $b \rightarrow c\bar{c}s$ decays of the B^0 meson is caused by the interference between mixed and unmixed decay amplitudes.

A state initially prepared as a B^0 ($\overline{B}^0$) can directly decay to $J/\psi K^0_s$ or can oscillate into a $\overline{B}^0$ (B^0) and then decay to $J/\psi K^0_s$. With very little theoretical uncertainty, the phase difference between those two amplitudes is equal to the angle β of the CKM Unitarity Triangle.

In e^+e^- storage rings operating at the $\Upsilon(4S)$ resonance the produced $B^0\overline{B}^0$ system evolves in a coherent P-wave state until the B^0 mesons decay. If one of the B mesons can be ascertained to decay to a state of known flavor at a certain time (the so called tagging prozess), the other B is at that time known to be of the opposite flavor. By measuring the proper time interval between the to B-meson decay vertizes, it is possible to determine the time evolution of the initially pure B^0 or $\overline{B}^0$ state.

The final measured quantity is the time dependent asymmetry ($\mathcal{A}_{CP}(\Delta t)$) between the number of B^0 or $\overline{B}^0$ mesons, decaying to the same final state

$$\mathcal{A}_{CP}(\Delta t) \sim \mathcal{D} \sin 2\beta \times \sin \Delta m_d \, \Delta t$$

[1] Work supported by Department of Energy contract DE–AC03–76SF00515.

[2] Carsten Hast, LAL Orsay, BP 34, F-91898 Orsay, CEDEX, FRANCE.
work address: Stanford Linear Accelerator Center, Stanford University, Stanford, California 94309

The Role of Neutrinos, Strings, Gravity, and Variable Cosmological Constant in Elementary Particle Physics, Edited by Kursunoglu *et al.*, Kluwer Academic/Plenum Publishers, New York, 2001

The dilution factor $\mathcal{D}$ is given by $\mathcal{D} = 1 - 2\omega$, where ω is the probability that the flavour of the tagging B was identified incorrectly. Δm_d describes the mixing between the B-mesons.

2. MACHINE AND EXPERIMENT

The asymetric e^+e^- accelerator PEP II at SLAC had its first collisions on May 26th 1999. Since then the beam currents and luminosity could be increased in a unprecedented way. The design luminosity of $3.0 \times 10^{33} cm^{-2} s^{-1}$ was reached and exceeded by 5 % at the end of the first running period in October 2000. Figure 1.1 shows the dayly recorded and figure 1.2 the integrated luminosity during this time. BA$\overline{B}$AR could record an integrated luminosity in excess of 24 fb^{-1} comprising the

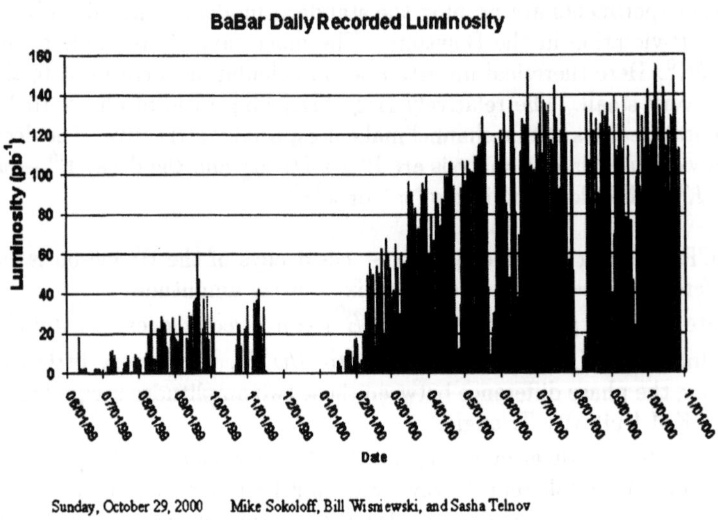

Figure 1.1: Daily recorded luminosity during the 1999/2000 running period

largest B-meson sample recorded by an experiment so far.
During the last year many improvements in the reconstruction of the data were achieved. Particle identification with DIRC (Detection of Internally Reflecting Cherencov Light) could be improved to a $3.3\,\sigma$ separation between pions and kaons up to momenta of 3.3 GeV. Figures 1.3 and 1.4 show for the decay of neutral D-mesons into a charged kaon and pion the Cherencov angle θ_C after demanding a well identified pion (1.3) or kaon (1.4). The excellent separation is clearly visible. The best achieved resolution so far is $\theta_C = 2.5$ for muons from the process $e^+e^- \rightarrow \mu^+\mu^-(\gamma)$.

The improved reconstruction of photons in the CsI electro-magnetic calorimeter leads to a π^0 mass resolution of 6.9 MeV as can be seen in figure 1.5.

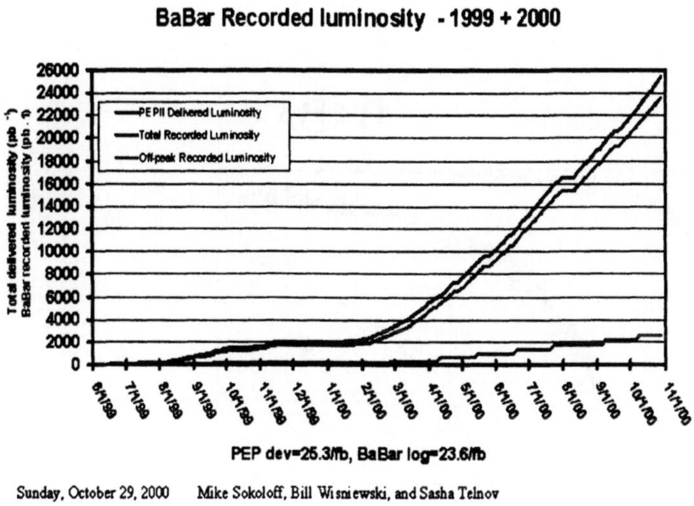

Figure 1.2: Integrated luminosity during the 1999/2000 running period

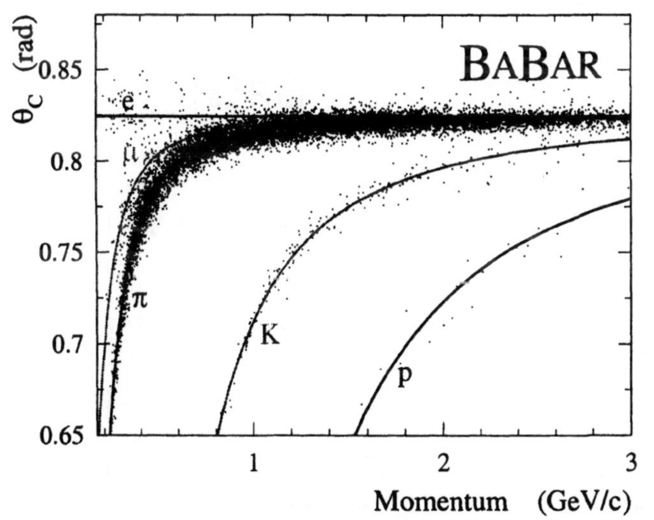

Figure 1.3: Cherencov angle for $D^0 \to K^+\pi^-$ decays after demanding a well identified pion

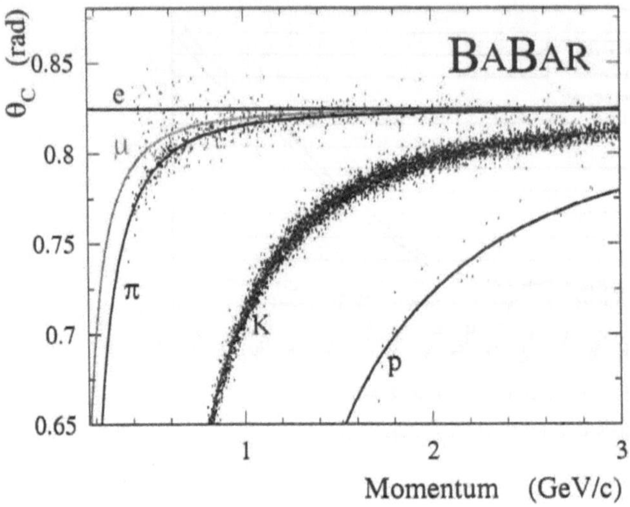

Figure 1.4: Cherencov angle for $D^0 \rightarrow K^+\pi^-$ decays after demanding a well identified kaon

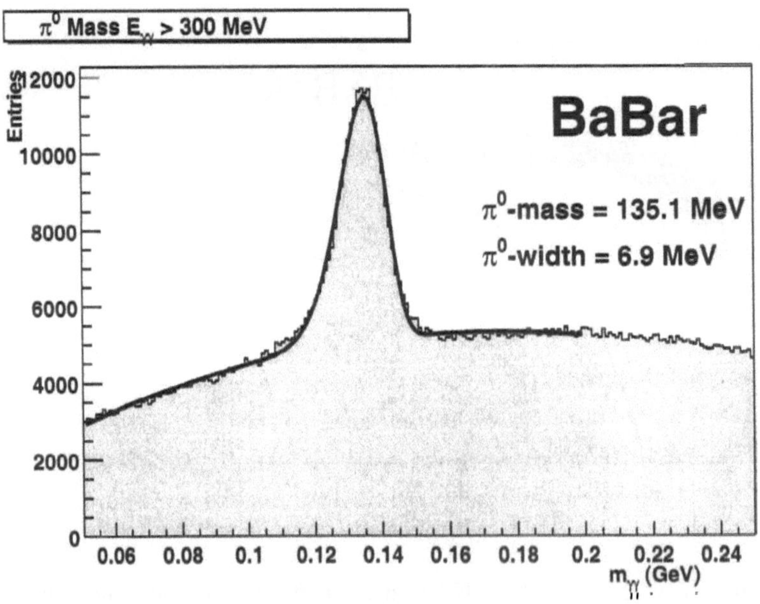

Figure 1.5: π^0 resolution for $E_{\gamma\gamma} > 300\,\text{MeV}$ and $E_\gamma > 100\,\text{Mev}$

3. OVERVIEW OF THE ANALYSIS

The measurement of the CP-violating asymmetry has five main components:

- Selection of the signal $B^0/\overline{B}^0 \rightarrow J/\psi K_s^0$ and $B^0/\overline{B}^0 \rightarrow \psi' K_s^0$ events.
- Measurement of the distance Δz between the two B^0 decay vertices along the $\Upsilon(4S)$ boost axis.
- Determination of the flavor of the B_{tag}.
- Measurement of the dilution factor $\mathcal{D}$ from the data.
- Extraction of the amplitude of the *CP* asymmetry and the value of sin2β with an unbinned maximum likelihood fit.

Whenever possible, we determine time and mass resolutions, efficiencies and mistag fractions from the data.

For this analysis we use a sample of 9.8 fb^{-1} of data recorded by the BABAR detector between October 1999 and the end of June 2000, of which 0.8 fb^{-1} were recorded 40 MeV below the $\Upsilon(4S)$ resonance (off-resonance data).

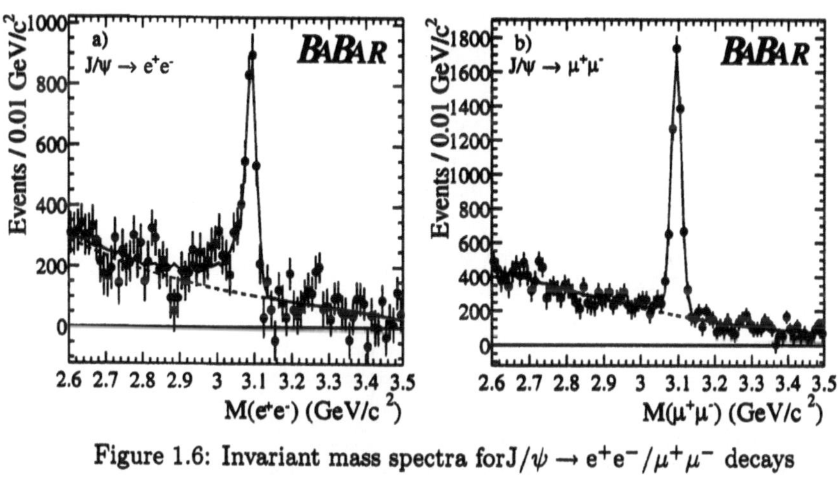

Figure 1.6: Invariant mass spectra forJ$/\psi \rightarrow e^+e^-/\mu^+\mu^-$ decays

4. K_S^0 AND J/ψ RECONSTRUCTION

Using a tight selection we obtain a quite pure sample of K_S candidates. They are identified as pairs of oppositely signed charged tracks from a common vertex and with mass consistent with the K_S^0 mass. Furthermore we require that the K_S^0 candidates point back to the interaction point and a minimum transverse momentum in the decay (to suppress contributions from combinatoric and Λ decays). We reconstruct J/ψ candidates in their decays modes into e^+e^- and $\mu^+\mu^-$, which have a 12% combined branching ratio. Electrons are identified using the pattern of energy deposited in the calorimeter and by requiring the ratio, E/p, of the energy measured in the calorimeter to the measured track momentum to be close to one. This channel presents a significant tail in the invariant mass distribution due to bremsstrahlung. Muons are identified using their minimum-ionizing signature in the calorimeter. Figure 1.6 shows the mass spectra for electrons and muons, respectively.

5. MEASUREMENT OF $\sin 2\beta$

Combining the selected J/ψ and K_S^0 mesons in the various decay channels involving e^+e^- or $\mu^+\mu^-$ and $\pi^+\pi^-$, or $\pi^0\pi^0$ pairs leads to the following results: The following 3 figures 1.7, 1.8, and 1.9 show the reconstructed B-meson mass for events who's reconstructed B-meson energy lies close within the beam energy in the center of mass system.

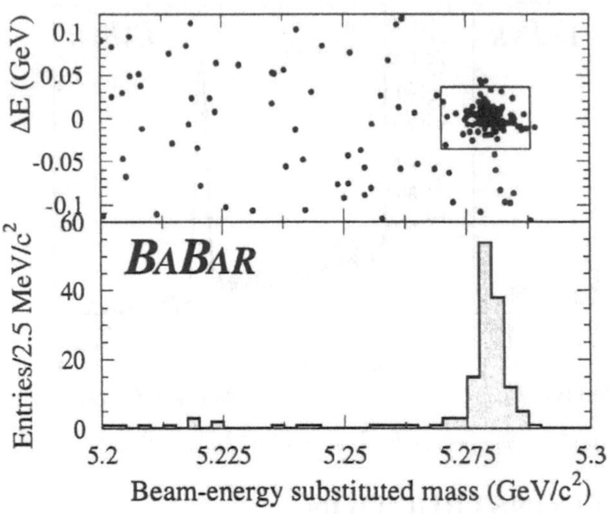

Figure 1.7: $B^0/\overline{B}^0 \to J/\psi K_S^0$ ($K_S^0 \to \pi^+\pi^-$) signal

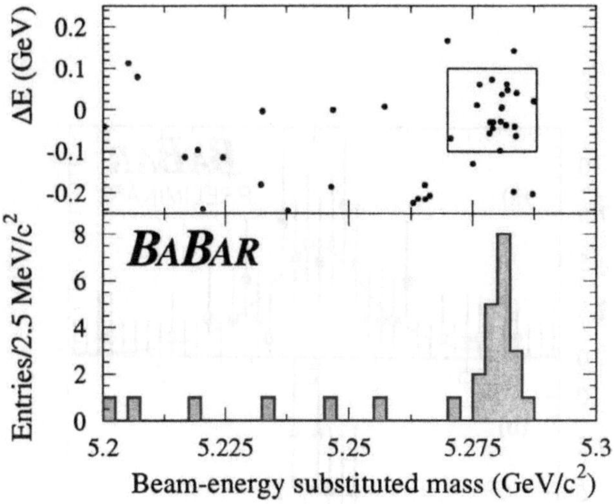

Figure 1.8: $B^0/\overline{B}^0 \to J/\psi\, K^0_S$ ($K^0_S \to \pi^0\pi^0$) signal

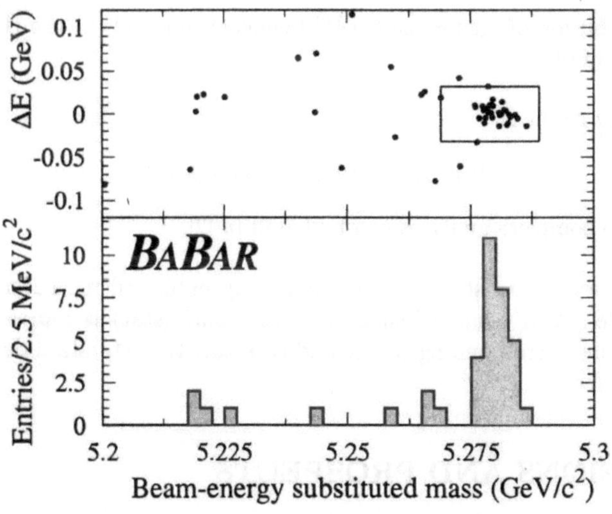

Figure 1.9: $B^0/\overline{B}^0 \to \psi(2S)\, K^0_S$ ($K^0_S \to \pi^+\pi^-$) signal

The time dependent fit to these events is shown in figure 1.10.

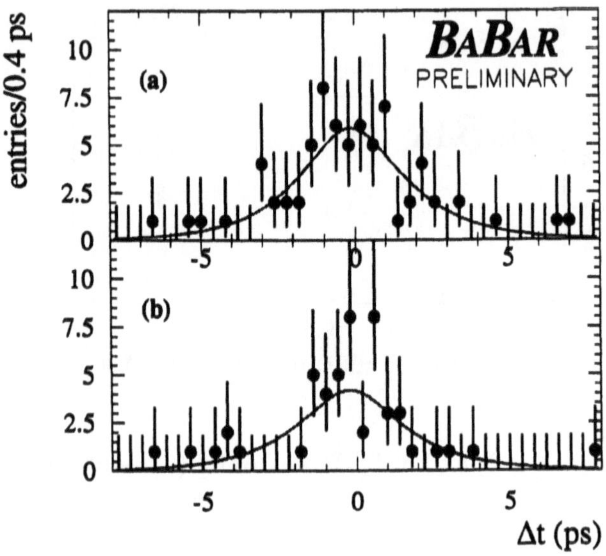

Figure 1.10: Distribution of Δt for the (a) B^0 tagged events and (b) the $\overline{B}^0$ tagged events in the CP sample.

Our preliminary result is:

$$\sin 2\beta = 0.12 \pm 0.37 \text{ (stat) } \pm 0.09 \text{ (syst)}$$

More details of the complete analyses can be found in [2].

Figure 1.11 shows the constraints of the currently published measurements on the standard model. Our result of $\sin 2\beta = 0.12 \pm 0.37$ (stat) is represented by cross-hatched regions corresponding to one and two statistical standard deviations.

6. CONCLUSIONS AND PROSPECTS

We have presented BA$\overline{\text{B}}$AR 's first preliminary measurement of the CP-violating parameter $\sin 2\beta$ in the B-meson system: $\sin 2\beta = 0.12 \pm 0.37$ (stat) ± 0.09 (syst). Our measurement is consistent with the world average of $\sin 2\beta = 0.9 \pm 0.4$ [3], and is currently limited by the size of our CP sample. We have more than doubled our statistics and are currently analyzing the data. A first journal publication is planned for mid February 2001.

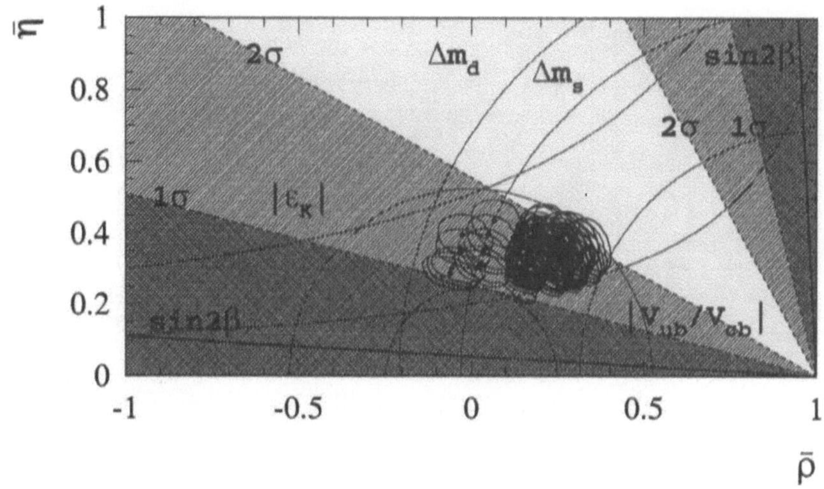

Figure 1.11: Present constraint on the position of the apex of the Unitary Triangle in the $\overline{\rho}\,\overline{\eta}$ plane

PEP II operation is expected to start in the beginning of February 2001 and we hope to collect an additional data sample of another 40 to $50\,\text{fb}^{-1}$. This would reduce our statistical error on $\sin 2\beta$ by a factor 4.

7. ACKNOWLEDGMENTS

I would like to thank the organizers of the Orbis Scientiae conference and in particular Prof. Behram Kursunoglu for inviting me to talk about the present and future of the BA$\overline{B}$AR experiment. I'd like to thank the BA$\overline{B}$AR collaboration and the staff of the SLAC accelerator department for their outstanding efforts.
Work supported by Department of Energy contract DE-AC03-76SF00515.

REFERENCES

[1] The BA$\overline{B}$AR physics book, P.H. Harrison and H.R. Quinn editors, SLAC-R-504 (1998)

[2] "A Study of time-dependent CP-violating asymmetries in $B^0 \to J\psi K^0_S$ and $B^0 \to \psi(2S)\,K^0_S$ decays"
B.Aubert et.al, The BA$\overline{B}$AR Collaboration. Contributed paper to ICHEP2000 conference in Osaca, Japan. hep-ex/0008048

[3] Particle Data Group, C. Caso *et al.*, Review of Particle Properties, The European Phsyical Journal **C3**, (1998).

NEW EXPERIMENT ON SPINNING PROTONS' VIOLENT COLLISIONS

Alan D. Krisch*

This lecture reviewed a recent proposal to the Institute for High Energy Physics (IHEP) in Protvino, Russia for the SPIN@U-70 Experiment[1]. SPIN@U-70 plans to measure the Analyzing Power A_n in 70 GeV proton-proton elastic scattering at Very-High-$P_\perp^2$. I first gave a brief summary of the large and unexpected spin effects found in *pp* elastic and inclusive scattering.[2] In particular, the *pp* analyzing power A_n was found to be unexpectedly large and still rising at $P_\perp^2$ of 7 $(GeV/c)^2$ in our 24 and 28 GeV Brookhaven AGS experiments during 1982 to 1990.[3] SPIN@U-70 should extend this AGS data to higher energy and to much larger $P_\perp^2$. The new experiment will use our state-of-the-art solid polarized proton target containing radiation-doped frozen NH_3 beads at 5 T and 1 K which provides up to 95% proton polarization even when used with very high intensity beams. SPIN@U-70 will use the very intense (10^{12}/pulse) extracted proton beam available at IHEP-Protvino's 70 GeV U-70 accelerator; it will detect the recoil protons using a 32-meter-long spectrometer with a large solid angle and a momentum precision of better than ±0.1%. More details about SPIN@U-70 are accessible on the Spin Physics Center website *http://spinbud.physics.lsa.umich.edu*, by clicking the *SPIN@U-70* button.

References

[1] The SPIN@U-70 Collaboration consists of scientists from University of Michigan, University of Virginia, IHEP-Protvino, JINR-Dubna, and TRIUMF.

[2] See in A.D. Krisch, Z. Phys. C - Particles and Fields **46**, S133-147 (1990).

[3] D.G. Crabb *et al.*, Phys. Rev. Lett. **65**, 3241 (1990); and references therein.

*Spin Physics Center, University of Michigan, Ann Arbor, MI 48109-1120; *krisch@umich.edu*.

The Role of Neutrinos, Strings, Gravity, and Variable Cosmological Constant in Elementary Particle Physics, Edited by Kursunoglu *et al.*, Kluwer Academic/Plenum Publishers, New York, 2001

THE GRAND UNIFIED THEORY OF CLASSICAL QUANTUM MECHANICS

Randell L. Mills*

1. INTRODUCTION

A theory of classical quantum mechanics (CQM), derived from first principles,[1] successfully applies physical laws on all scales. The classical wave equation is solved with the constraint that a bound electron cannot radiate energy. The mathematical formulation for zero radiation based on Maxwell's equations follows from a derivation by Haus.[2] The function that describes the motion of the electron must not possess spacetime Fourier components that are synchronous with waves traveling at the speed of light. CQM gives closed form solutions for the atom, including the stability of the n=1 state and the instability of the excited states, the equation of the photon and electron in excited states, the equation of the free electron, and photon which predict the wave particle duality behavior of particles and light. The current and charge density functions of the electron may be directly physically interpreted. For example, spin angular momentum results from the motion of negatively charged mass moving systematically, and the equation for angular momentum, **r x p**, can be applied directly to the wave function, called an orbitsphere (a current density function), that describes the electron. The magnetic moment of a Bohr magneton, Stern Gerlach experiment, g factor, Lamb shift, resonant line width and shape, selection rules, correspondence principle, wave particle duality, excited states, reduced mass, rotational energies, and momenta, orbital and spin splitting, spin-orbital coupling, Knight shift, and spin-nuclear coupling are derived in closed form equations based on Maxwell's equations. The calculations agree with experimental observations.

For or any kind of wave advancing with limiting velocity and capable of transmitting signals, the equation of front propagation is the same as the equation for the front of a light wave. By applying this condition to electromagnetic and gravitational fields at

* Randell L Mills, President, BlackLight Power, Inc., 493 Old Trenton Road, Cranbury, NJ 08512, Phone: 609-490-1090, e-mail: rmills@blacklightpower.com; www.blacklightpower.com

The Role of Neutrinos, Strings, Gravity, and Variable Cosmological Constant in Elementary Particle Physics, Edited by Kursunoglu *et al.*, Kluwer Academic/Plenum Publishers, New York, 2001

particle production, the Schwarzschild metric (SM) is derived from the classical wave equation which modifies general relativity to include conservation of spacetime in addition to momentum and matter/energy. The result gives a natural relationship between Maxwell's equations, special relativity, and general relativity. It gives gravitation from the atom to the cosmos. The Universe is time harmonically oscillatory in matter energy and spacetime expansion and contraction with a minimum radius that is the gravitational radius. In closed form equations with fundamental constants only, CQM gives the deflection of light by stars, the precession of the perihelion of Mercury, the particle masses, the Hubble constant, the age of the Universe, the observed acceleration of the expansion, the power of the Universe, the power spectrum of the Universe, the microwave background temperature, the uniformity of the microwave background radiation, the microkelvin spatial variation of the microwave background radiation, the observed violation of the GZK cutoff, the mass density, the large scale structure of the Universe, and the identity of dark matter which matches the criteria for the structure of galaxies. In a special case wherein the gravitational potential energy density of a blackhole equals that of the Plank mass, matter converts to energy and spacetime expands with the release of a gamma-ray burst. The singularity in the SM is eliminated.

2. COSMOLOGICAL THEORY BASED ON MAXWELL'S EQUATIONS

Maxwell's equations and special relativity are based on the law of propagation of a electromagnetic wave front in the form

$$1/c^2(\delta\omega/\delta t)^2 - \left[(\delta\omega/\delta x)^2 + (\delta\omega/\delta y)^2 + (\delta\omega/\delta\omega)^2\right] = 0 \quad (1)$$

For any kind of wave advancing with limiting velocity and capable of transmitting signals, the equation of front propagation is the same as the equation for the front of a light wave. Thus, the equation $1/c^2(\delta\omega/\delta t)^2 - (grad\omega)^2 = 0$ acquires a general character; it is more general than Maxwell's equations from which Maxwell originally derived it.

A discovery of the present work is that the classical wave equation governs: (1) the motion of bound electrons, (2) the propagation of any form of energy, (3) measurements between inertial frames of reference such as time, mass, momentum, and length (Minkowski tensor), (4) fundamental particle production and the conversion of matter to energy, (5) a relativistic correction of spacetime due to particle production or annihilation (Schwarzschild metric), (6) the expansion and contraction of the Universe, (7) the basis of the relationship between Maxwell's equations, Planck's equation, the de Broglie equation, Newton's laws, and special, and general relativity.

The relationship between the time interval between ticks t of a clock in motion with velocity v relative to an observer and the time interval t_0 between ticks on a clock at rest relative to an observer[24] is

$$(ct)^2 = (ct_0)^2 + (vt)^2. \quad (2)$$

Thus, the time dilation relationship based on the constant maximum speed of light c in any inertial frame is $t = t_0\big/\sqrt{1-\left(v^2/c^2\right)}$. The metric $g_{\mu\nu}$ for Euclidean space is the

Minkowski tensor $\eta_{\mu\nu}$. In this case, the separation of proper time between two events x^μ and $x^\mu + dx^\mu$ is $d\tau^2 = -\eta_{\mu\nu} dx^\mu dx^\nu$.

3. THE EQUIVALENCE OF THE GRAVITATIONAL MASS AND THE INERTIAL MASS

The equivalence of the gravitational mass and the inertial mass m_g/m_i = universal constant which is predicted by Newton's law of mechanics and gravitation is experimentally confirmed to less 1 X 10^{-11}.[3] In physics, the discovery of a universal constant often leads to the development of an entirely new theory. From the universal constancy of the velocity of light c, the special theory of relativity was derived; and from Planck's constant h, the quantum theory was deduced. Therefore, the universal constant m_g/m_i should be the key to the gravitational problem. The energy equation of Newtonian gravitation is

$$E = \frac{1}{2}mv^2 - \frac{GMm}{r} = \frac{1}{2}mv_0^2 - \frac{GMm}{r_0} = \text{constant} \tag{3}$$

Since h, the angular momentum per unit mass, is $h = L/m = |\mathbf{r} \times \mathbf{v}| = r_0 v_0 \sin\phi$, the eccentricity e may be written as

$$e = [1 + \left(v_0^2 - \frac{2GM}{r_0} \right) \frac{r_0^2 v_0^2 \sin^2\phi}{G^2 M^2}]^{1/2}, \tag{4}$$

where m is the inertial mass of a particle, v_0 is the speed of the particle, r_0 is the distance of the particle from a massive object, ϕ is the angle between the direction of motion of the particle and the radius vector from the object, and M is the total mass of the object (including a particle). The eccentricity e given by Newton's differential equations of motion in the case of the central field permits the classification of the orbits according to the total energy E [4] (column 1) and the orbital velocity squared, v_0^2, relative to the gravitational velocity squared, $2GM/r_0$ [4] (column 2):

$E < 0$	$v_0^2 < 2GM/r_0$	$e < 1$	ellipse
$E < 0$	$v_0^2 < 2GM/r_0$	$e = 0$	circle (special case of ellipse)
$E = 0$	$v_0^2 = 2GM/r_0$	$e = 1$	parabolic orbit
$E > 0$	$v_0^2 > 2GM/r_0$	$e > 1$	hyperbolic orbit

4. CONTINUITY CONDITIONS FOR THE PRODUCTION OF A PARTICLE FROM A PHOTON TRAVELING AT LIGHT SPEED

A photon traveling at the speed of light gives rise to a particle with an initial radius equal to its Compton wavelength bar.

$$r = \lambda_C = \frac{\hbar}{m_0 c} = r_\alpha^*, \tag{5}$$

The particle must have an orbital velocity equal to Newtonian gravitational escape velocity v_g of the antiparticle.

$$v_g = \sqrt{\frac{2Gm}{r}} = \sqrt{\frac{2Gm_0}{\lambda_C}}. \tag{6}$$

The eccentricity is one. The orbital energy is zero. The particle production trajectory is a parabola relative to the center of mass of the antiparticle.

4.1 A Gravitational Field as a Front Equivalent to Light Wave Front

The particle with a finite gravitational mass gives rise to a gravitational field that travels out as a front equivalent to a light wave front. The form of the outgoing gravitational field front traveling at the speed of light is $f(t - r/c)$ and $d\tau^2$ is given by

$$d\tau^2 = f(r)dt^2 - \frac{1}{c^2}\left[f(r)^{-1}dr^2 + r^2 d\theta^2 + r^2 \sin^2\theta d\phi^2\right] \tag{7}$$

The speed of light as a constant maximum as well as phase matching and continuity conditions of the electromagnetic and gravitational waves require the following form of the squared displacements:

$$(c\tau)^2 + (v_g t)^2 = (ct)^2, \tag{8}$$

$$f(r) = \left(1 - \left(\frac{v_g}{c}\right)^2\right). \tag{9}$$

In order that the wave front velocity does not exceed c in any frame, spacetime must undergo time dilation and length contraction due to the particle production event. *The derivation and result of spacetime time dilation is analogous to the derivation and result of special relativistic time dilation* wherein the relative velocity of two inertial frames replaces the gravitational velocity.

The general form of the metric due to the relativistic effect on spacetime due to mass m_0 with v_g given by Eq. (6) is

$$d\tau^2 = \left(1 - \left(\frac{v_g}{c}\right)^2\right)dt^2 - \frac{1}{c^2}\left[\left(1 - \left(\frac{v_g}{c}\right)^2\right)^{-1} dr^2 + r^2 d\theta^2 + r^2 \sin^2\theta d\phi^2\right]. \tag{10}$$

The gravitational radius, r_g, of each orbitsphere of the particle production event, each of mass m_0 and the corresponding general form of the metric are respectively

$$r_g = \frac{2Gm_0}{c^2}, \tag{11}$$

$$d\tau^2 = \left(1 - \frac{r_g}{r}\right)dt^2 - \frac{1}{c^2}\left[\left(1 - \frac{r_g}{r}\right)^{-1} dr^2 + r^2 d\theta^2 + r^2 \sin^2\theta d\phi^2\right]. \tag{12}$$

Masses and their effects on spacetime *superimpose*. The separation of proper time between two events x^μ and $x^\mu + dx^\mu$ is

$$d\tau^2 = \left(1 - \frac{2GM}{c^2 r}\right)dt^2 - \frac{1}{c^2}\left[\left(1 - \frac{2GM}{c^2 r}\right)^{-1} dr^2 + r^2 d\theta^2 + r^2 \sin^2\theta d\phi^2\right]. \tag{13}$$

The metric $g_{\mu\nu}$ for non-Euclidean space due to the relativistic effect on spacetime due to mass M is *the Schwarzschild metric which gives the relationship whereby matter causes relativistic corrections to spacetime that determines the curvature of spacetime and is the origin of gravity.*

4.2. Particle Production Continuity Conditions from Maxwell's Equations, and the Schwarzschild Metric

The photon to particle event requires a transition state that is continuous wherein the velocity of a transition state orbitsphere is the speed of light. The radius, r, is the Compton wavelength bar, $\lambda\!\!\!^{-}_C$, given by Eq. (5). At production, the Planck equation energy, the electric potential energy, and the magnetic energy are equal to m_0c^2.

The Schwarzschild metric gives the relationship whereby matter causes relativistic corrections to spacetime that determines the masses of fundamental particles. Substitution of $r = \lambda\!\!\!^{-}_C$; $dr = 0$; $d\theta = 0$; $\sin^2\theta = 1$ into the Schwarzschild metric gives

$$d\tau = dt\left(1 - \frac{2Gm_0}{c^2 r_\alpha^*} - \frac{v^2}{c^2}\right)^{\frac{1}{2}}, \tag{14}$$

with $v^2 = c^2$, the relationship between the proper time and the coordinate time is

$$\tau = ti\sqrt{\frac{2GM}{c^2 r_\alpha^*}} = ti\sqrt{\frac{2GM}{c^2 \lambda\!\!\!^{-}_c}} = ti\frac{v_g}{c}. \tag{15}$$

When the orbitsphere velocity is the speed of light, continuity conditions based on the constant maximum speed of light given by Maxwell's equations are mass energy = Planck equation energy = electric potential energy = magnetic energy = mass/spacetime metric energy. Therefore, $m_0c^2 = \hbar\omega^* = V = E_{mag} = E_{spacetime}$

$$m_0c^2 = \hbar\omega^* = \frac{\hbar^2}{m_0\lambda_C^2} = \alpha^{-1}\frac{e^2}{4\pi\varepsilon_0\lambda_C} = \alpha^{-1}\frac{\pi\mu_0 e^2\hbar^2}{(2\pi m_0)^2\lambda_C^3} = \frac{\alpha h}{1\,\text{sec}}\sqrt{\frac{\lambda_C c^2}{2Gm}} \; . \tag{16}$$

The continuity conditions based on the constant maximum speed of light given by the Schwarzschild metric are:

$$\frac{\text{proper time}}{\text{coordinate time}} = \frac{\text{gravitational wave condition}}{\text{electromagnetic wave condition}} = \frac{\text{gravitational mass phase matching}}{\text{charge/inertial mass phase matching}} , \tag{17}$$

$$\frac{\text{proper time}}{\text{coordinate time}} = i\frac{\sqrt{\frac{2Gm}{c^2\lambda_C}}}{\alpha} = i\frac{v_g}{\alpha c} \; . \tag{18}$$

5. MASSES OF FUNDAMENTAL PARTICLES

Each of the Planck equation energy, electric energy, and magnetic energy corresponds to a particle given by the relationship between the proper time and the coordinate time. The electron and down-down-up neutron correspond to the Planck equation energy. The muon and strange-strange-charmed neutron correspond to the electric energy. The tau and bottom-bottom-top neutron correspond to the magnetic energy. The particle must possess the escape velocity v_g relative to the antiparticle where $v_g < c$. According to Newton's law of gravitation, the eccentricity is one and the particle production trajectory is a parabola relative to the center of mass of the antiparticle.

5.1. The Electron-Antielectron Lepton Pair

A clock is defined in terms of a self consistent system of units used to measure the particle mass. The proper time of the particle is equated with the coordinate time according to the Schwarzschild metric corresponding to light speed. The special relativistic condition corresponding to the Planck energy gives the mass of the electron.

$$2\pi\frac{\hbar}{mc^2} = sec\sqrt{\frac{2Gm^2}{c\alpha^2\hbar}} \; , \tag{19}$$

$$m_e = \left(\frac{h\alpha}{sec\, c^2}\right)^{\frac{1}{2}}\left(\frac{c\hbar}{2G}\right)^{\frac{1}{4}} = 9.1097\, X\, 10^{-31}\, kg \; , \tag{20}$$

$$m_e = 9.1097\, X\, 10^{-31}\, kg - 18\, eV/c^2\,(\nu_e) = 9.1094\, X\, 10^{-31}\, kg \; , \tag{21}$$

$$m_{e\,\text{experimental}} = 9.1095\, X\, 10^{-31}\, kg \; . \tag{22}$$

5.2. Down-Down-Up Neutron (DDU)

The corresponding equation for production of the neutron is

$$2\pi \frac{2\pi\hbar}{\frac{m_N}{3}\left[\frac{1}{2\pi}-\frac{\alpha}{2\pi}\right]c^2} = sec\sqrt{\frac{2G\left[\frac{m_N}{3}\left[\frac{1}{2\pi}-\frac{\alpha}{2\pi}\right]\right]^2}{3c(2\pi)^2\hbar}}, \tag{23}$$

$$m_{N\,calculated} = (3)(2\pi)\left(\frac{1}{1-\alpha}\right)\left(\frac{2\pi h}{sec\, c^2}\right)^{\frac{1}{2}}\left(\frac{2\pi(3)ch}{2G}\right)^{\frac{1}{4}} = 1.6744X10^{-27}\, kg, \tag{24}$$

$$m_{N\,experimental} = 1.6749X10^{-27}\, kg. \tag{25}$$

6. GRAVITATIONAL POTENTIAL ENERGY

The gravitational radius, α_G or r_G, of an orbitsphere of mass m_0 is defined as

$$\alpha_G = r_G = \frac{Gm_0}{c^2}, \tag{26}$$

when the $r_G = r_\alpha^* = \lambda_C$, the gravitational potential energy equals m_0c^2

$$r_G = \frac{Gm_0}{c^2} = \lambda_c = \frac{\hbar}{m_0c}, \tag{27}$$

$$E_{grav} = \frac{Gm_0^2}{r} = \frac{Gm_0^2}{\lambda_C} = \frac{Gm_0^2}{r_\alpha^*} = \hbar\omega^* = m_0c^2. \tag{28}$$

The mass m_0 is the Planck mass, m_u,

$$m_u = m_0 = \sqrt{\frac{\hbar c}{G}}, \tag{29}$$

the corresponding gravitational velocity, v_G, is defined as

$$v_G = \sqrt{\frac{Gm_0}{r}} = \sqrt{\frac{Gm_0}{\lambda_C}} = \sqrt{\frac{Gm_u}{\lambda_C}}. \tag{30}$$

6.1. Relationship of the Equivalent Planck Mass Particle Production Energies

For the Plank mass particle, the relationships corresponding to Eq. (16) are: (mass energy = Planck equation energy = electric potential energy = magnetic energy = gravitational potential energy = mass/spacetime metric energy)

$$m_oc^2 = \hbar\omega^* = V = E_{mag} = E_{grav} = E_{spacetime}, \tag{31}$$

$$m_0 c^2 = \hbar\omega^* = \frac{\hbar^2}{m_0 \lambda_C^2} = \alpha^{-1}\frac{e^2}{4\pi\varepsilon_0 \lambda_C} = \alpha^{-1}\frac{\pi\mu_0 e^2 \hbar^2}{(2\pi m_0)^2 \lambda_C^3} = \alpha^{-1}\frac{\mu_0 e^2 c^2}{2h}\sqrt{\frac{Gm_0}{\lambda_C}}\sqrt{\frac{\hbar c}{G}} = \frac{\alpha h}{1\,\mathrm{sec}}\sqrt{\frac{\lambda_C c^2}{2Gm}} \quad (32)$$

These equivalent energies give the particle masses in terms of the gravitational velocity, v_G, and the Planck mass, m_u

$$m_0 = \alpha^{-1}\frac{\mu_0 e^2 c}{2h}\frac{\sqrt{\frac{Gm_0}{\lambda_C}}}{c} m_u = \alpha^{-1}\frac{\mu_0 e^2 c}{2h}\sqrt{\frac{Gm_0}{c^2 \lambda_C}} m_u = \alpha^{-1}\frac{\mu_0 e^2 c}{2h}\frac{v_G}{c} m_u = \frac{v_G}{c} m_u . \quad (33)$$

6.2. Planck Mass Particles

A pair of particles each of the Planck mass corresponding to the gravitational potential energy is not observed since the velocity of each transition state orbitsphere is the gravitational velocity, v_G that in this case is the speed of light; whereas, the Newtonian gravitational escape velocity v_g is twice the speed of light. In this case, an electromagnetic wave of mass energy equivalent to the Planck mass travels in a circular orbit about the center of mass of another electromagnetic wave of mass energy equivalent to the Planck mass wherein the eccentricity is equal to zero and the escape velocity can never be reached. The Planck mass is a "measuring stick." The extraordinarily high Planck mass ($\sqrt{\hbar c/G} = 2.18 X 10^{-8}\ kg$) is the unobtainable mass bound imposed by the angular momentum and speed of the photon relative to the gravitational constant. It is analogous to the unattainable bound of the speed of light for a particle possessing finite rest mass imposed by the Minkowski tensor.

6.3. Astrophysical Implications of Planck Mass Particles

The limiting speed of light eliminates the singularity problem of Einstein's equation that arises as the radius of a blackhole equals the Schwarzschild radius. General relativity with the singularity eliminated resolves the paradox of the infinite propagation velocity required for the gravitational force in order to explain why the angular momentum of objects orbiting a gravitating body does not increase due to the finite propagation delay of the gravitational force according to special relativity.[5] When the gravitational potential energy density of a massive body such as a blackhole equals that of a particle having the Planck mass, the matter may transition to photons of the Planck mass. Even light from a blackhole will escape when the decay rate of the trapped matter with the concomitant spacetime expansion is greater than the effects of gravity which oppose this expansion. Gamma-ray bursts are the most energetic phenomenon known that can release an explosion of gamma rays packing 100 times more energy than a supernova explosion.[6] The annihilation of a blackhole may be the source of *γ-ray bursts*. The source may be due to conversion of matter to photons of the Planck mass/energy which may also give rise to cosmic rays which are the most energetic particles known, and their origin is also a mystery.[7] According to the GZK cutoff, the cosmic spectrum cannot extend beyond 5 X $10^{19} eV$, but AGASA, the world's largest air shower array, has shown that the spectrum is extending beyond $10^{20} eV$ without any clear sign of cutoff.[8] Photons each of the Planck mass may be the source of these inexplicably energetic cosmic rays.

7. RELATIONSHIP OF MATTER TO ENERGY AND SPACETIME EXPANSION

The Schwarzschild metric gives the relationship whereby matter causes relativistic corrections to spacetime. The limiting velocity c results in the contraction of spacetime due to particle production, which is given by $2\pi r_g$ where r_g is the gravitational radius of the particle. This has implications for the expansion of spacetime when matter converts to energy. Q the mass/energy to expansion/contraction quotient of spacetime is given by the ratio of the mass of a particle at production divided by T the period of production.

$$Q = \frac{m_0}{T} = \frac{m_0}{\frac{2\pi r_g}{c}} = \frac{m_0}{\frac{2\pi \frac{2Gm_0}{c^2}}{c}} = \frac{c^3}{4\pi G} = 3.22 X 10^{34} \frac{kg}{sec}. \quad (34)$$

The gravitational equations with the equivalence of the particle production energies [Eq. (16)] permit the conservation of mass/energy ($E=mc^2$) and spacetime ($c^3/4\pi G=3.22$ $X\ 10^{34}\ kg/sec$). With the conversion of $3.22\ X\ 10^{34}\ kg$ of matter to energy, spacetime expands by 1 sec. The photon has inertial mass and angular momentum, but due to Maxwell's equations and the implicit special relativity it does not have a gravitational mass.

7.1. Cosmological Consequences

The Universe is closed (it is *finite but with no boundary*). It is a 3-sphere Universe-Riemannian three dimensional hyperspace plus time of constant positive curvature at each r-sphere. The *Universe is oscillatory in matter/energy and spacetime* with a finite minimum radius, the gravitational radius. Spacetime expands as mass is released as energy which provides the basis of the atomic, thermodynamic, and cosmological arrows of time. Different regions of space are isothermal even though they are separated by greater distances than that over which light could travel during the time of the expansion of the Universe.[9] Presently, stars and large scale structures exist which are older than the elapsed time of the present expansion as stellar, galaxy, and supercluster evolution occurred during the contraction phase.[10–16] The maximum power radiated by the Universe which occurs at the beginning of the expansion phase is $P_U=c^5/4\pi G = 2.89\ X\ 10^{51} W$. Observations beyond the beginning of the expansion phase are not possible since the Universe is entirely matter filled.

7.2. The Period of Oscillation of the Universe Based on Closed Propagation of Light

Mass/energy is conserved during harmonic expansion and contraction. The gravitational potential energy E_{grav} given by Eq. (28) with $m_0 = m_U$ is equal to $m_U c^2$ when the radius of the Universe r is the gravitational radius r_G. The gravitational velocity v_G [Eq. (30) with $r=r_G$ and $m_0 = m_U$] is the speed of light in a circular orbit wherein the eccentricity is equal to zero and the escape velocity from the Universe can never be reached. The period of the oscillation of the Universe and the period for light to transverse the Universe corresponding to the gravitational radius r_G must be equal. The harmonic oscillation period, T, is

$$T = \frac{2\pi r_G}{c} = \frac{2\pi G m_U}{c^3} = \frac{2\pi G\left(2\,X\,10^{54}\,kg\right)}{c^3} = 3.10\,X\,10^{19}\,sec = 9.83\,X\,10^{11}\,years\,, \tag{35}$$

where the mass of the Universe, m_U, is approximately 2 X 10^{54}*kg*. (The initial mass of the Universe of 2 X 10^{54}*kg* is based on internal consistency with the size, age, Hubble constant, temperature, density of matter, and power spectrum.) Thus, the observed Universe will expand as mass is released as photons for 4.92 X 10^{11}*years*. At this point in its world line, the Universe will obtain its maximum size and begin to contract.

8. THE DIFFERENTIAL EQUATION OF THE RADIUS OF THE UNIVERSE

Based on conservation of mass/energy ($E=mc^2$) and spacetime ($c^3/4\pi G=3.22\ X\ 10^{34}$ *kg/sec*). The Universe behaves as a simple harmonic oscillator having a restoring force, **F**, which is proportional to the radius. The proportionality constant, k, is given in terms of the potential energy, E, gained as the radius decreases from the maximum expansion to the minimum contraction.

$$\frac{E}{\aleph^2} = k\,. \tag{36}$$

Since the gravitational potential energy E_{grav} is equal to $m_U c^2$ when the radius of the Universe r is the gravitational radius r_G

$$F = -k\aleph = -\frac{m_U c^2}{r_G^2}\aleph = -\frac{m_U c^2}{\left(\frac{G m_U}{c^2}\right)^2}\aleph\,. \tag{37}$$

And, the differential equation of the radius of the Universe, $\aleph$ is

$$m_U\,\ddot{\aleph} + \frac{m_U c^2}{r_G^2}\aleph = m_U\,\ddot{\aleph} + \frac{m_U c^2}{\left(\frac{G m_U}{c^2}\right)^2}\aleph = 0\,. \tag{38}$$

The *maximum radius of the Universe*, the amplitude, r_0, of the time harmonic variation in the radius of the Universe, is given by the quotient of the total mass of the Universe and Q, the mass/energy to expansion/contraction quotient.

$$r_0 = \frac{m_U}{Q} = \frac{m_U}{\frac{c^3}{4\pi G}} = \frac{2\,X\,10^{54}\,kg}{\frac{c^3}{4\pi G}} = 1.97\,X\,10^{12}\,light\ years\,. \tag{39}$$

The *minimum radius* which corresponds to the gravitational radius r_g, given by Eq. (11) with $m_0=m_U$ is 3.12 X 10^{11} *light years*. When the radius of the Universe is the gravitational radius, r_g, the proper time is equal to the coordinate time by Eq. (15) and the gravitational escape velocity v_g of the Universe is the speed of light. The radius of the Universe as a function of time is

$$\aleph = \left(r_g + \frac{cm_U}{Q}\right) - \frac{cm_U}{Q}\cos\left(\frac{2\pi t}{\frac{2\pi r_G}{c}}\right) = \left(\frac{2Gm_U}{c^2} + \frac{\frac{cm_U}{c^3}}{4\pi G}\right) - \frac{\frac{cm_U}{c^3}}{4\pi G}\cos\left(\frac{2\pi t}{\frac{2\pi G m_U}{c^3}}\right). \quad (40)$$

The expansion/contraction rate, $\dot{\aleph}$, is given by time derivative of Eq. (40)

$$\dot{\aleph} = 4\pi c X 10^{-3} \sin\left(\frac{2\pi t}{\frac{2\pi G m_U}{c^3}}\right)\frac{km}{sec}. \quad (41)$$

9. THE HUBBLE CONSTANT

The *Hubble constant* is given by the ratio of the expansion rate given in units of *km/sec* divided by the radius of the expansion in *Mpc*. The radius of expansion is equivalent to the radius of the light sphere with an origin at the time point when the Universe stopped contracting and started to expand.

$$H = \frac{\dot{\aleph}}{t\,Mpc} = \frac{4\pi c X 10^{-3} \sin\left(\frac{2\pi t}{\frac{2\pi G m_U}{c^3}}\right)\frac{km}{sec}}{t\,Mpc}, \quad (42)$$

for $t = 10^{10}\ \mathit{light\ years} = 3.069\ X\ 10^{3}\ Mpc$, the Hubble constant, H_0, is 78.6 *km/sec·Mpc*. The experimental value is[17] H_0=80 ± 17 *km/sec·Mpc.*

10. THE DENSITY OF THE UNIVERSE AS A FUNCTION OF TIME

The density of the Universe as a function of time $\rho_U(t)$ is given by the ratio of the mass as a function of time and the volume as a function of time.

$$\rho_U(t) = \frac{m_U(t)}{V(t)} = \frac{m_U(t)}{\frac{4}{3}\pi\aleph(t)^3} = \frac{\frac{m_U}{2}\left(1 + \cos\left(\frac{2\pi t}{\frac{2\pi G m_U}{c^3}}\right)\right)}{\frac{4}{3}\pi\left(\left(\frac{2Gm_U}{c^2} + \frac{\frac{cm_U}{c^3}}{4\pi G}\right) - \frac{\frac{cm_U}{c^3}}{4\pi G}\cos\left(\frac{2\pi t}{\frac{2\pi G m_U}{c^3}}\right)\right)^3}, \quad (43)$$

for $t = 10^{10}\ \mathit{light\ years}$, $\rho_U = 1.7\ X\ 10^{-32}\ g/cm^3$ The density of luminous matter of stars and gas of galaxies is about $\rho_U = 2\ X\ 10^{-31}\ g/cm^3$.[18–19]

11. THE POWER OF THE UNIVERSE AS A FUNCTION OF TIME, $P_U(t)$

From $E = mc^2$ and Eq. (34),

$$P_U(t) = \frac{c^5}{8\pi G}\left(1 + \cos\left(\frac{2\pi t}{\frac{2\pi r_G}{c}}\right)\right) \tag{44}$$

For $t = 10^{10}$ *light years*, $P_U(t) = 2.88 X 10^{51} W$. The observed power is consistent with that predicted.

12. THE TEMPERATURE OF THE UNIVERSE AS A FUNCTION OF TIME

The temperature of the Universe as a function of time, $T_U(t)$, follows from the Stefan-Boltzmann law.

$$T_U(t) = \left(\frac{1}{1 + \frac{Gm_U(t)}{c^2 \aleph(t)}}\right)\left[\frac{R_U(t)}{e\sigma}\right]^{\frac{1}{4}} = \left(\frac{1}{1 + \frac{Gm_U(t)}{c^2 \aleph(t)}}\right)\left[\frac{\frac{P_U(t)}{4\pi \aleph(t)^2}}{e\sigma}\right]^{\frac{1}{4}}. \tag{45}$$

The calculated uniform temperature is about 2.7 K which is in agreement with the observed microwave background temperature.[9]

13. POWER SPECTRUM OF THE COSMOS

The power spectrum of the cosmos, as measured by the Las Campanas Survey, generally follows the prediction of cold dark matter on the scales of 200 million to 600 million light-years. However, the power increases dramatically on scales of 600 million to 900 million light-years. The infinitesimal temporal displacement, $d\tau^2$, is given by Eq. (13).

The relationship between the proper time and the coordinate time is

$$\tau = t\sqrt{1 - \frac{2Gm_U}{c^2 r}} = t\sqrt{1 - \frac{r_g}{r}}. \tag{46}$$

The power maximum in the proper frame occurs at

$$\tau = 5 X 10^9 \text{ light years}\sqrt{1 - \frac{3.12 X 10^{11} \text{ light years}}{3.22 X 10^{11} \text{ light years}}} = 880 X 10^6 \text{ light years}. \tag{47}$$

The power maximum of the current observable Universe is predicted to occur on the scale of 880 $X\ 10^6$ *light years*. There is excellent agreement between the predicted value and the experimental value of 600–900 $X\ 10^6$ *light years*.[16]

14. THE EXPANSION/CONTRACTION ACCELERATION, $\ddot{\aleph}$

The expansion/contraction acceleration rate, $\ddot{\aleph}$, is given by the time derivative of Eq. (41).

$$\ddot{\aleph} = 2\pi\frac{c^4}{Gm_U}\cos\left(\frac{2\pi t}{\frac{2\pi Gm_U}{c^3}sec}\right) = \ddot{\aleph} = H_o = 78.7\cos\left(\frac{2\pi t}{3.01\,X\,10^5\ Mpc}\right)\frac{km}{sec\cdot Mpc} \tag{48}$$

The differential in the radius of the Universe, $\Delta\aleph$, due to its acceleration is given by $\Delta\aleph = 1/2\ddot{\aleph}t^2$. The differential in expanded radius for the elapsed time of expansion, $t = 10^{10}$ *light years* corresponds to a decease in brightness of a supernovae standard candle of about an order of magnitude of that expected where the distance is taken as $\Delta\aleph$. This result based on the predicted rate of acceleration of the expansion is consistent with the experimental observation.[20–22]

Furthermore, the microwave background radiation image obtained by the Boomerang telescope[23] is consistent with a Universe of nearly flat geometry since the commencement of its expansion. The data is consistent with a large offset radius of the Universe with a fractional increase in size since the commencement of expansion about 10 billion years ago.

15. THE PERIODS OF SPACETIME EXPANSION/CONTRACTION AND PARTICLE DECAY/PRODUCTION FOR THE UNIVERSE ARE EQUAL

The period of the expansion/contraction cycle of the radius of the Universe, T, is given by Eq. (35). It follows from the Poynting power theorem with spherical radiation that the transition lifetimes are given by the ratio of energy and the power of the transition.

$$\tau = \frac{energy}{power} = \frac{[\hbar\omega]}{\left[\frac{2\pi c}{[(2l+1)!!]^2}\left(\frac{l+1}{l}\right)k^{2l+1}\left|Q_{lm} + Q'_{lm}\right|^2\right]}$$

$$= \frac{1}{2\pi}\left(\frac{h}{e^2}\right)\sqrt{\frac{\mu_o}{\varepsilon_o}}\frac{[(2l+1)!!]^2}{2\pi}\left(\frac{l}{l+1}\right)\left(\frac{l+3}{3}\right)^2\frac{1}{(kr_n)^{2l}\omega} \tag{49}$$

Exponential decay applies to electromagnetic energy decay $h(t) = e^{-\frac{1}{T}t}u(t)$. The coordinate time is imaginary because energy transitions are spacelike due spacetime expansion from matter to energy conversion. For example, the mass of the electron (a fundamental particle) is given by

$$\frac{2\pi \lambda_C}{\sqrt{\frac{2Gm_e}{\lambda_C}}} = \frac{2\pi \lambda_C}{v_g} = i\alpha^{-1} sec , \tag{50}$$

where v_g is Newtonian gravitational velocity [Eq. (6)]. When the gravitational radius r_g is the radius of the Universe, the proper time is equal to the coordinate time by Eq. (15), and the gravitational escape velocity v_g of the Universe is the speed of light. Replacement of the coordinate time, t, by the spacelike time, it, gives

$$h(t) = Re\left[e^{-i\frac{1}{T}t}\right] = cos\frac{2\pi}{T}t , \tag{51}$$

where the period is T [Eq. (35)]. The continuity conditions based on the constant maximum speed of light (Maxwell's equations) are given by Eqs. (16). The continuity conditions based on the constant maximum speed of light (Schwarzschild metric) are given by Eqs. (17–18). The periods of spacetime expansion/contraction and particle decay/production for the Universe are equal because only the particles which satisfy Maxwell's equations and the relationship between proper time and coordinate time imposed by the Schwarzschild metric may exist.

16. WAVE EQUATION

The general form of the light front wave equations is given by Eq. (1). The equation of the radius of the Universe, $\aleph$, may be written as

$$\aleph = \left(\frac{2Gm_U}{c^2} + \frac{\frac{cm_U}{c^3}}{4\pi G}\right) - \frac{\frac{cm_U}{c^3}}{4\pi G} cos\left(\frac{2\pi}{\frac{2\pi G m_U}{c^3} sec}\left(t - \frac{\aleph}{c}\right)\right) m , \tag{52}$$

which is a solution of the wave equation for a light wave front.

17. CONCLUSION

Maxwell's equations, Planck's equation, the de Broglie equation, Newton's laws, and special, and general relativity are unified.

REFERENCES

1. R. Mills, *The Grand Unified Theory of Classical Quantum Mechanics*, January 2000 Edition, BlackLight Power, Inc., Cranbury, New Jersey, Distributed by Amazon.com.
2. H. A. Haus, On the radiation from point charges, *American Journal of Physics*, **54**, 1126–1129 (1986).
3. E. G. Adelberger, C. W. Stubbs, B. R. Heckel, Y. Su, H. E.Swanson, G. Smith, J. H. Gundlach, *Phys. Rev. D*, **42**(10), 3267-3292 (1990).
4. G. R. Fowles, *Analytical Mechanics*, Third Edition, Holt, Rinehart, and Winston, New York, (1977), pp. 154–155.
5. T. Van Flandern, The Speed of Gravity—What the Experiments Say, *Physics Letters A*, **250**, 1–11 (1998).
6. R Cowen, Gamma-ray burst makes quite a bang, *Science News*,**153**(19), 292(1998).
7. M. Chown, The ultimate free lunch, *New Scientist*, **154**(2081), 50–51 (1997)/
8. B. Schwarzschild, Giant air shower array shows cosmic-ray spectrum violating greisen cutoff, *Physics Today*, **51**(10), 19–21 (1998).
9. J. C. Mather, E. S. Cheng, A preliminary measurement of the cosmic microwave background spectrum by the Cosmic Background Explorer (COBE) satellite, *Astrophysical Journal Letters*, **354**, L37–L40 (May 10, 1990).
10. W. Saunders, C. Frenk; et al; The density field of the local universe; *Nature*, **349**(6304), 32–38 (1991).
11. R. P. Kirshner, A. Oemler, Jr., P.L. Schechter, S. A. Shectman, A deep survey of galaxies, *Astronomical Journal*, **88**, 1285–1300 (September 1983).
12. V. de Lapparent, M.J. Geller, J.P. Huchra, The mean density and two-point correlation function for the CfA redshift survey slices, *Astrophysical Journal*, **332**(9) 44–56 (September 1, 1988).
13. A. Dressler, D. Lynden-Bell, D. Burstein, et. al., Spectroscopy and photometry of elliptical galaxies. I–A new distance estimator, *Astrophysical Journal*, **313**(2), 42–58 (1987).
14. S. Flamsteed, Crisis in the Cosmos, *Discover*, **16**(3), 66(1995).
15. J. Glanz, CO in the early universe clouds cosmologists' views, *Science*, **273**(5275), 581 (1996)
16. S. D. Landy, Mapping the Universe, *Scientific American*, **280**(6), 38–45 (1999).
17. W.L. Freeman, et al., Distance to the Virgo cluster galaxy M100 from Hubble Space, *Nature*, **371**(6500), 757–762 (1994).
18. R. M. Wald, *General Relativity*, University of Chicago Press, Chicago, (1984), pp. 114-116.
19. P. J. E. Peebles, J. Silk, Joseph, A cosmic book of phenomena, *Nature*, **346**(6281), 233 (1990)
20. J. Glanz; Astronomers see a cosmic antigravity force at work, *Science*, **279**(5355), 1298–1299 (1998).
21. R. Cowen, Living with lambda, *Science News*, **153**(22), 344 (1998).
22. R. Cowen, Studies support an accelerating Universe, *Science News*, **154**(18), 277 (1998).
23. P. de Bernardis et al., A flat Universe from high-resolution maps of the cosmic microwave background radiation, *Nature*, Vol. 404, (2000), p. 955; http://www.physics.ucsb.edu/~boomerang.
24. A. Beiser., *Concepts of Modern Physics*, 4th Edition, McGraw-Hill Book Company, New York, (1978), 2–10.

INDEX

MIX
Papier aus verantwortungsvollen Quellen
Paper from responsible sources
FSC® C105338

If you have any concerns about our products,
you can contact us on
ProductSafety@springernature.com

In case Publisher is established outside the EU,
the EU authorized representative is:
Springer Nature Customer Service Center GmbH
Europaplatz 3, 69115 Heidelberg, Germany

Printed by Libri Plureos GmbH
in Hamburg, Germany